DARK FORCES

Also by Kenneth R. Timmerman

Nonfiction

Shadow Warriors: Traitors, Saboteurs, and the Party of Surrender

Countdown to Crisis: The Coming Nuclear Showdown with Iran

Preachers of Hate: Islam and the War on America

Shakedown: Exposing the Real Jesse Jackson

The Death Lobby: How the West Armed Iraq

Fiction

St. Peter's Bones

Honor Killing

The Wren Hunt

DARK FORCES

THE **TRUTH** ABOUT WHAT HAPPENED IN **BENGHAZI**

KENNETH R. TIMMERMAN

BROADSIDE BOOKS
An Imprint of HarperCollins*Publishers*
www.broadsidebooks.net

HarperCollins books may be purchased for educational, business, or sales promotional use. For information, please e-mail the Special Markets Department at Spsales@harpercollins.com.

Broadside Books™ and the Broadside logo are trademarks of HarperCollins Publishers.

FIRST EDITION

Library of Congress Cataloging-in-Publication Data has been applied for.

ISBN: 978-0-06-232119-0

14 15 16 17 18 DIX/RRD 10 9 8 7 6 5 4 3 2 1

FOR TYRONE WOODS, GLEN DOHERTY, SEAN SMITH, AND CHRIS STEVENS;
AND FOR THOSE WHO SURVIVED, NAMED AND UNNAMED:
THIS IS YOUR STORY.

"Greater love hath no man than this, that a
man lay down his life for his friends."

—**John 15:13 (KJV)**

"The global network of espionage is a dark underworld,
full of ruthless individuals, a moral vacuum where
ego and self-gratification generally rule."

—**Former CIA officer Kevin Shipp,** *Company of Shadows*

Contents

Contents

Acknowledgments

This book relies extensively on sources developed over the past two decades, including present and former U.S. government officials, trusted sources within the intelligence community, and members of Congress. I have also been aided by extensive access to defectors from Iranian intelligence organizations who agreed to assist my investigations at great risk to themselves and to their networks inside Iran.

I was particularly blessed to receive assistance from a network of Special Forces operators, both U.S. and foreign, including individuals who worked on the ground in Benghazi and who wanted to share their experience and insights to ensure that I got this story right.

Some of the sources I relied on for information and analysis I can name. Representatives Darrell Issa, Jason Chaffitz, Ed Royce, James Lankford, Martha Roby, Louis Gohmert, Trent Franks, and their staffs were particularly helpful. So were Representative Frank Wolf, who led the charge to create a select committee to investigate Benghazi; Senator Jim Inhofe; and former House Intelligence Committee chairman Representative Pete Hoekstra. Senator Dianne Feinstein and her Republican colleague, Senator Saxby Chambliss, produced the most far-reaching report on Benghazi to date. Many other members of Congress from both parties also worked hard to get at the truth and deserve the appreciation of all Americans.

A special thanks to Lieutenant Colonel Andrew Wood, who headed the security detail at the U.S Embassy in Tripoli; to Rear

Admiral Richard Landolt, Director of Operations (J3) at AFRICOM; to former CIA officers John Maguire, Dewey Clarridge, Bob Baer, Clare Lopez, Larry Johnson, Wayne Simmons, Kevin Shipp, and Gary Bernsten; to Charles Woods, father of Ty Woods; to Colonel Dick Brauer, Captain Larry Bailey, and other members of the organization Special Operations Speaks; to Charles and Mary Ann Strange and to Billy and Karen Vaughn, parents of SEAL Team 6 members who perished on *Extortion 17* in Afghanistan, and to Larry Klayman and Dina James, who introduced me to them; to Chris Farrell of Judicial Watch; Roger Aronoff of Accuracy in Media; Major General Paul Vallely, Lieutenant General Tom McInerney, and Admiral James ("Ace") Lyons of the Citizens Commission on Benghazi; Victoria Toensing; former Libyan ambassador Ali Aujali; Simon Henderson of the Washington Institute for Near East Policy; researcher Tom Anderson with the National Legal and Policy Center; Sebastian Gorka, Walid Phares, Patrick Sookhdeo, and Tawfiq Hamid; former UN official Salim Raad; Maurice Botbol, Bernard Lugan, Jean-Louis Bruguiere, and Michel Garfunkiel in France; and Steve and Shoshana Bryen, always a source of insight and wisdom. And a special hat tip to the extraordinary reporting of Catherine Herridge at Fox News, Sharyl Attkisson at her former employer CBS News, Jerome Corsi at WorldNetDaily.com, and to bloggers Walid Shoebat, Cynthia Farahat, and Raymond Ibrahim.

Others I cannot name. You know who you are. It is my honor to say that, together, we serve the cause of freedom.

My family as ever have provided a finely tuned sounding board, while sharing the frustration of confronting frequent lies, obfuscation, and false leads that surround this story. I can only guess what our next adventure will be. I am blessed to be able to share your lives.

DARK FORCES

Prologue

Four dead Americans, one missing general, and thirty missing Special Operations and intelligence personnel from a black site in Libya . . .

A congressman so terrified of compromising classified U.S. government operations that he interrupts a public hearing three times in an attempt to prevent witnesses from talking . . .

Family members of SEAL Team 6 outraged at Pentagon stonewalling over an ill-fated operation that cost the lives of seventeen of their loved ones in Afghanistan, where U.S. military helicopters were getting shot down by Stinger missiles provided by the CIA to Islamic groups in Libya . . .

What really happened in Benghazi?

We know one thing for sure: The story concocted by the Obama administration on the night of the attack, of a demonstration caused by an Internet movie that went out of control, bears no resemblance to the truth. It's not even close.

So, why all the lies?

No Americans died during Watergate or as a result of Iran-Contra. That is what makes the attacks of September 11, 2012, in Libya the deepest, the darkest, and the dirtiest political scandal of recent American history.

It's a story of clandestine arms deliveries by the United States and its allies to Libya that wound up in the hands of Islamist guerillas allied with al Qaeda. It's a story of a romantic diplomat, in love with the

Middle East and with a mystical version of Islam, who gets caught up in a whirlwind beyond his comprehension or control. It's a story of bald-faced lies, heroic acts, and the deepest corruption.

This is the true story of what happened in Benghazi.

MISSING MISSILES

On August 25, 2012, a Libyan fishing boat, *Al Entisar*, docked in the southern Turkish port of Iskenderun, where Libyan Islamists unloaded 400 tons of weapons and military supplies they had brought from Benghazi.

They had purchased the weapons from stockpiles left over from the fight against Muammar Qaddafi, with funds provided by U.S. allies Saudi Arabia and Qatar.

The arrival of so many weapons created havoc between a local Islamist supposed charity, which claimed control over the goods, and brigades of the Free Syrian Army, which said it needed them on the battlefield.

"Everyone wanted a piece of the ship," said Suleiman Hawari, an Australian-Syrian working with the ship's captain from Benghazi. "Certain groups wanted to get involved and claim the cargo for themselves. It took a long time to work through the logistics."

The infighting among rival rebel commanders reached a feverish pitch, and caught the attention of the Turkish authorities and their allies. Because, as the rebels argued over who would get what, word leaked out that weapons were on the ground, and foreign journalists started asking questions.

On September 2, CIA director David Petraeus made an unannounced trip to Ankara hoping to straighten out the mess. He was worried by reports that the shipment included portable surface-to-air missiles, known as MANPADS, deadly weapons the United States was desperately trying to collect because of the threat they posed to civilian airliners.

Down in Iskenderun, the Libyans were boasting to journalists that they were going to be the kingmakers in Syria and would shoot down Syrian Air Force jets and helicopters.

Hillary Clinton was just as worried as Petraeus, since she had been a big supporter of the secret arms pipeline to the Libyan rebels that helped them get rid of Qaddafi. Among those weapons, my sources reveal, were 400 CIA-supplied Stingers and 50 launchers (see chapter 5). These were an upgraded version of the deadly missiles the Reagan administration supplied to the Afghan mujahideen that helped them to defeat the Soviet army in the late 1980s.

If word got out that U.S. MANPADS—or even Russian-made missiles from Qaddafi's looted stockpiles—were on the loose in Syria, there would be hell to pay. The Libyans were just too chaotic, too disorganized, and too damn talkative. Someone had to get them back on the reservation.

The man who knew the militias and tribal elders in Benghazi best was not a spy, but a diplomat: U.S. Ambassador to Libya J. Christopher Stevens.

Stevens knew the rebel leaders. He had broken bread with them as the State Department's special envoy to Benghazi in the heady days of the anti-Qaddafi uprising in 2011. So, in early September 2012, Hillary Clinton instructed him to travel from Tripoli to Benghazi to see what he could do.

This book is a narrative of the events that led up to that deadly decision, as well as its consequences. It sets out the facts as we have come to learn them, reveals new information, and dispels quite a number of rumors that have inevitably clouded the picture of what happened that night, and why.

THE ATTACKS

Ambassador Stevens arrived in Benghazi on September 10, 2012. Everyone knew the situation on the ground was dicey. Indeed, just a few days earlier, Stevens himself had cabled to Washington that the Benghazi Mission was on "maximum alert," because of an intensifying series of anti-Western terrorist attacks (see chapter 12).

As soon as he arrived at the Special Mission Compound (often, but inaccurately, referred to as the "Consulate"), he was briefed by the State Department's Regional Security Officer on new evacuation procedures in case of emergency.

His next stop was to visit a building known as "the Annex," the top-secret headquarters of the CIA team that had been training the Libyan rebels and now was assisting the MANPADS collection effort. There he was briefed about the weapons shipment to Turkey that was giving everyone such headaches.

That evening, he dined with tribal elders at a local hotel. Despite efforts by his security detail to keep the meeting quiet, the local press showed up. The word was out: The U.S. ambassador was in town.

The next day was the eleventh anniversary of the September 11, 2001, terror attacks on America. It began with an eerie warning that went unheeded.

At around 6:45 AM, one of the local unarmed security guards at the mission noticed that a Libyan policeman had parked his car by the front gate and climbed to the roof of a building under construction just across the street, from which he was taking photographs inside the mission's thirteen-acre walled compound.

The information so alarmed the State Department Regional Security Officer that he drafted letters of complaint to the Benghazi police chief and to the Libyan Ministry of Foreign Affairs, to report the troubling surveillance by someone in police uniform in an official police car.

However, Ambassador Stevens pursued his mission undeterred, meeting with local political contacts, the head of a Turkish shipping company in Benghazi, and, ultimately, with the Turkish consul general in Benghazi, Ali Sait Akin. According to the timeline provided by the State Department, Stevens escorted Akin to the front gate of the walled compound at around 7:40 PM.

No one noticed anything out of the ordinary. However, unbeknownst to Stevens or his meager security detail, he was being watched. His emergence on the street gave a crucial—and unexpected—"eyes-on" confirmation of his presence in the compound to the man who masterminded the attacks that night.

I will name that mastermind and his accomplices. Their presence in Benghazi that night, and for several months beforehand, has never been publicly revealed.

The first wave of attacks began at 9:42 PM and happened so quickly that the Mission complex was a burning wreck and the ambassador missing within twenty minutes. Reinforcements from the CIA Annex arrived roughly forty-five minutes after the attack began. It took them another fifteen minutes to fight their way into the compound and rescue the State Department security officers who had hunkered down without ever firing a shot. Together, they began the search for the missing ambassador and State Department Communications Officer Sean Smith.

It's been said repeatedly that the initial attack occurred so fast that nothing could have been done to save Ambassador Stevens or Sean Smith. However, my sources, and a careful examination of the official record, show that that is not true. Tyrone Woods, who died later that night defending the Annex, was told three times to stand down by his CIA chief of base during the first critical twenty-two minutes of the attack, even though he had assembled an immediate reaction force of heavily armed Special Forces troops who were ready to roll.

They were a five-minute drive from the diplomatic compound. Had they arrived at the beginning of the attack, just as the fires were being lit, there's a good chance that they could have saved the ambassador and Sean Smith.

Seven hours later, a precision mortar attack on the CIA Annex killed former Navy SEALs Tyrone (Ty) Woods and Glen Doherty in less than a minute. As I will reveal in chapter 14, the men who fired those mortars were trained and commanded by officers from the Quds Force, the overseas terrorist battalion of Iran's Revolutionary Guards Corps (IRGC).

Benghazi was a state-sponsored terrorist attack, carried out on orders from the leadership of the Islamic Republic of Iran.

QUESTIONS

Many questions about the Benghazi debacle remain unanswered, despite a multitude of congressional hearings and media investigations:

Why did Washington turn down repeated requests for additional security at the Benghazi diplomatic compound? Why did Secretary of State Hillary Clinton demand that it remain open, despite multiple, specific threats to the compound—even as other countries and international organizations were fleeing Benghazi because of intensifying, jihadi violence, in addition to specific threats to the U.S. compound?

Why weren't reinforcements sent that night from Croatia, where a fifty-man U.S. Army C-110 counterterrorism/hostage rescue team was engaged in a military training mission? Or from Naval Air Station Sigonella in Sicily, a 450-mile flight away?

Why wasn't President Obama personally engaged in coordinating the response?

Why did Hillary Clinton and John Brennan decide to stand down the State Department–led Foreign Emergency Support Team (FEST), an extraordinary operational unit, which was standing by and whose main purpose was to rescue U.S. diplomats under attack?

Why were no precautionary measures taken anywhere in the world to protect U.S. assets and facilities on the anniversary of the September 11 terrorist attacks?

Who concocted the outrageous cover story that the attack was a spontaneous protest over an Internet movie, when, in fact, there were no reports of a protest, but of a massive and well-coordinated attack?

As Congress continues its investigation into the Benghazi attacks, we will learn a lot more about these issues.

However, the big question not being asked is this: Why even bother sending a top diplomat to Benghazi when the State Department and the CIA knew how dangerous it was? The answer, I believe, reveals the utter cynicism of Hillary Clinton and her reckless disregard for the men and women who served under her command.

In Watergate, the big questions were: What did the president know, and when did he know it?

In Benghazi, this became: What did Hillary Clinton and Barack Obama know about the covert arms transfers to the Syrian rebels, and when did they know it?

In chapter 12, I reveal for the first time evidence that an illegal covert action was being run out of the White House.

U.S. POLICY SHIFT

The Benghazi attacks were the culmination of a dramatic shift in U.S. policy, set into motion by President Obama in the weeks and months after he took office in 2009.

His famous tilt to the Muslim world, aimed at convincing Muslim leaders and public opinion that the United States was not at war with Islam, achieved results far beyond the president's wildest expectations.

It convinced Muslim leaders that the United States had lost its resolve and was not to be taken seriously. It convinced them that the United States was weak.

It began with President Obama's overture to the Islamic Republic of Iran in 2009, and his stubborn refusal to hear the calls from millions of pro-freedom demonstrators in the streets of Iranian cities, begging for U.S. support. Rather than help the forces of freedom, President Obama sought a deal with the radical leadership of Iran's Islamist regime, a deal that is continuing to play out today.

As the Arab Spring took off in mid-January 2011, spreading from Tunisia to Egypt and finally a month later to Libya, the United States quickly discarded longtime allies to embrace forces aligned with the Muslim Brotherhood.

In Tunisia, we dropped the pro-Western Zine al-Abidine Ben Ali in favor of exiled Muslim Brotherhood leader Rachid Ghannouchi.

In Egypt, we ditched long-term U.S. ally Hosni Mubarak, who had kept the peace with Israel for thirty years, in favor of a volatile coalition of Islamist organizations ultimately dominated by the Muslim Brotherhood.

In Libya, the United States joined with Muslim jihadis, many of whom had close ties to al Qaeda, to overthrow a Qaddafi who had voluntarily ended his WMD programs and cut off his support to international terrorist groups, in hopes of becoming a friend of the West.

Some of the jihadis we supported in Libya had been captured fighting against us on the battlefields of Afghanistan or Pakistan and sent to Guantánamo Bay.

To nurture the struggle against Qaddafi, the CIA set up shop in Benghazi in early 2011 to arm and train the rebels, both directly and through proxies. Once Qaddafi was overthrown in September 2011, this covert operation shifted gears to aid similar groups in Syria, the newest front in the Muslim Brotherhood war on secular Mideast regimes.

The Benghazi arms pipeline went awry from the very start. One of the partners the United States engaged was Qatar, a tiny emirate in the Persian Gulf best known for its sponsorship of Al Jazeera, which critics refer to as Jihad TV.

My sources relate an astonishing incident in the desert of northern Chad, where a French military patrol confronted a convoy led by Qatari special forces officers that was bringing Stingers and other advanced weapons to the Libyan rebels at the start of the fight against Qaddafi. When the French officers sought to intercept it, they were told by Paris to stand down, because the shipment had been approved by Washington.

Once it became clear that those weapons were making their way to al Qaeda–related groups in Libya, the Obama White House desperately sought to clean up the mess and keep it from becoming public.

To do so, they sent members of the National Security Staff (formerly known as the National Security Council) to Libya on operational missions to negotiate arms buybacks from Libyan rebel leaders, in direct violation of the National Security Act of 1947. (Congress threatened to impeach President Ronald Reagan for similar abuse of his authority during the Iran-Contra scandal twenty-five years earlier.)

Dark Forces will examine in detail the Obama administration's covert operation to supply weapons to the Libyan rebels during the civil war, knowing full well that many of the rebel leaders had ties to al Qaeda.

It will show how White House counterterrorism advisor John Brennan (now CIA director) was personally in charge of U.S. covert operations in Libya, including a Fast and Furious–style gun-walking operation that allowed 800 Russian-made surface-to-air missiles from Qaddafi's arsenal to reach al Qaeda groups in Africa and beyond.

Some of these missiles were used to shoot down U.S. combat helicopters in Afghanistan. Others made their way into the Sudan and

Gaza, prompting Israel to launch air strikes to take them out. Still others were walked into the Sinai Peninsula under the control of al Qaeda groups close to deposed Egyptian president Mohamed Morsi. Photographic evidence I will present in this book shows that some wound up in the hands of Syrian rebels.

I believe Congress needs to investigate these missile-walking operations before more American lives are lost.

THE IRANIAN GAMBIT

The United States and Iran have been at war since 1979. Americans often are lulled in believing that the war is over, or that some cease-fire has been reached when the bodies stop piling up. That happened for a few years in the early 1990s, once Iran released U.S. hostages in Lebanon and stopped attacking our embassies and our military.

However, the Iranians take a longer view, especially the Quds Force, the overseas terror battalions of Iran's Revolutionary Guards Corps. They wait until the next opportunity to strike a deadly blow, preferably in such a way as to leave no trail back to Tehran. I detailed many of these covert Iranian terror attacks in an earlier book, *Countdown to Crisis: The Coming Nuclear Showdown with Iran.*

So, I was not surprised when I began hearing anecdotal evidence of an Iranian involvement in Libya.

Americans on the ground in Benghazi during the early days of the anti-Qaddafi rebellion were already reporting on the Iranian presence at that time (see chapter 6). Once Benghazi became the hotbed for arming the Syrian rebels—Iran's deadly enemies—they expanded their activity dramatically.

By June 2012, the CIA was briefing the U.S. Embassy in Tripoli on Iran's efforts to fund, train, and equip jihadi militias in Benghazi, including Ansar al-Sharia, the group most frequently blamed for carrying out the attacks. Those briefings—with redacted headers—were incorporated into a damning report released by the Senate Select Committee on Intelligence in January 2014.

But the CIA fatally underestimated Iranian resolve. In chapter 12, I reveal the details of the ingenious ploy the Iranians devised to lull the CIA chief of base into believing the danger they posed was over, along with the names of the Iranian operatives in charge of the attacks and the mechanism they used to finance them.

Put simply, the CIA got played. The story of how the Iranians outsmarted them will not be remembered as one of the agency's finest hours.

Benghazi will "go down in history as the greatest cover-up. And I'm talking about the Pentagon Papers, Iran-Contra, Watergate and the rest of them," predicted the top Republican on the Senate Armed Services Committee, Oklahoma Senator James Inhofe.[1]

A bipartisan report from the Senate Select Committee on Intelligence concluded in January 2014 that the Benghazi attacks "were likely preventable." The intelligence community "produced hundreds of analytic reports in the months preceding the September 11–12, 2012, attacks, providing strategic warning that militias and terrorist and affiliated groups had the capability and intent to strike U.S. and Western facilities and personnel in Libya," the senators concluded.[2]

The real scandal of Benghazi did not begin on September 11, 2012, but years earlier. This book will tell that story.

1

FROM TERRORIST TO FRIEND

It's hard to believe that a U.S. president could once look to Libya as a success story. To most Americans, Libya has become synonymous with chaos, a wild and dangerous place where, as in Iraq, American dreams of democracy went to die. President Obama's hesitation to use U.S. military might against Qaddafi in early 2011 prompted his political opponents to accuse him of leading from behind, even though U.S. aircraft, U.S. airmen, and U.S. taxpayers bore the brunt of the NATO-led no-fly zone over Libya during the first few months of the conflict, at a cost to taxpayers of $550 million for the first two weeks alone.

The debacle in Benghazi and the image of U.S. weakness it projected to the world only heightened this sense of futility. It prompted at least one prospective Republican presidential hopeful, Senator Rand Paul, to argue in favor of a broad American pullback from around the world and a major military downsizing. It also will undoubtedly become a campaign issue should former secretary of state Hillary Clinton enter the 2016 race.

As they contemplated a similar U.S. military involvement in Syria's bloody civil war over the summer of 2013, politicians of both parties became increasingly worried that U.S. military aid could fall into the hands of jihadi terrorist groups, such as the ones who benefited from the U.S.-Qatari arms pipeline to the Libyan rebels who ousted Qaddafi.

But one U.S. president *could* look to Qaddafi's Libya as a success story. And it's a story that has never been fully told.

THE MAD DOG OF TRIPOLI

For several weeks in March 1986, U.S. and Libyan warships and combat jets had been dancing toward war in the Gulf of Sidra, the giant bay stretching from Misrata, just outside of Tripoli, all the way to Benghazi. Qaddafi drew a straight line across the Mediterranean between those two points and claimed everything south of it as Libyan territorial waters. He dared anyone—meaning the United States—to cross this "line of death."

Qaddafi's exclusion zone included waters seventy miles from the nearest Libyan coastline, far beyond the twelve-nautical-mile limit recognized as the international standard. President Ronald Reagan asserted the right of the United States and its NATO allies to conduct naval operations in international waters and on March 23, 1986, ordered three U.S. carrier battle groups—USS *America,* USS *Coral Sea*, and USS *Saratoga*—with 225 aircraft and some thirty warships, to cross Qaddafi's double-dare line. It was a formidable armada only a madman would try to oppose.

U.S. warships crossed the line of death twice that year without incident. However, on March 24, 1986, the Libyans responded, sending missile boats and MiG-23 Flogger and MiG-25 Foxbat fighters jets to counter the Americans. In every engagement, the Americans blew away their Libyan counterparts or forced them to flee before the shooting began. The Americans sunk two of Qaddafi's French-built Combattante II missile boats, a Soviet-built corvette, and killed thirty-five

Libyan sailors. Qaddafi was humiliated and vowed revenge, publicly calling on Arabs everywhere to kill Americans.[1]

Operation Prairie Fire appeared to be a resounding success, projecting precisely the image of a strong America that President Reagan had worked so hard to build after the "malaise" of the Carter years.

Just one week later, Qaddafi took his revenge. America was still vulnerable, and he proved it with cowardly skill.

On April 2, 1986, a member of the Abu Nidal terrorist group, which was then based in Libya and armed by Qaddafi, placed a bomb made with Semtex H plastic explosive under the seat of TWA flight 840 as it was on approach to the Athens airport on the short flight from Rome. Because of the relatively low altitude at the time of the explosion, the plane did not explode. But four passengers—all Americans—were sucked out of the hole in the fuselage. Pilot Pete Peterson was welcomed as a hero for his skill in safely landing his badly damaged aircraft at the Athens airport. While Qaddafi quickly announced he had nothing to do with the attack, it was well known that the Abu Nidal organization were his protégés.

Just three days after TWA 840, a bomb exploded in the early morning hours at La Belle discotheque in West Berlin, Germany, a favorite haunt of American soldiers. Sergeant Kenneth T. Ford, twenty-one, and a twenty-nine-year-old Turkish woman, Nermin Hannay, were sitting near the disc jockey's booth and died instantly. Sergeant James E. Goins, twenty-five, died two months later of his injuries. Another 230 people were wounded, including seventy-nine American servicemen, many of whom lost limbs or were permanently disabled. The terrorist had placed a bomb filled with shrapnel and two kilograms of Semtex beneath a table by the dance floor, then left the scene before it went off. That was ten times the amount of the deadly plastic explosive used on TWA 840.

President Ronald Reagan pointed the finger at Colonel Muammar Qaddafi, the outlandish Libyan dictator who portrayed himself as a one-man army out to defeat American and Zionist "imperialism." This was the same Qaddafi who often elicited smiles—even smirks—because of his flair for the exotic, dressing alternately in designer capes and Bedouin hats, or in outlandish military uniforms that bore a greater resemblance to Sergeant Pepper than to Sergeant Shaft.

Reagan wasn't amused when he took the podium at a White House press conference on April 9, 1986. Notorious left-wing reporter Helen Thomas, who was born in Lebanon and later in life revealed herself to be a rabid anti-Semite, asked Reagan if U.S. policies weren't to blame for the attacks on America.

"Well, we know that this mad dog of the Middle East has a goal of a world revolution," Reagan said. "And where we figure in that, I don't know. Maybe we're just the enemy because—it's a little like climbing Mount Everest—because we're here. But there's no question but that he has singled us out more and more for attack, and we're aware of that."[2]

What Reagan couldn't say was that the NSA had intercepted communications between Qaddafi himself and intelligence officers working out of the Libyan embassy in East Berlin, ordering them to carry out the disco attack in a manner "to cause maximum and indiscriminate casualties."[3]

Less than one week later, Reagan ordered air strikes on Libya, code-named Operation El Dorado Canyon, widely seen as an attempt to assassinate Qaddafi. In addition to Libyan military barracks, air defense sites and air bases in Tripoli and Benghazi, U.S. jets hit a residential compound used by Qaddafi and his family, killing his three-year-old adopted daughter, Hana, and wounding his youngest son, Khamis. The boy grew up to command the notorious Khamis Brigade, the best-trained and best-equipped unit in the Libyan armed forces, responsible for several military victories and atrocities against rebel forces in the 2011 civil war.[4]

Contrary to Reagan's expectation, the air strikes didn't put the "mad dog of the Middle East" out of the terrorism business. Just four days before Christmas 1988, as Reagan was preparing to hand over the White House to President-elect George H. W. Bush, Pan Am Flight 103 departed Heathrow Airport headed for New York. Thirty-eight minutes after takeoff, a block of Semtex plastic explosive hidden in a suitcase ripped a giant hole out of the Boeing 747, splitting the plane into pieces. The cockpit of the *Maid of the Seas* landed virtually intact near a churchyard in Lockerbie, Scotland, and became an iconic image. One wing of the aircraft, filled with jet fuel, burst into flames on impact, killing eleven people on the ground. With the 259 passengers and crew, 189 of them American, it was the deadliest terrorist attack on America since Iran hit the U.S. Marine barracks in Lebanon in October 1983.

While the CIA and other U.S. government agencies initially suspected Iran of carrying out the Lockerbie attack, the discovery of a fragment of the detonator used in the bomb ultimately allowed prosecutors to trace it through the Swiss manufacturer to a batch sold to a Libyan intelligence operative.

With the fall of the Berlin Wall in November 1989 and the collapse of the Soviet Empire, a new pro-Western government led by playwright Václav Havel came to power in Czechoslovakia. Havel startled the world in March 1990 by announcing that his communist predecessors had exported one thousand tons of Semtex plastic explosive to Libya. Here was an opportunity to unlock some of the Soviet Bloc's best-kept secrets, so I traveled to Prague to find out more. I wasn't disappointed.

Although the actual amount turned out to be closer to seven hundred tons, it was still enough to keep terrorists busy for the next one hundred fifty years, as Havel said. Officials at Omnipol, the arms export emporium of the former communist regime, told me that Libya accounted for 98 percent of all their Semtex sales. Qaddafi may have used some of the explosives to blast huge tunnels for his Great Man-Made River project, the ostensible end use for the sales. But he also re-exported the deadly plastic explosive to every terrorist group imaginable, including the Irish Republican Army (IRA), Ahmed Jibril's PFLP-GC (a prime suspect in Pan Am 103), the Lebanese Armed Revolutionary Faction (LARF) of George Ibrahim Abdallah, the Secret Army for the Liberation of Armenia (ASALA), the Organization of 15 May-Abu Ibrahim (based in Iraq), and Yasser Arafat's praetorian guard, Force 17.[5]

Crippling international sanctions and a travel ban imposed by the United Nations in reprisal for the Pan Am 103 attack kept ordinary Libyans in a box throughout the 1990s, but they did little to keep Colonel Qaddafi at bay. He continued to provide arms, plastic explosives, money, and training to terrorist groups around the world and, in the late 1990s, turned his sights to acquiring a nuclear weapon.

For three successive American presidents—Reagan, Bush 41, and Clinton—Qaddafi remained a deadly pariah, who seemed to revel in his "mad dog" image.

After the September 11, 2001, attacks on America, all that began to change.

THE FALCONER

In a way, it was Osama bin Laden who first pushed Qaddafi toward the West. But it took a fellow falconer to close the deal.

Libyans who had gone to Afghanistan to fight the great jihad against the Soviet Union came home to wage jihad against Qaddafi in the 1990s. In 1994, they stormed a prison in Benghazi to liberate fellow Islamists and declared their allegiance to bin Laden. After nearly eighteen months of running gun battles with regime forces, they announced the formation of the Libyan Islamic Fighting Group (LIFG) in September 1995. Their initial communiqué, issued by Libyan exiles granted political asylum in London, called Qaddafi's rule "an apostate regime that has blasphemed against the faith of God Almighty" and declared its overthrow to be "the foremost duty after faith in God."[6]

In February 1996, an LIFG member threw a bomb beneath Qaddafi's motorcade that killed several of his bodyguards. A former MI5 officer, David Shayler, later told Britain's *Observer* newspaper that British intelligence financed the assassination attempt "to the tune of $160,000."[7] In November, another LIFG operative tossed a grenade at Qaddafi while he was visiting the desert town of Brak.

Qaddafi the terrorist had become a target of bin Laden's terrorist gang. So when the 9/11 attacks hit America, Qaddafi condemned bin Laden publicly, asked Libyans to donate blood, and said the United States was justified to retaliate.

In his account of Qaddafi's turnaround, former CIA director George Tenet called Qaddafi's 9/11 statement "an interesting sign," and felt it was a good time to revive an intelligence back channel established two years earlier by the second in command of the CIA's counterterrorism center, Ben Bonk.[8]

To get the ball rolling, the White House quietly designated the LIFG as an international terrorist organization on September 25, 2001, and froze their assets in the United States. In mid-October, Tenet dispatched Bonk to London for a face-to-face meeting with Qaddafi's intelligence chief, Musa Kusa, at the home of Saudi ambassador Bandar bin Sultan. The Libyan handed over information on LIFG members that helped the United States identify several top deputies to Osama bin Laden when they were picked up in counterterrorism raids in Pakistan and Egypt later that year.[9]

However, when Bonk moved on to another CIA job a few months later, the back channel languished. It fell to a lifelong British spy named Mark Allen to revive it.

Allen was head of the counterterrorism division of MI6 (formally known as the Special Intelligence Service), putting him a notch higher in the bureaucracy than Bonk. Allen had learned Arabic at Oxford decades earlier and spent a summer as a young man crisscrossing the Jordanian desert on a camel he had purchased at a local souk. Sitting on his haunches sipping bitter coffee with Bedouins, he fell in love with their simple lifestyle. At twenty-eight, he published a book on falconry with a preface by Wilfred Thesiger, a modern-day Lawrence of Arabia. He later went to hone his language skills at Britain's fabled spy school in the mountains above Beirut, the Middle East Centre for Arabic Studies. How many American spies could boast of such training?[10]

On what he claimed was a personal initiative, Allen began talking to Musa Kusa separately after that initial meeting at Prince Bandar's London mansion.

Allen knew that Qaddafi was seeking an exit from sanctions, which had been suspended, but not removed, two years earlier when Libya handed over two intelligence agents found guilty in a Scottish court for their role in the Pan Am 103 attack. And he knew that Musa Kusa had Qaddafi's ear. So, he invited him one afternoon in late 2001 to the Travellers Club, a posh London watering hole frequented by diplomats, millionaires, and spies.

The 9/11 attacks have changed the world, Allen began. That's why my American colleague sought you out. It's no longer possible to conduct murder and mayhem and think you can retreat back home and no one will find you. Look what's happening to the Taliban in Afghanistan.

If you want this relationship to develop, you've got to put Lockerbie behind you, he said. That's the only way to bring Libya back into the community of nations. You'd be surprised how quickly we could discover a mutual interest in fighting al Qaeda and their Islamist friends.

We have more files on these guys, Kusa said. Libyans who went to Afghanistan to fight the great jihad. I could see about making them available.

A few weeks later, he brought files on hundreds of LIFG terrorists—and more. Why not come to Tripoli and speak with the Guide [Qaddafi] directly, Kusa suggested.

So it was in early 2002 that Mark Allen, top British spy and falconer, the sport of Arab royals, traveled to Libya to meet with Colonel Qaddafi, the self-proclaimed champion of the common man. Qaddafi received him in his giant tent complex in Tripoli, the Bab al-Azizia Barracks. Allen was a good pitchman. He knew from long experience to appreciate the bitter Bedouin coffee Qaddafi offered him, letting the *boab* refill the tiny cup three times then gently shaking it to signal he was satisfied. He nodded appreciatively at the lavish cushions on the couch, embroidered with sayings from Qaddafi's Green Book.

The two men spoke mostly in Arabic, although Qaddafi understood much more English than he liked most visitors to know. Allen made the same pitch he had made in London to Qaddafi's spy chief, adding that if they could put Lockerbie behind them, Britain would help get the United Nations sanctions removed for good.

I hear that some of your advisors want to open Libya up to Western investors and technology, Allen said.

Have they told you this? Qaddafi gave his intelligence chief a sharp look.

Allen laughed. Nothing more than they have been quoted saying in your own newspapers.

The key to getting the sanctions lifted permanently was compensation for the Pan Am 103 victims, Allen went on. I understand that you have agreed to this. But the Americans are insisting that you go a step beyond that and publicly renounce terrorism as well.

Qaddafi played along, as though he had expected this. We are victims of terrorism just like you, he said. Al Qaeda tried to kill me twice! If we agree to the principle of paying compensation for Lockerbie, it is because we want to put this file behind us.

The diplomat who recounted this exchange to me summarized it bluntly. "In other words, Qaddafi wasn't admitting guilt. He was being practical." [11]

That August, British foreign minister Mike O'Brien followed in Allen's footsteps, traveling to Qaddafi's summer residence in a luxuriously appointed white tent complex in the desert outside Sirte, Qaddafi's home town. O'Brien became the point man in the Lockerbie negotiations for the Foreign and Commonwealth Office, along with Assistant Secretary of State William Burns on the American side. Qaddafi appointed Foreign Minister Mohamed Abdelrahman Shalgam as the official head of his negotiating team, even though to the Americans and the Brits it was clear that his eldest son, Saif al-Islam, was really in charge. "We called the Lockerbie talks 'the London channel,' " one participant said. [12]

In September 2002, British Prime Minister Tony Blair sent a personal letter to Qaddafi hectoring him for supporting Robert Mugabe in Zimbabwe, just as international pressure was mounting on the aging dictator to resign. He also summoned Qaddafi to abandon his WMD programs, a new demand put on the table at the urging of U.S. President George W. Bush.

In December, Qaddafi sent a rambling eleven-page reply, "essentially saying, why are you picking on me?" a senior British official told me. The established nuclear powers had thirty-thousand warheads, and Israel had more. Anything Libya might have would be a drop in the bucket, Qaddafi wrote.

Blair was frustrated, and shared it with Bush. Diplomacy alone was failing to get Qaddafi to break with his deadly past. More was needed to bring him around.

On March 19, 2003, the United States, Britain, and a coalition of some forty countries launched hostilities against Iraq, beginning with a shock-and-awe air campaign that set Baghdad aglow like the land of the midnight sun. Thanks to satellite broadcasting, the air strikes were visible for all the world to see. "Qaddafi watched Baghdad burn on CNN," one of his close advisors told me. The next day, he instructed Musa Kusa to head back to London, this time without Shalgam. Qaddafi wanted to talk to the CIA.

Qaddafi was convinced the Americans would find no active nuclear weapons program in Iraq. But, if they looked, he knew they would find one in Libya. He was worried that after Saddam Hussein, he would be next.[13]

CIA Director George Tenet thought it was time to take another pass at Qaddafi. But, with the war in Iraq, he wanted more than just Qaddafi's personal enemies list. He wanted Qaddafi's WMD programs, not just the chemical weapons, which everyone knew about. He wanted the nukes, the missiles, the whole works.

So, Tenet sent his deputy director of operations, Steve Kappes, to London to meet with the Libyans a few days after the Iraq War began. Kappes later told his favorite journalists that he single-handedly won Qaddafi over, and reported back to President Bush in person, to prevent news of the Qaddafi opening from "leaking" to neoconservatives who weren't keen on rehabilitating the Libyan strongman. After one of his encounters with Qaddafi, Kappes said that he found himself actually starting to like the man, even though he knew he had slaughtered so many innocents.[14]

Tenet reported in his memoir that Kappes was the messenger of the Bush administration policy, not the architect. After speaking to the president about the opening to Qaddafi, Tenet says he briefed Kappes and "put the project in his hands and got back to worrying about Iraq."[15]

For several months the two sides played poker, as Qaddafi tried to figure out how much the Americans really knew about his programs.

Behind the scenes, Qaddafi's son and his advisors in Tripoli were bending his ear, arguing that Libya's security would be enhanced, not reduced, by giving up the nuclear program. "We had no delivery system," a top Qaddafi advisor explained to me. "I told the Guide, if Libya were to start a nuclear war, our missiles won't even reach Malta. If the U.S. starts it, Libya will be erased from the map."[16]

Qaddafi still held back. "At the same time they carried on negotiations with us, they continued with their WMD programs," a British official privy to the negotiations told me.

Two months had gone by since that first meeting with the Americans. Kappes flew back to London to confer with Allen, then the two met with Kusa and Qaddafi's son on May 14, 2003. Saif al-Islam al-Qaddafi was clearly in the driver's seat, and he was far more direct and thoughtful than the files on him suggested. He spoke rapid-fire English, honed while writing (some said, plagiarizing) his graduate thesis at the London School of Economics. Gone were the days when he would keep a pet tiger at his palatial apartment in Vienna. Tutored by Musa Kusa in the affairs of state, he was now his father's negotiator-in-chief.[17]

The younger Qaddafi asked for a written guarantee of what the United States would offer if Libya gave up as-yet-unspecified WMD programs. Allen's reply, sent one month later, was a remarkable mixture of official diplomacy and the familiar.

"Greetings to Musa from Steve and Mark. We hope that you are well," the cable began. "Condoleez[z]a Rice has told CIA that if Libya proceeds with its proposal to destroy its WMD programmes, Libya would be welcomed back to rejoin the community of nations and all that goes with that . . ." He then outlined all the good things Britain and the United States would do for Libya to make that happen, including a visit to Qaddafi's tent by British Prime Minister Tony Blair, something Saif al-Islam made clear his father wanted.[18]

Kappes traveled to Libya with Allen for his first meeting with Qaddafi later that summer, where they pressed the Libyan leader for details on his weapons programs. Qaddafi hemmed and hawed. No matter how many times they asked for specifics about his chemical weapons stockpiles, his delivery systems, and his nuclear program, Qaddafi changed the subject. Who's going to guarantee my security if I announce that I am giving up these weapons? he asked.

Meanwhile, Libyan nuclear weapons scientists were rushing to complete their secret bomb program. It was almost as if Qaddafi wanted to have the bomb before he gave it up, so that he could get a better price.

For several years, the Libyans had been working with Pakistani bomb-maker A. Q. Khan and his far-flung Bombs R Us network of middlemen and suppliers. At the very moment Qaddafi was talking with the British and American spies, Ma'atouq Mohamed Ma'atouq (Qaddafi's "Minister of Bad Things," as the Americans later called him) placed an order with Khan's business partner in Dubai, a Sri Lankan named B. S. A. Tahir. The Libyan wanted parts for several thousand advanced P-2 centrifuges, so that they could make highly enriched uranium needed for the bomb.

Tahir sent the order to a machine shop in Malaysia, Scomi Precision Engineering, which he had set up for the clandestine bomb supply line a few years earlier. What none of them knew was that an employee at Scomi, a young Swiss engineer named Urs Tinner, had dual loyalties. As soon as he saw the order come in, he phoned a contact at the CIA.

The tip-off from Tinner allowed the CIA to track production of the Libyan centrifuges, and have eyes on as they were transported from the machine shop to a shipping company, packed into containers, and loaded onto a German-owned cargo ship, *BBC China*. U.S. spy satellites tracked the ship to Dubai, and later, through the Suez Canal. Just as it steamed out into the Mediterranean, a U.S. warship, working with Italian customs, intercepted the *BBC China* and forced it into the port of Taranto, Italy. They seized the five containers of Libyan centrifuge components, labeled "used machinery," on October 4, 2003. President George W. Bush outed the Malaysian company and others in the A. Q. Khan network in a dramatic speech at the National Defense University a year later.[19]

"It was a big shipment—the guts of what he needed," a U.S. official involved in the Bush administration's Proliferation Security Initiative told me. PSI was a newly minted program, promoted by Undersecretary of State for Arms Control John R. Bolton, to enhance cooperation among friends and allies of the United States to interdict shipments of WMD gear on the high seas. "That seizure broke the back of Qaddafi's nuclear weapons program. Without it, he would have had to go back to square one."

Kappes had a moment of panic when news of the *BBC China* seizure became public. He told everyone on Capitol Hill (and his preferred journalists) that it risked scaring the Libyans off, blowing the whole operation. But, in fact, the opposite occurred. As with the invasion of Iraq six months earlier, the additional pressure unlocked a door in Tripoli. Suddenly, Kusa told Allen they should come to Libya immediately to make the final arrangements for the inspection team.

On October 19, 2003, Kappes and Allen returned to Tripoli along with a fifteen-member technical team of U.S. and British weapons experts to map out the scope of the Libyan program. They flew into

Wheelus Air Base outside Tripoli on a CIA black aircraft, a Boeing 737 that was being used to render terrorists captured overseas to secret CIA prisons around the world.[20]

Saif al-Islam later told reporter Judith Miller that the initiative to give up WMD was his, not his father's, "an astonishing assertion that no diplomat believes."

And yet, that is very likely the case. Every step of the way, Qaddafi hesitated, stalled for time, played coy on the equipment and weapons he possessed. Saif al-Islam and other key advisors I interviewed later in Tripoli convinced Qaddafi that the game was up and the United States had him dead to rights.

Just to make sure the Libyans followed through on their pledges to open up their nuclear weapons facilities to the technical team, Kappes and Allen gave Musa Kusa a CD-ROM when they arrived in October that contained an audio recording of a February 28, 2002, discussion between Ma'atouq Ma'atouq, the Minister of Bad Things, and A. Q. Khan. It contained a candid review of Libya's entire nuclear weapons program. All the cards were now on the table.[21]

Before making his historic renunciation of his WMD programs on December 19, 2003, Qaddafi sought counsel from an unusual source— Ukrainian president Leonid Kuchma. During a private meeting at Qaddafi's tent, in Tripoli, he asked Kuchma how America had treated him when he gave up his nuclear weapons after the fall of the Soviet Union.[22]

The final event that sealed the fate of Qaddafi's nuclear weapons program took place on December 13, 2003, along the borders of the Tigris River just south of Tikrit, Iraq when soldiers of the 4th Infantry Division pulled a bearded, long-haired Saddam Hussein out of a spider hole hidden on a small farm. The Americans broadcast footage of Saddam's capture the next day. "When Qaddafi watched a U.S. medic probe Saddam's hair for lice and poke around his mouth, he turned white," a Qaddafi confidant told me in Sirte.

Until Saddam's capture, "We were still negotiating. Both sides were sparring back and forth," a British official involved in the talks over Qaddafi's WMD programs told me. "Things radically changed course after that." Just six days later, Qaddafi made his public announcement that Libya was giving up its WMD programs and had invited U.S. and British experts into his country to verify the dismantling of his weapons plants.

Leading Democrats, including Senator John Kerry, later argued that the Libyan case showed that diplomacy worked better in the war on terror than force. "If diplomacy was so effective," a Bush administration official involved in the weapons cleanup effort told me, "why did Colonel Qaddafi continue to procure equipment at the same time our diplomats were talking?"

The timeline of Qaddafi's concessions is crystal clear. The Lockerbie negotiations, which dragged on for a dozen years, were all about his economy. The WMD talks, which took nine months, were about his survival. Getting Qaddafi to renounce terrorism and give up his WMD programs was a huge victory for George W. Bush. Even today, very few Americans know just how successful power diplomacy can be.

THE VERIFIER

The woman put in charge of verifying that Qaddafi really came clean was an old hand at arms control. Paula DeSutter likes to tell the story of coming home from college to her native Alabama one summer and telling her grandmother that she was studying arms control.

"What's a matter, honey? Don't you like guns?"

"No, Grams. I don't like *big* guns," she said.

During the Cold War, DeSutter worked at the Arms Control and Disarmament Agency (ACDA) and helped draft the State Department's annual verification and compliance report. President George W. Bush brought her back in 2002 and made her assistant secretary of state for verification and compliance. She was sworn in that August and was thrilled to be back. They were going to kick butt.

Over the Christmas holiday, she and her team of verifiers put together a conceptual plan of the sites, the people, and the equipment they needed to see in Libya, in order to confirm that Qaddafi's stated intention of giving up his WMD programs was for real.

On New Year's Day 2004, she flew with Undersecretary of State John Bolton to London to resolve outstanding issues with their British counterparts before they sent the action team down to Libya.

There were moments of comedy mixed with the drama. Because U.S. laws prohibited any economic exchanges with Libya—even by U.S. government officials—one of her lawyers had to "bust a piggy-bank" at the U.S. embassy in London to get cash to buy plane tickets for the group. Once in Libya, American Express refused to cash their Travelers Cheques. Although they had gone to Libya to dismantle Qaddafi's secret nuclear weapons program and to take possession of his missiles and chemical weapons, AmEx wanted a special license from the Treasury Department's Office of Foreign Assets Control so they could spend money.

Two weeks later, team members had packed Libya's top-secret nuclear bomb designs into an oversized briefcase and flew back to Dulles Airport in Washington, where they had expected to discreetly deliver their precious cargo to DeSutter and her verifiers out in the parking lot.

Instead, Energy Secretary Spencer Abraham sent plainclothes guards wearing bomber jackets and packing heat to the international arrivals area, who greeted them in front of incoming passengers. It was definitely not discreet, DeSutter thought. The big black case was festooned with IAEA and Energy Department high-security seals. As they frog-marched her men out to the parking lot, they looked like a bunch of kooks out of *Dr. Strangelove*, she remembered thinking.

The sophisticated bomb designs were written in English and in Chinese. They were part of the nuclear bomb package that the inimitable Dr. Khan had sold the Libyans. As the verifiers went over the files, they realized that Dr. Khan had most likely sold the same package to the Iranians as well. It was so obvious that they hadn't seen it until now.

Iran and Libya had been feeding from the same trough, one of DeSutter's top aides told me. The Iranians would be guilty of proliferation malpractice if they didn't get the bomb design, too.[23]

As DeSutter told Congress that March:

> We removed several containers of uranium hexa-
> fluoride and centrifuges purchased from [the A. Q.
> Khan] network for the purpose of enriching uranium.
> We received detailed descriptions of the Libyan mis-
> sile research and development activities. We removed
> five Scud-C missile guidance sets, including their
> gyroscopes. And we began assisting the Libyans in
> the preparation of their Chemical Weapons Conven-
> tion declaration, and witnessed the first destruction of
> chemical munitions when three chemical munitions
> were destroyed.[24]

But was Qaddafi's conversion the real thing? Or was he just playing with the United States, trying to save himself from Saddam's fate? I had the opportunity to accompany a congressional delegation led by Representative Curt Weldon (R-PA) to Libya in March 2004 to find out. There I met with Qaddafi, his senior aides, and top British, U.S., and European diplomats and learned how the United States ultimately convinced Qaddafi not just to renounce terrorism, but also to give up his clandestine weapons program.

THE STIFF

Qaddafi had summoned his American visitors to the new capital in Sirte, his tribal homeland, where he planned to address the People's Congress. He liked to pretend that he held no position in the Libyan government; that he was just the "Guide" of the revolution, a behind-the-scenes grey eminence. That allowed him to disown his own government when it was convenient; and to surprise them, by taking them in a whole new direction no one had expected. As we soon learned, it could be quite a spectacle.

The seven American representatives were seated in the front row of the giant hall of the Libyan National People's Congress, just to the right of the rostrum. As I surveyed the six hundred members of his rubber-stamp parliament, I also saw some unusual faces: an Iranian cleric with his retinue of trim-bearded intelligence agents, come no doubt to make sure Qaddafi didn't spill the beans on their involvement in the Lockerbie attack; visitors from black Africa in their colorful tribal costumes; Central Asians in a variety of headgear; and Chinese in expensive European business suits.

Suddenly the rustling in the great hall stopped. From out of nowhere, a half-dozen Qaddafi girls appeared and took up position at all the exits. They were the Guide's famous personal security detail, trained in martial arts, and they alone would have been worth the trip. Gone were the blond East Germans Qaddafi used to employ as his personal bodyguard. Today's Qaddafi girls were Arab and black African, and wore camo and red berets. As I looked closer, I saw that all of them had fingernails at least an inch long, coated in a deep purple gloss that looked like congealed blood.

As a foreign diplomat remarked to me later, there was a design behind the ghoulishness. Because everyone in the hall had turned their attention to the unusual spectacle of the Qaddafi girls, no one saw the Guide himself whisk in from the wings. The next thing we knew, he was seated at the table up on the stage and was talking to us in a halting whisper.

The performance, I learned later, was vintage Qaddafi, but with a twist. Instead of a long, rambling diatribe denouncing America and the West, Qaddafi embarked on a lengthy justification of his decision to abandon terrorism and embrace the West. And his message was unequivocal: Yesterday's enemies were about to become Libya's friends—or so he hoped.

"At first, I was just listening to the speech," California Democrat Susan Davis told me afterward, "but what he was saying was so amazing that I started writing it down so I could report to my constituents. I took twenty-four pages of notes."

It was a bizarre performance by any standard. For the first thirty minutes or so, Qaddafi hardly spoke above a whisper. The words we heard through our headsets appeared to be rambling, almost incoherent. He talked about Rousseau, Tolstoy, and Socrates, comparing himself to each. After he had warmed up, he launched into a brutally self-critical account of Libya's past support for terrorist movements, and announced that those days were gone.

Libya had helped the Irish Republican Army, the Palestine Liberation Organization, and the African National Congress. Now all of them had made their separate peace with their former enemies, leaving Libya behind to fight on. "Are we more Irish than the Irish?" he said. "Are we more Palestinian than the Palestinians . . . ? How can

[Arafat] enter the White House and we not improve our relations with the United States?"

It was time to turn the page, he said. And, most remarkably, at least to all the foreigners present, was the Guide's apparent willingness to take the blame on himself. "No one separated Libya from the world community," Qaddafi insisted. "Libya voluntarily separated itself from others" by its actions. "No one has imposed sanctions on us or punished us. We have punished ourselves." The irony, Qaddafi stated, was that "all these things were done for the sake of others."

And on it went for another half hour or more.

That evening, he gave the first detailed public account of the reasons behind the surprise announcement he had made on December 19, 2003, that Libya was abandoning its previously secret nuclear weapons program. "Yes, there was such a program," he said, to the astonishment of many people in the room. Libya chose to declare it to the United States and Britain and seek their help in dismantling it "because it is in our own interest and for our own security." At another point, he said, "We got rid of it. It was a waste of time, it cost too much money."

After the speech, Representative Weldon was ecstatic. He had promised his congressional colleagues a show and, by golly, they got their money's worth. "We were part of history tonight," he told reporters who were there. "Colonel Qaddafi's statements were unequivocal. There were no ifs, ands, or buts. It reminds me of the sea change that occurred when the Berlin Wall came down, or when Yeltsin stood on top of a tank in front of the Russian White House. As startling as it is to us, we'd better take advantage of it."

After the speech, Weldon and his colleagues were shown into a private reception hall to greet the Libyan leader. Qaddafi was wearing one of his typical outlandish uniforms: black silk shirt, white suit with wide lapels, festooned with U.S.-style battle decorations across the chest. He whispered something to his chief of staff, Nouri al-Mismari, who then pointed to the pin Weldon was wearing with the American and green Libyan flag intertwined. Curt took it off and was about to hand it to him when Mismari whispered in his ear. "Pin it on him," he said. As he did so, everyone started taking pictures.

Qaddafi was beaming like a six-year-old with his first bicycle.

THE SENATOR

The next morning Weldon and his delegation were joined by Senator Joe Biden, who flew into Sirte on his own Gulfstream 5. Curt met him on the tarmac, then briefed him on the extraordinary events of the previous twenty-four hours in the airport VIP lounge. Biden was accompanied by a reporter, Daniel Klaidman. Qaddafi's men called him, "Mr. *Newsweek*."

Shortly after 1:30 PM, we were all brought to Qaddafi's tent complex. This time, Qaddafi was wearing his Bedouin outfit: a thick dark brown cape made of stiff wool, and a black sheepskin cap. He grunted as he greeted Biden and the rest of us in the tiny room. He seemed much taller than he had been last night. Elevator shoes? It was also clear that he understood English, as he responded to Biden's small talk before his translator had finished the Arabic. I dubbed him "The Stiff" that afternoon because of his demeanor.

Biden came in his preachy mode, prepared to talk about the requirements of American-style democracy, not just Qaddafi's apparent new opening to the West. At one point, he asked Qaddafi if the People's Congress had the power to overthrow him. The translator gulped, but Biden insisted that he render it word for word. Qaddafi didn't flinch. "Of course not, because I am the Founder of the Revolution."

"Gee, I'd like to have your job security," Biden said. For the first time, Qaddafi laughed.

After Qaddafi emerged from his one-on-one with Biden, a *Washington Post* reporter asked him if he was planning to allow Islamic organizations to operate in Libya. The answer was not just no, but hell, no.

"We don't want to involve Allah, or God, in material affairs like infrastructure and sewage. He has nothing to do with that," Qaddafi said. "We need technology for infrastructure. We are talking about sewage, water, electricity, housing. Allah is another thing. How can we involve Allah in such things of daily life?"

I asked him about Lockerbie. "Last night you said you were going to tell the Libyan people the whole truth. Were there others involved—other countries involved—besides the two people taken to the Scottish court?"

"Lockerbie is buried and we don't want to dig it out," he said. "We're finished with Lockerbie. It's something of the past."

Later that evening, at a dinner hosted for Biden by the governor of Sirte and local notables, the future vice president related more of his exchange with Qaddafi, adding a comment no one paid much attention to at the time.

"Qaddafi raised the issue of the double standard. He could understand the United States getting rid of Saddam Hussein, who had killed hundreds of thousands of his own people. 'But you will not do that with the Saudis,' he said. 'And you should.'

"I think that is a legitimate criticism, that we selectively support democracy. And I say to my friends in Saudi Arabia, you are becoming too heavy a burden for us to carry. I'm ready to cut a deal with Russia for the six percent of our oil supplies that we import from you. I'm not going to be blamed for your lack of openness. It just breeds anger and hatred of America," Biden added.

For all the changes Qaddafi put in place over the next seven years, he never held elections, which, in Biden's view, were the hallmark of a democracy. The breathtaking arrogance and naïveté of that approach would contribute heavily to the decision made by the Obama White House to cast Qaddafi aside despite all of the benefits his newfound alliance with America had brought.

INSIDE THE WIRE

The U.S. officials in charge of eliminating Qaddafi's WMD factories and materials were stunned by the cooperation the Libyans offered them. Some of the verifiers reported back to Paula DeSutter that their experience on the ground had been sobering. "Many of the places they showed us we had never heard of before," one of them told me. "It shows you the limits of international arms inspections. Without Qaddafi's cooperation, we never would have found most of this stuff." It also showed the limits of U.S. intelligence collection against hard targets such as Libya or Iran.

The minister of bad things, Ma'atouq Ma'atouq, took DeSutter's people to warehouses and factories, machine shops and ammo dumps, where they loaded up thousands of tons of equipment to produce ballistic missiles and nuclear weapons. They set up a chemical weapons disposal facility, similar to what the United Nations did in Iraq, and rapidly eliminated Qaddafi's stockpiles and the leftover chemicals he had imported to make more of the deadly brew.

Before I left Libya later that week, DeSutter's teams loaded a thousand tons of uranium centrifuge enrichment gear onto a U.S. ship in Tripoli harbor, and sailed for the United States. Also on the ship were five complete SCUD-C missiles and their bug-eyed launch vehicles, which Qaddafi had imported from North Korea. DeSutter proudly hung an access panel from one of them in her State Department office as a reminder of what they had accomplished. The centrifuges went to America's first bomb plant in Oak Ridge, Tennessee, where U.S. experts are later believed to have reassembled them to test the Stuxnet computer virus they introduced into Iran's centrifuge plant.

Before I left, I paid a visit to the two American diplomats who had recently arrived in Tripoli to pave the way for reopening a U.S. embassy there. I was curious as to why they hadn't bothered to join the rest of the diplomatic corps in Sirte to witness the Qaddafi speech.

"Frankly, we didn't even think about it," the woman who was chief of mission told me. "We're stuck in day-to-day logistics. We don't have an embassy. We're operating out of a room in this hotel while we look for property. Look, it took two of our employees three days just to receive a DHL package."

In other words, because of security concerns, these diplomats were instructed not to venture outside the wire, even though the Qaddafi regime clearly was friendly and effusively cooperating with the United States, which was not the case when Chris Stevens went to Benghazi on September 10, 2012.

Although they had just arrived in Tripoli, they had used their first meetings with Libyan officials to make clear that "human rights concerns" would affect the tenor of U.S.–Libyan relations going forward. But they also gave Qaddafi high marks for following through on his promised disarmament. "When they've made a commitment, they've followed through. We've done the same," they told me.

I was curious how they saw the anarchic Libyan system, where Qaddafi seemed to rule by fits and starts—and sometimes with great theatrics—as when he personally took to the controls of a bulldozer to demolish the Tripoli office of his prime minister, who had refused to move to the new capital in Sirte.

"The more we look at it, the less we understand," the chief of mission said.

"They do seem to have a budget process," her deputy added. "We just don't know what it is."

At the time, I wasn't sure whether their candor was refreshing or pathetic.

MR. SUSPENDERS OF THE CIA

The CIA's deputy director of operations, Steve Kappes, regaled members of the Senate and House intelligence committees in closed-door briefings about his one-on-one meetings with Qaddafi. Known as "Mr. Suspenders" to many DO officers because of the effete dress habits that made him look more like a banker than a clandestine operator, Kappes was a smooth talker who knew when to put forward his tough-guy persona as a former U.S. Marine and when to play the cosmopolitan spy.[25]

But, in fact, despite all his boasting that he spoke fluent Arabic, Kappes was accompanied by a translator in his rare meetings with Qaddafi and top Libyan officials.[26] And the timeline I have presented above shows clearly that it was Mark Allen—not Kappes—who opened the door to Qaddafi.

"Kappes's philosophy was that liaison is the most important part of intelligence," said Dewey Clarridge, a CIA legend who founded the counterterrorism center in the mid-1980s and has remained in the intelligence game as a private contractor to United States Special Operations Command (SOCOM). "Kappes believed the U.S. shouldn't do unilateral operations. That is pure lunacy! That's how we get fed lies by other services for their own reasons," Clarridge told me.[27]

Kappes put his liaison skills to good use in Libya, turning to Qaddafi and his subordinates for key assistance in the global war on terror.

"Dear Musa," he wrote on March 17, 2004. "I am glad to propose that our services take an additional step in cooperation with the establishment of a permanent CIA presence in Libya. We have talked about this move for quite some time, and Libya's cooperation on WMD and other issues, as well as our nascent intelligence cooperation mean that now is the right moment to move ahead." [28]

Formalizing the CIA presence in Tripoli, more than two years before the United States would open a full-fledged embassy there, showed just how important an ally Qaddafi had become. With Qaddafi's nuclear program now shipped off to Oak Ridge, Tennessee, Kappes felt it was time to press on to more urgent matters.

TRIPOLI STATION

The first order of business for the CIA in Tripoli was to secure the rendition to Libya of a top al Qaeda operative, Abdullah al-Sadiq, the head of the Libyan Islamic Fighting Group. After fighting alongside the Taliban for several years, he fled Afghanistan and eventually surfaced in Malaysia in early March 2004 with his wife, Fatima Bouchar, who was four and a half months pregnant.

The Libyans wanted him, and the CIA was only too happy to oblige. They knew that Qaddafi's intelligence services could elicit much more information from their prisoners than the CIA could ever hope to obtain. A series of faxes discovered after the revolution by Human Rights Watch shows the almost hour-by-hour communications between Kappes, Mark Allen, and Musa Kusa over this critical terrorist rendition.

Once al-Sadiq surfaced, the British and U.S. governments moved rapidly with the Malaysian government to detain the couple, who were seeking political asylum in Britain. Five days later, the CIA sent an aircraft to Kuala Lumpur and flew them to Tripoli, separated from each other and bound head to toe in duct tape. Husband and wife later claimed that they had been tortured by the Americans and ultimately filed a lawsuit in 2012 against the British government for complicity in their rendition to Libya.

Qaddafi kept al-Sadiq in the infamous Abu Salim prison for the next six years, where he was repeatedly interrogated by "foreigners." Sadiq claims in his lawsuit that his torturers were CIA and MI6 officers. He was finally released in March 2010 along with thirty-three other Islamist operatives in a "deradicalization" program put in place by Qaddafi's son, Saif al-Islam.[29]

Al-Sadiq never forgave the Americans and the Brits for his rendition, and was not grateful to the Qaddafis for his release. He went underground almost immediately, only to surface less than one year later during the Libya civil war under the real name, Abdelhakim Belhaj.

In May 2011, Belhaj, now a powerful military leader of the anti-Qaddafi rebels in Benghazi, found himself working with the same Americans who had committed the original sin of handing him over to torture and imprisonment. Chief among them was the U.S. special envoy to the Transitional National Council, J. Christopher Stevens.

2

THE MAKING OF AN AMBASSADOR

I got to know Chris Stevens in 1997, when he served an eighteen-month stint as the State Department's Iran desk officer in Washington. As the president of the nonprofit Foundation for Democracy in Iran, I spoke with him regularly about political developments inside Iran, U.S. policy, and about the needs of Iranian "refugees with special knowledge"—a euphemism for defectors. In our initial discussions, he professed his lack of area expertise in Iran, but was a quick study. By the time he left that position in 1999, he probably knew as much about Iran as anyone in the State Department.

As a twenty-three-year-old Peace Corps volunteer in Morocco in 1983, Chris Stevens learned Arabic and polished his college French. For this young history graduate from Berkeley, it was easy to fall in love with the softer side of the Middle East that Morocco presented. After two years teaching English to young Moroccans, he returned to his native California and got a law degree in 1989, eventually passing the Foreign Service Officer Test (FSOT) and entering the State Department two years later at the age of thirty-one. Reporters and bloggers have speculated that, in the meantime, he had been recruited by the CIA and was assigned to the State Department in a cover position. One reason for that speculation was his subsequent posting to Benghazi as special representative to the National Transitional Council in late April 2011, at a time when there was zero security, al Qaeda groups and Islamist militias ruled the streets, and the place was awash with hostile intelligence services running guns. "You would never send an ordinary diplomat into such a situation," a retired two-star general told me.

And, yet, that seems to be what happened. I have seen no evidence to suggest that Chris Stevens was a clandestine CIA officer. Indeed, his State Department career followed a pretty standard trajectory, alternating two-year postings in Damascus, Riyadh, Tunis, and Cairo, where he honed his Arabic, and stints back in Washington, D.C.

In 2006, Stevens was posted the deputy chief of mission at the U.S. Consulate in East Jerusalem, putting him in daily contact with senior Palestinian Authority officials and NGOs. His reporting back to the State Department showed him regularly backing up Palestinian requests for additional U.S. aid, and detailing the negative impact on the Palestinian population of U.S. aid restrictions put in place by the Bush administration as a result of the Hamas takeover of Gaza.[1] While most Americans agreed with the aid restrictions, Stevens painted them as harmful to Palestinian children, going out of his way to find a United Nations study that showed a rising rate of infection among newborn babies because of contaminated tap water, supposedly caused by decreased U.S. aid to sewage treatment projects. This bleeding-heart approach toward the Palestinians and, later, toward Islamist groups, was the norm, not the exception, among State Department Arabists.

MISSION: TRIPOLI

The United States resumed full diplomatic relations with Qaddafi's Libya in 2006, after taking Libya off the State Department's list of governments supporting international terrorist organizations. Chris Stevens arrived in Tripoli the following year as the deputy chief of mission (DCM), a significant promotion, and clearly did not take well to Qaddafi or to his regime. In one of his earliest cables, reporting on the state visit of French President Nicolas Sarkozy to Tripoli in July 2007, Stevens went out of his way to point out—twice—that Qaddafi had never repaired damage from the 1986 U.S. bombing of his residence in Tripoli and had erected a statue "displaying a fist crushing a US fighter plane" in front of it.[2]

I have read through scores of cables Stevens sent back to Washington from this period and they reveal an engaging diplomat, who regularly went outside the embassy to meet with sources and who enjoyed traveling around Libya without a security escort. However, they also reveal an activist who clearly was not happy with his government's rapprochement with the Qaddafi regime and constantly sought to undermine it.

Early in his tenure as DCM, Stevens became enmeshed in the top-secret negotiations over U.S. access to Guantánamo detainees who had been returned to Libya. It was clear that he suspected the Libyan government of torturing many of these terrorist detainees. His reporting betrayed a remarkable sympathy for the detainees and for the Libyan Islamic Fighting Group (LIFG), to which most of them belonged.

In one cable dated November 7, 2007, Stevens reported on the recently announced merger between the LIFG and al Qaeda. This was a significant event, since the two groups were now united in calling for the overthrow of the Qaddafi regime. While regime insiders were worried that political violence might hurt their personal economic interests, many average Libyans showed enthusiasm for the merger, he wrote.[3]

In another, dated December 13, 2007, Stevens intervened with the Libyan authorities to convey the State Department's concern over the welfare of two former Guantánamo detainees, Muhammad Mansur al-Rimi (ISN 194), and bin Qumu Abu Sufian Ibrahim Ahmed Hamuda (ISN 557).[4] Bin Qumu later founded the group known as Ansar al-Sharia, which initially claimed responsibility for the attack on Benghazi in which Stevens perished. In repeated meetings with the Libyans, the embassy insisted that the Libyan government treat terrorist detainees "humanely," and provide medical care and "timely access" so the State Department could visit them in jail to monitor potential abuse. Humane treatment of detainees transferred back to Libya became a condition for Libyan intelligence officers to interrogate other Libyan terrorist detainees still held in Guantánamo, so it gave Stevens and his colleagues real leverage with the Libyan government.[5]

In a December 18, 2007, cable, Stevens reported that a pair of U.S. companies had just won civilian construction contracts worth between $2 billion and $3 billion, a windfall by any measure, especially as the U.S. recession began to kick in. And yet, Stevens remained peevish: "The agreements underscore that the [government of Libya] continues to directly link commercial contracts to political relationships. The potential return to Libya of large numbers of non-official Americans has positive potential for people-to-people and cultural exchange, but could further anger Islamic militants already troubled by U.S.–Libyan reengagement."[6]

This type of walk-on-eggshells deference to Islamic militants would later make Stevens the key interlocutor for the Obama administration with the anti-Qaddafi rebels. And it would get him killed.

With the retirement of his predecessor, Charles O. Cecil, Stevens was promoted in December 2007 to chargé d'affaires, the top U.S.

diplomat in Tripoli. He was well liked and competent, and was complimented by the State Department's inspector general for providing "a steady hand at the helm."[7] Stevens encouraged his political officers to seek out unofficial sources of information and opinion, so the embassy's reporting wouldn't just mirror what the Qaddafi regime was putting out in its propaganda. This was a lesson the State Department had learned from its own failure to predict the downfall of the Shah of Iran in the late 1970s.

In one cable, for example, they talked to the director of Libya's newly formed National Security Council, who lifted the veil on inside-regime rivalries among Qaddafi's sons.[8] In another, they picked up gossip at a dinner party attended by Qaddafi's second wife, Safia Farkash, to report her concern with the economic reforms being promoted by Saif al-Islam, her eldest son, because they exacerbated resentment against the Qaddafi family and their cronies who had profited disproportionately from contracts with the West. Indeed, one of Stevens' reporting officers wrote, the perception of corruption was widespread: "[T]he level of dissatisfaction with Qadhafi's family and regime is such that some Libyans are willing to support any alternative perceived to be viable in the hope that the next regime will be less oppressive."[9]

In 2008, Stevens worked incessantly to secure the freedom of jailed dissident Fathi El-Jahmi, whose case had attracted international attention. He sent more than a dozen cables back to Washington about El-Jahmi between February and June 2008.

El-Jahmi was a government bureaucrat jailed in 2002 after he called for free speech and democracy. President George W. Bush mentioned his plight in a speech on March 12, 2003, just as the nuclear negotiations with Qaddafi got started. Qaddafi released him as a gesture to Senator Joe Biden, following their meeting in Sirte in March 2004, only to send him back to prison a few weeks later after El-Jahmi gave a television interview reiterating his call for democratic reforms.[10]

The case became more complicated because of El-Jahmi's health . . . and his family, who refused offers by the regime to take him back.

A mixture of heart problems and prostate cancer landed El-Jahmi in a state-run medical center, where, apparently, he was receiving exceptional care. Stevens reported in minute detail on El-Jahmi's medical exams, met repeatedly with his attendant physician, and even tried to get the State Department to retain a U.S. doctor to provide a second opinion on his test results! But as Stevens reported, the problem remained that his family didn't want him back, apparently daunted by the cost of private medical care.

The Qaddafi Development Foundation, run by Saif al-Islam, and the Libyan government "were now at a loss as to what to do with him, for if they released him now and his family did not accept their responsibility to care for him, he could become destitute," Stevens wrote. Saif asked Stevens to convey that message to Senator Biden, since he had been involved in El-Jahmi's initial release, with a request that he weigh in with the family.[11]

One of Stevens' reporting officers was "ambushed" by a European television network while visiting El-Jahmi at the hospital in May, leading El-Jahmi's son to send the embassy a letter of complaint. "Your frequent visits to him in the hospital are causing us problems; I urge you stop visiting our father." Stevens asked the State Department to check with El-Jahmi's brother in Boston to see whether the letter was written under duress.[12] Stevens believed that the family was being pressed into signing a written agreement guaranteeing that El-Jahmi would not talk about his detention or make any other political statement, as a condition for his release.[13] In the end, Stevens' approach backfired, and the embassy was forced to step back from the case. In early 2009, without U.S. help, the family negotiated a deal by which the Qaddafi Foundation picked up the tab for sending him overseas for medical treatment.[14]

The ultimate irony of the El-Jahmi case is that the families of the Navy SEALs who tried to rescue Stevens in Benghazi four years later, and those who died in Afghanistan on board *Extortion 17* following leaks by Vice President Joe Biden and Defense Secretary Leon Panetta of their identity, were offered the same kind of deal by the Obama administration that Stevens criticized Qaddafi for offering El-Jahmi: pension benefits in exchange for silence.

DIE HARD IN DERNA

By far the most remarkable aspect of Chris Stevens' first tour of duty in Tripoli was his reporting on Islamic extremism in eastern Libya. Here, he at times displayed an almost romantic attraction to Qaddafi's Islamist opponents and their willingness to die in so-called martyrdom operations, despite their open alliance with al Qaeda.

Learning more about the Libyan Islamic Fighting Group and their strongholds in Benghazi and Derna became a top priority for the embassy in Tripoli in September 2007, when the U.S. 3rd Ranger Battalion of the 75th Ranger Regiment hit Objective Massey along the Syria-Iraq border and seized an important cache of documents on the foreign fighters flowing into Iraq. In their analysis of the ledgers, the CIA Counterterrorism Center "found that Derna, Libya was nearly tied with Riyadh, Saudi Arabia for second place" as a source of foreign fighters in Iraq. "Third and fourth place belonged to Mecca and Benghazi, respectively."[15]

The Qaddafi regime had virtually ignored the Cyrenaica—Libya's eastern seacoast region—for many years. The Qaddafis came from Tripolitania, from different tribal groups with a different background. They treated Benghazi and Derna a bit like outliers of the empire: nominally part of Libya, source of tribute, but also potential trouble.

According to a U.S.-Libya dual national who regularly reported to embassy officials on his travels to visit family in Benghazi and Derna, the Qaddafi policy had caused widespread resentment. The fruits of Qaddafi's turn to the West never reached these parts of Libya, where at least half the young men were unemployed. "The situation reflects in part the Qadhafi regime's belief that if it keeps the east poor enough, it will be unable to mount any serious political opposition to the regime. Explaining the rationale, [the businessman] cited a Libyan proverb: 'If you treat them like dogs, they will follow you like dogs,'" Stevens wrote. He made it clear in the cable he thought that was a dangerous approach.

Family dinner parties were dominated by news of young men who had taken up the jihad against the occupying "Crusader forces in Iraq," where dinner guests offered the families of the martyrs a mix of condolences and congratulations. News reports about Objective Massey and the role of Benghazi and Derna as a big supplier of young men for the anti-American jihad had given local families a perverse sense of pride, Stevens wrote.[16]

Fueling the calls to jihad were small mosques outside the main cities, relics of the pseudo-secret Sanussi Brotherhood of the mid-nineteenth century. The fact that they didn't resemble ordinary mosques "[made] it harder for [government of Libya] security organizations to identify them and easier to hold unobserved meetings and sermons."

Pseudo-secret mosques, radical imams, closed-door family dinners with jihad on the menu: It was a romantic brew for the Arabist in Chris Stevens. And he sensed that it spelled trouble for Qaddafi. He ended his cable with a carefully worded warning: "[C]laims by senior GOL officials that the east is under control may be overstated." Perhaps he would go down in the annals of the State Department as the clear-eyed diplomat who convinced his government to pull back its support from a dictator before he fell?

Chris Stevens didn't see Qaddafi's Islamist opponents as the enemy:

he saw them as an opportunity. The fact that the Qaddafi regime was already seeking an accommodation with the Libyan Islamic Fighting Group didn't matter. Stevens and his embassy were getting high marks back in Washington for their extensive and groundbreaking reports on the Libyan Islamists.[17]

In a follow-up cable filed on June 2, 2008, Stevens reported on the extraordinary visit to eastern Libya by his chief political/economics officer, John Godfrey, who took advantage of a personal trip to Roman ruins in Benghazi to elude Qaddafi's minders, then rented a car and driver to visit contacts in nearby Derna. His adventure was all the more remarkable because of its serendipity: While asking directions to the city's old fort, Godfrey encountered a local resident named Nouri al-Mansuri, who happened to hail from the same tribe as his driver/guide. "In typical fashion, Mansuri promptly dropped what he was doing and spent the next several hours accompanying us around Derna," Stevens wrote, taking advantage of "what appeared to be a rare gap in surveillance by security organizations."

During an impromptu lunch at a local restaurant, Mansuri filled them in on the young jihadis who were fighting the Americans in Iraq. They were motivated by what they perceived was the determined U.S. support for Qaddafi in the wake of his renunciation of terrorism and WMD programs. "Fighting against U.S. and coalition forces in Iraq represented a way for frustrated young radicals to strike a blow against both Qadhafi and against his perceived American backers," Mansuri explained.

Derna was not "uniformly extremist," he added. Many people were not happy about the increasingly conservative religious atmosphere that had prevailed since returning jihadis introduced " 'unnatural foreign influences' on religious practices in Derna." But the Islamists were winning, and the Libyan Islamic Fighting Group was laying the groundwork to overthrow Qaddafi by whipping up "a heady mixture of violence, religious conservatism and hatred of U.S. policy in Iraq and Palestine."

Derna's reputation as a historical locus of resistance to occupation, and its increasingly conservative Islamist bent, were the main factors driving so many young men to leave Derna for Iraq to fight coalition forces. "It's jihad—it's our duty," Mansuri said. Then Stevens threw in the punch line. "Referring to actor Bruce Willis' character in the action picture *Die Hard*, [Mansuri] said many young men in Derna viewed resistance against Qadhafi's regime and against coalition forces in Iraq as an important last act of defiance." [18]

It's okay, Washington; Libya is just another Hollywood action film. If you don't like it, you can change the channel. Later, as ambassador, Chris Stevens would himself face the ire of the *Die Hard in Derna* types he had romanticized just four years earlier

3

"A NEW BEGINNING"

While the United States was preoccupied with the surge in Iraq to beat back al Qaeda, another war was simmering in Afghanistan. This was the "right war" that many in Congress, including then-Senator Barack Obama, thought the United States should be fighting, as opposed to what fellow Democrats called the unwinnable war in Iraq.

But, on the battlefields of Afghanistan, unsettling things were happening that received little attention in the press or in Congress. Taliban fighters were getting hold of man-portable air defense systems (MANPADS) and shooting down U.S. helicopters.

The Pentagon knew just how deadly these shoulder-fired, surface-to-air missiles could be. After all, it was the CIA supply of Stinger missiles to the Afghan mujahideen starting in 1986 that turned the tide against the Soviet occupation. The deadly strikes against Soviet attack helicopters and jets devastated the morale of the Russian troops and ultimately convinced Gorbachev to begin evacuating Afghanistan in 1988. The mujahideen reportedly succeeded in bringing down 269 Soviet helicopters and aircraft with the U.S. missiles.[1]

THE DOWNING OF FLIPPER 75

The wake-up call came on May 30, 2007, when a U.S. CH-47D Chinook helicopter, call sign Flipper 75, was hit by a surface-to-air missile in the upper Sangin Valley, killing a British combat photographer, a Canadian, and the five-man American flight crew. The helicopter had just landed and deployed ground troops when it came under fire. The pilot tried to clear the landing zone, quickly gaining altitude, when Flipper 75 was hit at about two hundred feet above the ground and went down. At the time of the shoot-down, a NATO spokesman denied that a heat-seeking missile had downed the aircraft, suggesting instead that the helicopter had been hit by a lucky strike from small arms fire or a rocket-propelled grenade (RPG). "It's not impossible for small arms fire to bring down a helicopter," spokesman Major John Thomas said.[2]

The initial incident report read quite differently. "The missile struck the aircraft in the left engine. The impact of the missile projected the aft end of the aircraft up as it burst into flames followed immediately by a nose dive into the crash site with no survivors."[3] For nearly three years, the Pentagon denied that report.

Flipper 75 was part of a much larger operation that had been going on for several days, code-named Kulang Hellion, aimed at clearing out a known Taliban stronghold in Helmand province. It has been called the largest air assault since the beginning of the Afghan war. The still-classified after-action reports reveal that the Taliban were well prepared. After downing the Chinook, they fired two more MAN-PADS against a pair of Apache AH-64 attack helicopters, Arrow 22 and Arrow 25, which were providing fire support for NATO troops into the Kajaki area. "Clearly, the Taliban were attempting to down an Apache after downing the CH-47," an intelligence officer assigned to Task Force Corsair commented.[4]

The Afghan war logs show forty separate incidents involving MANPADS from 2005 to 2009, including an attack on a C-130 refueling at 11,000 feet, requiring it to launch flares and engage in evasive maneuvers. Many of these were Chinese-made HN-5 missiles or their Soviet-era equivalent, the SA-7. Some theorized that they had been looted from Iraqi stockpiles after the fall of Saddam and transferred to the Taliban via Iran.[5]

However, Flipper 75 was the first known downing of a NATO aircraft in Afghanistan by a surface-to-air missile and the first helicopter lost to ground fire since June 28, 2005, when a rescue mission to save four Navy SEALs, whose dramatic story was told in the *Lone Survivor* memoir by Marcus Luttrell, was hit by an RPG. It would not be the last.

Former U.S. flag officers and intelligence officials involved in the Stinger supply effort to the Afghan mujahideen in the 1980s tell me that they are 100 percent certain that the Stingers they delivered were no longer operational. "For one, their batteries give out after about four or five years, and these aren't off-the-shelf items, but specially designed units that shoot an argon gas coolant into the system in addition to powering it up," one source said. In addition, the propellant of the actual missile decomposes after about fifteen years, making the missiles not only inoperable but dangerous to handle.

The Afghan war logs show twenty-two separate incidents that specifically mention Stingers. Some of these were inoperable 1980s-generation missiles found in Taliban weapons caches. But others were new.

Some of the new missiles came from an unusual supplier: Iran. According to one U.S. intelligence report, "The MANPADS delivered to the Taliban elements were made in Iran and very closely resemble U.S.-made Stinger MANPADS, so much so that the MANPADS appear to be direct Stinger replicas."[6]

More Stingers would soon be on the way.

MUSLIM OUTREACH

Candidate Obama promised to make dramatic changes to U.S. policy in the Middle East if elected. He pledged to end the wars in Iraq and Afghanistan, and to launch a new era of cooperation with the Muslim world. In effect, he was returning to a 1970s vision of the world, which failed to recognize the dangers of political Islam. Just like Jimmy Carter, who welcomed the advent of Ayatollah Khomeini in Iran and showered the Islamist mujahideen in Afghanistan with weapons and cash, President Obama seemed to believe there was an ironclad barrier that separated Islam the religion from Islam the terror-drenched ideology.

Emphasizing the importance he placed on these changes, he gave his first television interview as president to al-Arabiya, an Arabic network in Dubai, on January 26, 2009. "My job to the Muslim world is to communicate that the Americans are not your enemy," he said. Former Assistant United States Attorney Andrew C. McCarthy slammed Obama's comment as "implicitly slanderous, claiming that, until his arrival, America, despite freeing tens of millions of Muslims from tyranny in Bosnia, Afghanistan, and Iraq, had positioned itself as Islam's enemy."[7]

President Barack Obama embarked in April 2009 on his first overseas trip, a whirlwind tour aimed at presenting another side of the United States to friends and allies. At the G-20 summit in Britain, he bowed deeply to Saudi monarch King Abdullah Abd al-Aziz. The *Washington Times* called the video clip of Obama's greeting a "shocking display of fealty to a foreign potentate."[8] In Prague, he announced his goal of eliminating all nuclear weapons and spoke of the dark side of America's past. In Iraq, he staged photo ops with cheering troops and the combatant commander.[9]

At his final stop in Istanbul, he got down to more serious business.

After saluting Prime Minister Tayyip Recep Erdoğan, a man who openly called for the restoration of the Islamic caliphate, he announced his plan to embark on a dramatically different Middle East policy than his predecessors. Behind closed doors, Obama then test-marketed his ideas on Erdoğan, the foreign leader who soon became his closest confidant.

He also previewed his shift of America's alliances to a group of Islamic scholars, who paid a secret visit to the White House later that April. As it turned out, they included top leaders of Egypt's shadowy Muslim Brotherhood, who came to Washington on a mission that mixed curiosity with seduction. They reassured Obama and his advisors that they "favored democracy and would support the U.S.-led war on terror." [10]

Several of these same men were invited by the White House to stand just behind the president on June 4, 2009, at Cairo University and nodded approvingly when he unveiled his new approach for the first time in public. Asked specifically about the presence of known Muslim Brotherhood leaders at the speech, White House Deputy National Security Advisor for Strategic Communications Denis McDonough said Obama wanted to address the "full range of political representation in Egypt," and so tasked the embassy in Cairo to send invitations to Muslim organizations. White House advisors portrayed the president as working over the final drafts by hand, stamping the speech like a craftsman with his personal insignia. [11]

No one could claim Obama was merely reciting some speechwriter's words.

"A NEW BEGINNING" IN CAIRO

The Cairo speech was an extraordinary mixture of apology and concession to Muslim Sharia law. Obama painted a picture of America that most Americans would find unfamiliar, if not alarming. The first part of the speech read like Forrest Gump in Arabia, with unnamed Muslims popping up to play key roles at every stage of America's history. He went on to describe an America with "a mosque in every state in our union," where women's rights meant "the right of women and girls to wear the hijab [Islamic head scarves] and to punish those who would deny it."

He extolled the "Holy Koran" (five times), claiming that Islam was "not part of the problem in combating violent extremism [but] part of promoting peace." Then he made an extraordinary concession to America's sworn enemies in coded language meant to deceive Americans listening back home, by promising to change U.S. laws that had "made it harder for Muslims to fulfill their religious obligation." He was referring to a massive FBI and Department of Justice prosecution of the Holy Land Foundation that had revealed how Muslim charities were raising millions of dollars in the United States to finance terrorist groups such as Hamas and al Qaeda. In addition to exposing the vast network of Muslim Brotherhood organizations operating in the United States—many of which Obama invited to send top officials to join his administration—the 2008 trial had disrupted the Brotherhood's fund-raising apparatus by shutting down their "zakat committees," ostensibly devoted to social works. Obama pledged in Cairo to fix that, so Muslims could once again fulfill "zakat."

Obama also claimed that America had overreacted to the 9/11 attacks and pledged to fix that, too. No more Gitmo, and no more supposed torture of terrorist suspects, even though the interrogation techniques used by the United States paled in comparison to those commonly used in the Muslim world, both by government intelligence agencies and Muslim extremist groups.

If Egyptian President Hosni Mubarak wasn't worried by this point of the speech, he should have been. Egypt had been a staunch ally in the war on terror during the Bush administration. Mubarak's intelligence services had helped the CIA to identify top al Qaeda operatives and render them to their home countries. They interrogated them in ways off-limits to U.S. officials—much as Qaddafi had done. Mubarak had also kept the Muslim Brotherhood under wraps since they assassinated his predecessor in 1981, and here their leaders were seated right behind the president of the United States!

Just in case Mubarak (and other pro-U.S. "tyrants" in the region) didn't take the hint, Obama made his intentions explicit. "You must maintain your power through consent, not coercion," he said.

Not once in the whole speech did Obama mention Mubarak by name, or thank him for his hospitality. Neither did his top advisors mention Mubarak when they briefed the White House press corps on the upcoming trip. The oversight was purposefully rude, and a warning: In Obama's eyes, and the eyes of the Muslim brothers, Mubarak was on probation.

Obama was turning long-standing U.S. policies on their head, but Mubarak just couldn't let himself believe it. Eighteen months later, he would become yet another Obama victim.

4

TEHRAN SUMMER, ARAB SPRING

The first test of Obama's new approach to the Middle East came within days of the Cairo speech, when the Islamic Republic of Iran held the most raucous election in its thirty-year history. Was Obama's intention to use his bully pulpit to launch a new freedom agenda to overthrow tyrants across the Muslim world? Or was it something else?

Candidates for Iran's presidency traded outlandish accusations in the first one-on-one televised debates in Iran's history. Former prime minister Mir Hossein Mousavi became the darling of a fawning media in the United States and Europe, who conveniently forgot his past involvement in the regime's terrorist activities and created an image of him as a so-called reformer, prepared to transform the Islamic Republic into a modern country open for business to the world. Young people in Iran came out to vote in massive numbers. Even Iranian exiles in Europe and America flocked to polling places set up for them by the regime in hotel ballrooms and restaurants and mosques. They expected Mousavi to win in a landslide against the retrograde incumbent, Mahmoud Ahmadinejad, a former Revolutionary Guards officer who lived a simple lifestyle, boasted of Iran's nuclear progress, and spoke frequently of his devotion to the Twelfth Imam. He also kept his wife in a bag—the full black *niqab* favored by Sharia-observant clerics—whereas Mousavi trotted out *his* wife adorned with colorful headscarves and university degrees.

The sophisticates in Iran and abroad found Ahmadinejad an inconvenient reminder that the core ideology of the Islamic regime remained rooted in a revolutionary vision of the end times that spelled doom for their aspirations of normality. Ahmadinejad's defiance of the West had brought on international economic sanctions and made it harder for Iran's powerful *bazaari* merchant class to maintain their wealthy lifestyle, as the Revolutionary Guards Corps increasingly monopolized Iran's economy, crowding them out of the most lucrative contracts. They preferred to put a smiley face on the regime, as did the State Department, the European Union, and millions of ordinary Iranians. With a so-called moderate as president, it was easier to sweep the thuggish behavior of the Islamic Republic under the rug and get on with business as usual. It had happened before with Rafsanjani and Khatami; Mousavi was just the latest version of a well-known brand.

"OBAMA, ARE YOU WITH US?"

At two in the morning on June 13, 2009, six hours after the polls had closed, officials from the Interior Ministry visited Mousavi at his residence. They were elated. Mr. President, they said. You must prepare your victory speech for the morning. Not only had he won, but their projections showed that he had won by a comfortable margin, winning well over the 50 percent needed to avoid a runoff.

Just a few blocks away, at an underground command center beneath Beit Rahbari, the sprawling complex of Supreme Leader Ayatollah Ali Khamenei, another meeting was under way. Led by the leader's middle son, Mojtaba, a junior-grade mullah, it included Major General Morteza Rezai, former intelligence chief of the Islamic Revolution Guards Corps (IRGC), the architect of the regime's plan to overturn the elections. Mojtaba was shaking with rage. He swatted the air with a letter addressed to his father from the Interior Minister, showing the vote tally so far.

We have prepared for this, Rezai said calmly. You know what has to be done. With a single phone call, he shut down the Interior Ministry's vote counting operation. Then Rezai sent Mojtaba's father, the Supreme Leader, to an undisclosed secure location in north Tehran and gave orders to deploy the troops positioned on the outskirts of the capital to seal off the city by dawn, when they planned to announce Ahmadinejad's victory.[1]

For weeks, Rezai and Mojtaba Khamenei had been refining the details of their election coup, which they called "Sharayet-e Khakestari" (Condition Grey). It was a not-so-subtle mix of hard power and persuasion. They called on top IRGC officials to threaten voters into voting the right way. An IRGC spokesman, Reza Saraj, warned that the Guards would not tolerate an incorrect result from the polls. "In all those countries where they have had 'color' revolutions, they didn't have a Revolutionary Guards Corps," he told the local media well before the voting began. "We do."

The idea of a "color" revolution haunted the IRGC leaders as it did Ayatollah Khamenei. So did the knowledge that Hillary Clinton's State Department had given orders to the Voice of America's Persian Service to throw its full backing behind Mousavi, whose supporters all wore green scarves in public and dubbed themselves the "green movement."

To make sure that everyone was on the same page, the IRGC instructed a pro-Ahmadinejad member of Parliament, Fatemeh Rahbar, to announce the results of the election three days before voting even began. Standing before a parliamentary committee, she predicted that Ahmadinejad would win 24 million votes—precisely the number the Interior Ministry eventually announced he had won. But rather than deter the people, the heavy-handed intervention by the IRGC only encouraged a greater turnout.

A sizeable contingent of IRGC troops showed up at Mousavi's house shortly after the Interior Ministry officials who had informed him of his victory left. As Mousavi and his sons soon discovered, they hadn't come to protect the new president, but to seal him off from the public. They promptly cut his telephone lines and shut off the electricity, so he couldn't use the illegal satellite dishes on his rooftop to access the Internet. He was under house arrest.

In his address to the people announcing the election results, Ayatollah Khamenei called the phony election result a "divine assessment," and summoned the people to unite behind Ahmadinejad for a second term. Mousavi managed to get a message out to his supporters through his son, calling the results "treason" and a "dangerous manipulation." His advisors told me the former prime minister was in shock. He had run for the presidency not to become a martyr or the leader of the opposition, but to play his part in the game and become the smiley face of the regime. He had planned to propose some modest domestic reforms—enough to ease relations with the West and give Iran's youthful population a cause for hope, but nothing that would endanger the regime he had helped to establish. He had expected Khamenei to play fair. Instead, this was a coup.

Clashes between pro-Mousavi supporters and the IRGC antiriot squads erupted almost immediately in Tehran and in other cities. By Saturday afternoon, the regime cut the text-messaging system used by Mousavi supporters, hoping to stymie the protests. By eleven that night, they closed down Mousavi's campaign headquarters in Tehran and dispersed his supporters using pepper gas.

On Monday, June 15, the Mousavi camp organized a massive protest in Tehran, marching from Enghelab Avenue to Azadi Square, the exact same path anti-shah protesters had used in 1978 in a famous confrontation with the shah's police that led to a bloodbath and helped spark the revolution. Western news agency photographers captured roving motorcycle gangs of pro-regime thugs, brandishing pistols and knives as they drove into the crowds. Other photos surfaced of Basij militiamen in full riot gear firing assault rifles into the crowd, and of Lebanese Hezbollah thugs beating protesters. Two particularly gruesome amateur videos shot by Green Movement supporters showed masked policemen in body armor cutting the tongue out of one demonstrator and beating to death another.[2] Millions of people took to the streets of major cities all across Iran, denouncing the stolen elections. The mass protests were dramatic in a country that kept a tight lid on any expression of public discontent. The gruesome violence of the regime response gripped headlines around the world.

French President Nicolas Sarkozy called the elections a "fraud." Other Western leaders called on the regime to exercise constraint. But an eerie silence came from the Obama White House. The protesters began holding up signs in English to make sure the international media got their message. Many of them read, "Obama, are you with us?" But as the regime thugs continued their brutal crackdown, killing and wounding hundreds of protesters and tossing thousands in jail, Obama remained silent.

When a reporter finally badgered the president on Tuesday at a press conference, Obama was visibly uncomfortable. Instead of condemning the violence, he reiterated his intention to hold a "tough dialogue" with the regime. "[W]e respect Iranian sovereignty," Obama said. "It is not productive, given the history of U.S.-Iranian relations to be seen as meddling—the U.S. president, meddling in Iranian elections."

Obama's comments were greeted with dismay by the Green Movement. Mohsen Sazegara, a former presidential candidate who had been jailed and tortured by the regime, told me he listened to Obama's press conference with a sense of "deep, deep, deep regret. I never expected President Obama to say something like that"[3]

But Obama's words were music to the ayatollah's ears.

To explain why Obama was turning his back on what appeared to be a pro-American freedom movement in Iran, the White House eventually leaked his lofty geopolitical goal. The president was engaged in a delicate diplomatic dancing act with the ayatollahs, and was still waiting to hear the response of the Supreme Leader to a secret letter he had sent him weeks earlier offering direct talks. For the United States to back the street protests now would doom this attempt to achieve a lasting peace with the Iranian regime and resolve the nuclear standoff.[4]

Hundreds of deaths later, the Iranian Green Movement sputtered to end, as Iranians realize that help was not on the way.

Why did Obama fail to come to the aid of the Iranian people, who were crying out for American help? Part of it was pure politics. In the euphoria of the 2008 presidential campaign, when candidate Obama pledged that the power of his personality alone would launch an era where the "earth would begin to heal," no problem seemed unsolvable. And clearly Iran and its quest for nuclear weapons and its ongoing support for terrorism had eluded the best efforts of President George W. Bush, just as it had his predecessors. Obama pledged he would fix that, by embarking on "negotiations without preconditions" with the Tehran regime as soon as he took office. As he told *Newsweek* in May

2009, just before meeting Israeli Prime Minister Benjamin Netanyahu in the White House, he believed his approach was "superior" to what had been tried before.[5]

But part of it was also Obama's reluctance to actively pressure the Islamic regime, despite its barbaric track record. He was reinforced in this view by his consigliere, Valerie Jarrett, who was born in Shiraz, where her father worked at the Nemazee Hospital in the 1950s. "Every memory from Iran is a very happy memory," Jarrett told a National Public Radio reporter. "It is where my life-long love of rice and lamb began." She was personally involved in the outreach to the Supreme Leader, as I will show in subsequent chapters, and urged Obama to continue the effort to strike a deal with the regime, rather than promote the secular opposition. So did his two Iranian-American advisors, Vali Nasr and Ray Takeyh, who had contempt for the secular greens and the opposition more generally. "Obama had to make a choice, and he made it," wrote conservative analyst Robert Kagan in the *Washington Post*. "His strategy toward Iran places him objectively on the side of the government's efforts to return to normalcy as quickly as possible, not in league with the opposition's efforts to prolong the crisis."[6]

Up until the point that Obama acknowledged he was not going to "meddle" in Iran's domestic affairs by supporting the Green Movement, Ayatollah Khomeini had found the new American president an enigma. He had sent multiple emissaries to talk to the Americans since Obama was elected, starting during the transition. In February 2009, he sent a former intelligence officer, Ahmad Samavati, to Washington with a five-man delegation of Foreign Ministry officials, with an offer to render top al Qaeda members who were living in Iran, but no one in Washington seemed interested. Later that month, Khomeini sent a military advisor, Sadeq Mir-Hejaz, to meet with the deputy commander of U.S. forces in Iraq on Kish Island to discuss cooperation in

Afghanistan. Those talks became the subject of heated discussion in the Iranian parliament, with critics accusing the government of caving in to American demands.[7]

But now, with millions of protesters on the streets, Khamenei was worried. News that a State Department official, Jared Cohen, had asked the directors of Twitter to postpone a scheduled maintenance period, so that Iranians could use their Twitter accounts to organize new protests had reached Tehran. (What they didn't realize was that Cohen had gone out on a limb, without consulting the State Department hierarchy.) The Voice of America appeared to be supporting the Green Movement, at least initially. Just like his IRGC advisors, Khamenei was terrified that the United States would back a color revolution using soft power, precisely because he knew how vulnerable the regime was to popular protest. So Obama's hand-wringing apology that he was not going to meddle in Iran's domestic affairs came as a huge relief. Khamenei concluded that Obama was either weak, or a fool. Either way, he was taking America out of play in Iran.

Fear of American action can be an important motivator to Middle Eastern leaders, as Qaddafi's turnaround showed. Similarly, the lack of fear can be a kind of drug, making leaders such as Ayatollah Khamenei feel they can get away with murder and pay no price. Obama's acquiescence to the Islamic Republic's killing machine gave Khamenei the breathing space he needed to finish the job. Intentional or not, Obama sealed the fate of freedom in Iran for years to come. And he emboldened the ayatollah to attack the United States elsewhere—from Afghanistan to Benghazi.

OPEN SEASON ON CHRISTIANS

Hana Said, fifty-five, will never forget the last conversation she had with her twenty-seven-year-old son, who had gone down to Baghdad to see relatives. She was at home in Karakosh, capital of the predominantly Christian Nineveh Plain in northern Iraq, when he called her in hushed tones on his cell phone. He was huddled in a side chapel behind a steel door in Our Lady of Salvation Catholic Church in Baghdad on October 31, 2010, waiting for Muslim terrorists to murder him.

"Ayoub was telling us what was going on, minute by minute," she told me. "He had been wounded and was begging us to get someone to open the outside door to the chapel to rescue them."

For several hours, Hana says, she kept calling her son's cell phone every few minutes while her husband tried to reach the authorities to tell them about the people trapped in the side chapel. Then, finally, at around seven-thirty that evening, Ayoub stopped answering. The police told Hana later that the terrorists had lobbed a hand grenade over the door into the side chapel, killing her son and the other two people who had taken refuge with him. In all, fifty-eight worshippers died while attending church that day, and another ninety-eight were wounded.[8]

President Obama had already begun his promised disengagement from Iraq, and relied increasingly on Prime Minister Nouri al-Malaki and his security forces to fill the vacuum. And yet, many of the people I spoke to in Iraq about this incident believe al-Malaki's own security forces were complicit with the attackers. For starters, the Iraqi Special Forces assault team that eventually stormed the church could have opened the outside door to the side chapel where Ayoub and his companions were hiding. Instead, they burst in through the front doors of the church and shot everybody in sight, including wounded worshippers.

Then there were the elaborate preparations for the attack itself. For several days, the Interior Ministry troops guarding the church surreptitiously moved the Jersey barriers closer to the church, so that the terrorists could drive right up to the front doors. The SUVs they used in the attack had dark-tinted windows and no license plates, the kind of vehicle available only to officials with high-level security clearance. This allowed them to get through checkpoints without being stopped, even though they were carrying assault weapons and explosives.

In responding to the attack, the White House press office issued a generic statement condemning a "senseless act of hostage taking and violence by terrorists linked to al-Qaida in Iraq," but failed to mention that the victims were Christians or that they had been attending church. In response to the bloodiest single attack on an Iraqi Christian church ever, President Obama kept silent. The blood of a single Muslim could work him into a rant, but the blood of Christians apparently left him cold.

Attacks on Christians were accelerating all across the Muslim world, yet the Obama White House took two years to fill the vacant position of ambassador-at-large for religious freedom. For Nina Shea, a member of the U.S. Commission for International Religious Freedom and a Hudson Institute fellow, "The goal of the extremists is to drive Christians from Iraq. It is religious cleansing." The attacks had spread to Egypt, where Coptic Christians were being prosecuted "for praying inside their own homes without a permit," according to Dina Guirguis, a research fellow at the Washington Institute for Near East Policy.

In Congress, members of a bipartisan human rights panel named after the late Representative Tom Lantos, the only Holocaust survivor to serve in Congress, urged the administration to put pressure on the governments of Iraq and Egypt to protect religious minorities. "Going to the market and riding the bus, Iraqi Christians face death every day. There is no question that Christians are being targeted," said Representative Anna Eshoo (D-CA), the granddaughter of Assyrian Christians who fled the genocide carried out by the Ottoman Turks at the end of World War I. But the Obama White House ignored Congress, while giving preferential treatment to Iraqi Muslims seeking to immigrate to the United States. Meanwhile, tens of thousands of Iraqi Christians desperately waited in line for visas at United Nations offices in Jordan, Lebanon, and Syria, where they had fled for their lives from their jihadi Muslim assailants.[9]

When the Patriarch of the Maronite (Catholic) Church in Lebanon, Bechara Rahi, sought an audience with President Obama to warn about the ongoing persecution of Christians in the Levant, his request was rejected by Obama's religious affairs advisor, Dalia Mogahed, a prominent Muslim activist. But if you were a member or sympathizer of the Muslim Brotherhood, the White House rolled out the red carpet at Iftar dinners and private receptions, even if you had ties to known terrorist organizations. It was open season on Christians all across the Muslim world.[10]

Soon, it would also be open season on Americans as well.

THE STREET VENDOR

Act II of Obama's new policy toward the Muslim world began with a slap in the face and two trays of confiscated fruit.

Mohamed Bouazizi was a twenty-six-year-old street vendor in the sleepy town of Sidi Bouzid on the edge of the Tunisian desert. Since dropping out of school as a teenager, he eked out a modest living for himself and his family by selling fruits and vegetables from a cart in the local market. In a good month, he made around one hundred fifty dollars to support his mother, her incapacitated second husband, and six siblings.

Like many other street vendors, "Basboosa," as locals knew him, never had enough money to buy a proper license from the authorities, so he regularly ran afoul of the police, who would harass him on a whim. Sometimes, they would demand a small cash bribe to leave him alone. On other occasions, they would take a tray of fruit. Whenever he saw them coming, he knew he was about to lose his profit for the day, and sometimes for the week. The police were corrupt and everyone knew it, but no one did anything about it.

On December 17, 2010, Bouazizi set off to the market at 8 AM, as was his habit, with the cartload of produce he had purchased from the wholesalers the night before. His mother remembers that he was happy. The trouble began at 10:30 AM, when a female municipal official started her song and dance about him lacking a permit. It was still early in the day, and he didn't have enough to pay her the money she was demanding. And besides, he had had enough. He started shouting, complaining of the constant shakedowns. That was enough to provoke the forty-five-year-old official, Faida Hamdi, to slap him in the face and spit on him. According to some reports, she reached into his cart and took two trays of apples and ostentatiously put them in her nearby car. According to eyewitnesses, she ordered policemen with her to turn over Bouazizi's cart, scattering and kicking his produce on the ground. She confiscated his electronic scales, and told him he could get them back by showing up at the local governor's office to pay his fine. She later denied she had ever asked him for a bribe or roughed him up.

Angered and humiliated by the confrontation, he ran to the governor's office to ask for his scales back, but the local officials refused to see him. Later, people remembered hearing him shout to the guards, "If you won't see me, I'll burn myself."

And that's what he did. Going to a nearby gas station, he purchased a can of gas, ran back to the governor's office, and set himself on fire in the middle of the crowded street. "How do you expect me to make a living?" he shouted, as the flames licked at his clothes. Cars started honking and a crowd gathered in shock. Someone tried to pour water on him, but that only made it worse. By the time an ambulance arrived, he was nearly dead.

Organized protests against the way the local officials and the police had treated Mohamed Bouazizi began the next day and ripped through decades of resentment and suppressed rage like a wildfire. Within twenty-four hours, they transformed dusty Sidi Bouzid into the focal point of the Arab and international media. Soon, the protests spread to the capital, Tunis, and focused not on reform, but on an end to the regime of president Zine El Abidine Ben Ali. Within days, anti-government protests also erupted in Algeria. Within less than a month, President Ben Ali fled the country for exile in Saudi Arabia. Thus began the events now known as the Arab Spring, in a dusty oasis town in the dead of winter, born out of the humiliation of a street vendor named Mohamed Bouazizi.

The Obama White House and the State Department were slow to sense the winds of change. Top on the agenda that Christmas season was the repeal of the Pentagon's "Don't Ask, Don't Tell" policy on gays in the military, the controversial new START treaty then in the Senate, and administration efforts to prevent a repeat of the al Qaeda underwear bomber of 2009, who nearly succeeded in blowing up a U.S. passenger jet en route to Detroit on Christmas Day.

On New Year's Eve, two Muslim men set off a car bomb at a Coptic Christian church in Alexandria, Egypt, killing twenty-three parishioners and wounding ninety-six more during midnight mass. This time, President Obama had the White House press office issue a statement in his name, saying he deplored the casualties "from both the Christian and Muslim communities," even though only Christians were targeted. French President Nicolas Sarkozy, by contrast, understood the jihadi nature of the attack, and warned of "a particularly perverse program of cleansing in the Middle East, religious cleansing."

The very first mention of the momentous events in Tunisia by the administration was this exchange between an unnamed reporter and

State Department spokesman P. J. Crowley during his daily briefing on January 4, 2011, after nearly three weeks of daily protests including the shooting deaths of several protesters.

> **QUESTION:** On Tunisia, there's continued, sort of, civil unrest there, and I was just wondering—
>
> **MR. CROWLEY:** What country?
>
> **QUESTION:** Tunisia. Tunisia. And I was wondering what you made of the situation there.
>
> **MR. CROWLEY:** Actually, I didn't get updated on Tunisia today. So we'll save that question—
>
> **QUESTION:** When was the last time you did get updated on Tunisia? (Laughter.)
>
> **QUESTION:** How about next on Libya? [11]

And so it went, with the Washington press corps yucking it up with the officials they were responsible for covering as if together they were the only forces who made history. At the White House, it was total chaos. By January 13, 2011—the day before Ben Ali fled—White House spokesman Robert Gibbs had yet to mention Tunisia. As reports came over the wires the next day of Ben Ali's departure, a lone reporter asked Gibbs for comment. He deferred the question to Tom Donilon, Obama's recently appointed national security advisor.

> **MR. DONILON:** [O]n Tunisia, I was told literally as I was walking in, that President Ben Ali—it's been reported, at least—is leaving the . . .
>
> **MR. VIETOR:** We've not confirmed that.
>
> **MR. DONILON:** Have not confirmed that—okay. We've seen obviously the state media reporting that the president has dis-

missed the government and has called for legislative elections, and we're monitoring these developments, Jake. Obviously, in this case we would condemn the ongoing violence in Tunisia, call upon the Tunisian authorities to fulfill the important commitments made by President Ben Ali in his speech yesterday to the Tunisian people, including respect for basic human rights in a process of much needed political reform. . . .[12]

When the demonstrations moved to Egypt on January 25, 2011, the administration shifted gears, stung by allegations that they were asleep at the switch as momentous events were taking place in a critical part of the world, or simply incompetent when it came to defending America's interests. Obama was determined his administration would not repeat its pathetic performance in Tunisia. He sent former U.S. ambassador Frank Wisner, a close confidant of President Hosni Mubarak, to Cairo, with orders that the Egyptian president should step down or watch as the United States cut aid to his regime and forced him out. Seeing that his American ally had deserted him, just as Jimmy Carter had deserted the Shah of Iran before him, Mubarak caved in to the inevitable and turned over power to subordinates just three weeks later.

This apparent U.S. success in negotiating Mubarak's departure before a massive bloodbath had occurred led Obama strategist David Axelrod to boast that the White House had been "out in front" of the Arab Spring well before the protests had even begun. In fact, to listen to Axelrod, the removal of long-standing U.S. allies—now termed *dictators* by the White House and senior administration officials—was U.S. policy.

"The way [Obama has] confronted it, is he went to Cairo [in June 2009] and talked about the need, the universal human rights of people. He's on several occasions directly confronted President Mubarak on it. And pushed him on the need for political reform in his country," Axelrod told ABC's Jake Tapper Friday, on the advisor's last day of work at the White House.

"To get ahead of this?" Tapper asked.

"Exactly. To get ahead of this. This is a project he's been working on for two years, and today the president is working hard to encourage restraint and a cessation of violence against the people of Egypt," said Axelrod.

"Nice myth," said one human-rights advocate Tapper asked about Axelrod's description.[13]

But, in a way, Axelrod was right and was revealing the secret goal behind the seemingly chaotic response of the Obama administration to the fast pace of events in the Middle East. Rather than empower the forces of democratic change sought by the human-rights advocates and liberal democrats in those countries, the policies of President Obama and his administration enhanced the standing of the Muslim Brotherhood, which by then was just barely hiding its head in the shadows.

This apparent turnaround in U.S. policy caught establishment commentators by surprise. When White House spokesman Gibbs announced that the post-Mubarak government must "include a whole host of important non-secular actors," the *Washington Post* explained that the United States was "rapidly reassessing its tenuous relationship with the Muslim Brotherhood."[14] Of course, that was simply not true. Since Obama's secret meeting at the White House with Brotherhood leaders in April 2009 and his invitation to them to join him during his Cairo speech two months later, Obama's mind was made up: He preferred political Islam over secular regimes in the Middle East. "After all, Brotherhood operatives are in the American government and working closely with it, thanks to Barack Obama," wrote Robert Spencer. "Why shouldn't the same situation prevail in Egypt?"[15]

During the first stage of the anti-Mubarak movement, the Muslim Brotherhood worked behind the scenes, throwing their support behind former International Atomic Energy Agency chief Mohamed ElBaradei with backing from megabucks Obama ally, George Soros. "The Muslim Brotherhood's cooperation with Mohamed ElBaradei, the Nobel laureate who is seeking to run for president, is a hopeful sign that it intends to play a constructive role in a democratic political system," Soros wrote just as news that Mubarak would not stand for reelection hit the airwaves. "My foundations are prepared to contribute what they can" to their success.[16] ElBaradei was seen as a liberal democrat, acceptable to the West, despite the fact that as secretary general of the IAEA, he helped keep the lid on Iran's secret nuclear weapons program for well over a decade.

As soon as Mubarak actually left power on February 11, however, the Muslim Brothers shed the mask of secular outreach and moderation. Remember the thirty-year-old Google executive from Egypt who became the spokesman for the anti-Mubarak demonstrations in Cairo's Tahrir Square? Wael Ghonim was hip, he was young, he spoke fluent English. He contributed to the convenient falsehood that the Arab Spring was a "Facebook Revolution," organized by legions of anonymous youngsters online, where women were free, Christians were welcome, everybody was an Egyptian patriot, and nobody died. Just ten days after Ghonim catapulted to fame by demanding that Hosni Mubarak and his regime abandon power in Egypt, he himself was barred from addressing the crowds in Tahrir Square.

Who could have such power over him? Mubarak's military? The police? Some secret, black-hooded assassination squad? No. It was Sheikh Yusuf al-Qaradawi, the spiritual guide of the Muslim Brotherhood.

Qaradawi had left his comfortable home in Doha, Qatar, to remind the Egyptian masses at Tahrir Square of their duty as Muslims.

Sheikh Qaradawi wasn't even Egyptian, but a Palestinian who sub-scribed to the strict Wahhabi school of Islamic doctrine. He preached hatred of Jews, the eradication of the state of Israel, and openly encour-aged suicide bombings. Because of his ties to the 9/11 hijackers and al Qaeda, the Bush administration banned him from travel to the United States in 2002. In 2004, he issued a fatwa authorizing Muslims to kill Americans in Iraq. Just as Sheikh Rashid Ghannouchi, the exiled leader of the Muslim Brotherhood in Tunisia who returned home to a hero's welcome once the pro-Western, secular government of President Ben Ali was gone, Sheikh Qaradawi came to Cairo just days after Mubarak left power to give the signal to the Muslim Brotherhood in Egypt that it was time to come out of the shadows and push aside liberal Democrats such as Mohamed ElBaradei and Wael Ghonim. Sheikh Qaradawi's message was simple, but to the point: Islam's time had come.[17]

HILLARY'S PRIORITIES

As the Arab Spring leapt from country to country, engulfing pro-Western, secular regimes, you would think that the State Department would show some concern for the security of U.S. diplomats. Surely the extent of civil unrest and the possibility for armed confrontation between Islamists and supporters of the secular regimes they replaced was obvious enough that Hillary Clinton and her seventh-floor man-agement team understood they had to step up physical security at U.S. diplomatic outposts in the region.

In fact, just the opposite was the case, in particular as it concerns Libya. When the State Department released its budget on February 14, 2011, for the upcoming year, it included $455 million in aid to the Hamas-run Gaza Strip, close to $2 billion for United Nations opera-tions, $626 million on Family Planning and Reproductive Health—

much of it to promote abortion overseas—and just $250 million on regional security operations to secure our embassies in more than twenty-eight countries in the Middle East. Congress upped the security budget by $50 million over the objections of Hillary Clinton and her paper-pushers.[18]

So much for the myth, conveniently created after Benghazi went up in flames, that Republicans in Congress were responsible for the lack of security personnel—and therefore the attack—because they had supposedly slashed the State Department's budget.

5

QADDAFI, THE ENEMY

The wave of protests we now call the Arab Spring swept into Benghazi on February 17, 2011, smashing the old order like a tsunami. After two days of mainly peaceful demonstrations at the courthouse square overlooking the Mediterranean, opposition groups in exile called for a "day of rage" on February 17. Already, the Islamist influence was palpable. They chose the date to commemorate the Benghazi riots of 2006, when a large crowd stormed a European consulate to protest Danish newspaper cartoons portraying Muhammad, the founder of Islam, with a bomb in his turban. In Benghazi, the police opened fire on protesters, killing fourteen. The next day, Qaddafi loyalists in the al Fadeel Abu Omar barracks, known simply as "the Katiba," fired on the funeral procession for the initial victims, killing twenty-four more.

THE FIRST BATTLE OF BENGHAZI

Over the next three days, local youths stormed the Abu Omar barracks repeatedly, setting cars on fire, ramming the gates, and scaling the facade to tear down the Libyan flag. At one point, they even commandeered bulldozers in an attempt to breach the outer walls, as Qaddafi loyalists shot at them from inside. Muslim Brotherhood spiritual leader Sheikh Yusuf al-Qaradawi issued a fatwa on Al Jazeera TV, calling on the Libyan army to assassinate Qaddafi. As the death toll rose into the hundreds, Qaddafi sent his interior minister, General Abdelfattah Younis, to relieve the besieged troops with fresh soldiers from a nearby base. But instead of crushing the rebels, Younis announced he was joining them, and offered the remaining loyalists holed up in the barracks safe passage out of town.

After shooting comrades who had refused to fire on the protesters, Qaddafi's troops fled, and Benghazi quickly became the center of the anti-Qaddafi rebel movement.[1]

Al Jazeera, the Qatar-government satellite channel sometimes known as "Jihad TV," tried to paint the protests as Libya's version of the Arab Spring movement for "jobs, equality, and more political freedom."[2] However, the organized civil disobedience that had toppled governments in Tunisia and Egypt was largely missing in Libya. Forty-two years of Qaddafi's erratic but iron-fisted rule had stifled any form of political expression. With government informers seemingly everywhere, ordinary people were always looking over their shoulders, afraid to speak to foreigners, let alone voice dissent. There was no equivalent of the Muslim Brotherhood social organizations or secular NGOs that could morph into a political opposition overnight, as they had in Tunisia and Egypt. Instead, popular anger with Qaddafi simply boiled over; crowds formed in city after city, surging this way and that, storming government buildings, no longer afraid of the Qaddafi loyalists. Libya's civil war was a popular uprising from day one.

Driving the violence was loosely knit cells of jihadi fighters, forged in battle against the American infidels in Iraq and Afghanistan. They stormed government arms depots in Brega and Derna at the very outset of the protests, ensuring that the conflict turned into an armed insurrection from the very start. They quickly dominated the groups of activists in exile, such as the National Front for the Salvation of Libya, who used Facebook and Twitter from afar in an attempt to get people to take to the streets.

Qaddafi, too, responded differently from Ben Ali and Mubarak. For several days he disappeared from sight, giving rise to rumors he had fled the country as his diplomats and military units defected to the opposition. His son Saif al-Islam initially offered to discuss a package of reforms, and released 110 members of the Libyan Islamic Fighting Group even as the security forces clashed with armed protesters in Benghazi and Tripoli. Then on February 20, he gave a televised speech where he blamed the violence on "foreign agents," and warned of "rivers of blood" if the clashes didn't end. "We will fight to the last man and woman and bullet. We will not lose Libya. We will not let Al Jazeera, Al Arabiya and BBC trick us," he said. The threat to unleash the army from someone who held no government position only added fuel to the fire.[3]

The next day, Qaddafi himself made a brief appearance, telling a reporter that he was "still in Tripoli." In the brief film sequence he looks pathetic, a bewildered and cartoonish figure in his long desert robes and black cap, clutching an open umbrella in the backseat of a vehicle, looking for all the world like he had been surprised squatting in an outhouse. But he came roaring back in an hour-long speech the next night, calling the protesters "rats" and "cockroaches," vowing to hunt them down "inch by inch, house by house, room by room, alley way by alley way." An Israeli musician, Noy Alooshe, set the speech to rap music, creating a spoof called "Zenga Zenga" (Arabic for "alley way") that became an instant YouTube hit with over four million views. A sanitized version, without the scantily clothed dancing girls in the background, received an additional million views.[4]

Saif al-Islam al-Qaddafi had long believed he could negotiate his way out of confrontation with the LIFG. At the urging of the United States, following the resumption of diplomatic relations in 2004, Saif got his father to allow Amnesty International to visit the infamous Abu Selim prison, where they found 300 political prisoners, not 3,000 or 30,000 as some claimed.[5] Even as the United States and Britain began to "render" Libyan jihadists captured on the battlefields of the Global War on Terror, Saif al-Islam's Qadhafi Development Foundation devised a rehabilitation program similar to one pioneered in Saudi Arabia, aimed at taking the jihad out of the jihadi, so that he could return to society and live a normal life. By March 2008, U.S. Chargé d'Affaires Chris Stevens wrote that the Qadhafi Foundation was negotiating with the LIFG over "the possible release from prison of approximately one-third of the cohort of LIFG members currently imprisoned in Libya."

How many al Qaeda terrorists did the barbaric Qaddafi regime hold in its prisons? Around ninety-three, Stevens reported. Saif al-Islam al-Qaddafi told the Americans he believed they had undergone "a sincere ideological revision" and could be released with minimal risk, since "reconciliation among former enemies was a 'natural part' of Libya's tribal culture." Clearly, he was wrong.[6]

LEADING FROM BEHIND

CNN reporter Ben Wedeman, the first Western journalist to enter Libya once the fighting began, documented the total collapse of the Libyan government in the east of the country, just four days after the fighting of the 17th February Revolution started in earnest. Coming in from Egypt, he found no officials, no passport control, no customs. "I've seen this before," Wedeman wrote. "In Afghanistan after the rout of the Taliban, in Iraq after the fall of Saddam Hussein. Government authority suddenly evaporates. It's exhilarating on one level; its whiff of chaos disconcerting on another."[7]

The international response to the violence in Libya was as rapid and unexpected as the events inside Libya itself. Less than one week after the first protests, former British foreign secretary David Owen called for a NATO no-fly zone to protect the civilian population from government air strikes. On February 26, just nine days after the uprising began, the United Nations passed a resolution placing economic sanctions and an arms embargo on the Libyan regime and referring Qaddafi to the International Criminal Court. That same day, President Obama called for Qaddafi to leave power.

On February 28, the U.S. Navy began positioning warships off Libya's coast; British Prime Minister David Cameron joined the call for a NATO intervention. Al Qaeda now officially said it would back the Libyan rebels, who had organized themselves in Benghazi into an ad hoc Transitional National Council. It was total war.[8]

While the French and the British were clamoring for NATO action, Defense Secretary Robert Gates opposed it, "point[ing] out a fact that many people didn't seem to understand: the first step in creating a no-fly zone would be to bomb the Libyan air defense," a move that inevitably would engage the United States in its third simultaneous war in the Muslim world.[9] Pushing for intervention were Secretary of State Hillary Clinton and National Security Council aide Samantha Power, the in-house champion of the "responsibility to protect," a much-disputed doctrine to prevent the massacre of civilians by dictatorial regimes. "R2P" groups, as they came to be known, had sprung up around the world and were lobbying heavily for international military intervention in Libya. In his account of these heady days, Gates warned aides and subordinates to refer any calls from Power to him, for fear she would convey specific target coordinates she had obtained from anti-Qaddafi groups. "Don't give the White House staff and NSS too much information on the military options," he wrote. "They don't understand it, and 'experts' like Samantha Power will decide when we should move militarily." [10]

On March 15, Qaddafi's troops were poised to take Ajdabiyah, the logistics hub that controlled food and fuel supplies to the rebel stronghold in Benghazi. If Ajdabiyah fell, Benghazi was next, and everyone feared that a victorious Qaddafi would hunt down his opponents house by house, as he had promised, in a massacre of staggering proportions. At a meeting in the Situation Room early that evening, Obama asked his advisors if a no-fly zone would stop Qaddafi's advance. The question itself revealed the president's lack of understanding of modern warfare, since Qaddafi was using tanks, not his air force, to stamp out the rebellion. When they told him the truth, Obama asked his aides to come up with some more robust military options, and left for dinner. [11]

That was how UN Security Council Resolution 1973, which passed on March 17, 2011, became an authorization for NATO and consenting Arab League countries to launch a full-scale military intervention in Libya, not just a no-fly zone. The United States called it Operation Odyssey Dawn and mobilized active-duty and Air Force reserve units from across the country, moving them into theater within seventy-two hours. KC-10 tankers came from as far away as March Air Reserve Base in Southern California to refuel combat aircraft eventually supplied by nineteen participating nations. On the ground in Benghazi, rebel leaders were calling the new U.S. effort "Hillary's war," openly mocking what they perceived as Obama's indecisiveness. [12]

While the United States provided the bulk of the military assets for the UN-approved fighting, to the tune of $550 million of U.S. taxpayer dollars for the first two weeks alone, the White House continued to claim that the United States was just playing a "supportive" role in a NATO-led United Nations operation. This gave rise to the famous description of the Obama Doctrine by a White House official to *New Yorker* journalist Ryan Lizza as "leading from behind."

"That's not a slogan designed for signs at the 2012 Democratic Convention, but it does accurately describe the balance that Obama now seems to be finding," Lizza wrote. "It's a different definition of leadership than America is known for," he added—an understatement, if ever there were one.

While supporters of the administration later argued that the term was unfair, noting that without U.S. leadership the United Nations never would have approved UNSC Resolution 1973 authorizing war against Qaddafi, leading from behind gave the impression of a timid U.S. foreign policy, operating by stealth, by an administration eager to shift political responsibility onto others. And that was far more accurate a description of Obama's agenda than anyone outside the White House then understood.

And so was that other phrase: Hillary's war.

BEARING GRUDGES

It was not just the brutality of Qaddafi's response to the armed uprising that turned the West against him. The change of heart among leaders of Britain, France, Italy, and the United States had been in the works for some time.

Despite Qaddafi having warmly welcomed Barack Obama as a "brother" and "fellow Muslim" during the 2008 election campaign— or perhaps, precisely because of it—relations with the United States gradually soured once Obama took office.[13] U.S. Ambassador Gene Cretz sent back a slew of cables to Hillary Clinton's State Department detailing the debauchery and corruption of the Qaddafi family, and magnifying every dispute with Qaddafi regime officials. There were no more "Dear Musa" letters from top CIA officials to Qaddafi's intelligence chief. Indeed, a request by Musa Kusa to Ambassador Cretz to arrange a one-on-one meeting for Qaddafi with President Obama during a G-8 summit in Italy in May 2009 was politely turned down after the embassy criticized Qaddafi for pressuring U.S. oil companies to make supposedly voluntary contributions to his own foundation, which he then used to pay terrorism compensation claims stemming from the Lockerbie bombing.[14]

In addition, Libya's nonchalance about taking back remaining Libyan prisoners in Guantánamo was decidedly unhelpful to the president's campaign promise to shut down the terrorist prison camp. "We are not overly concerned about them," Foreign Minister Kusa told Ambassador Cretz.[15]

Beyond that, the Qaddafi clan had big appetites. They felt they had not been adequately compensated for their decision to give up Libya's WMD programs and abandon terrorism in 2003, and wanted "security assurances in the form of a defensive alliance with the United States," Cretz cabled Washington in July 2009.[16] President Obama and his advisors found such a close embrace of Qaddafi distasteful—and potentially embarrassing—given the ongoing detention of political prisoners and his repression of the Islamist opposition. As Chris Stevens reported when the Lockerbie compensation arrangement was finalized, there were "high expectations . . . that the U.S. will seek to capitalize on the new tenor of the relationship to press Muammar al-Qadhafi to open further political space—particularly with respect to respect for human rights, freedom of the press and an expanded role for civil society—in what remains a tightly-controlled society." Clearly that never happened.[17]

Similarly, the British government was publicly embarrassed when a secret deal to release convicted Lockerbie bomber Abdulbasset Megrahi from a Scottish jail leaked to the press. The government initially claimed it had agreed to Megrahi's release on humanitarian grounds, because he had terminal cancer and doctors said he had just months to live. As it turned out, the doctors never made such a claim and Megrahi was released following extensive Libyan pressure, in order to secure a major new oil concession for British Petroleum. Making the scandal worse was the hero's welcome Qaddafi gave Megrahi when he returned to Libya on August 21, 2009, and the fact that the deal was reportedly arranged by Mark Allen, the former MI6 officer who brokered Qaddafi's return to the fold in 2003 and was now serving as a highly paid consultant to BP.[18]

Italian Prime Minister Silvio Berlusconi bore a grudge against Qaddafi ever since his flamboyant—and insulting—trip to Rome in June 2009. When Qaddafi emerged from one of the four aircraft ferrying the 446-member delegation to Rome, he had pinned on his favorite Sergeant Pepper uniform a picture of Libyan resistance leader Omar al-Mukhtar in chains during the Italian occupation of 1911–42. The gesture set Italian heads spinning, according to Italy's ambassador to Tripoli, Francesco Trupiano. Then, strutting before Berlusconi, who was writhing in pain from a recent back injury, Qaddafi went on to introduce "all of the individuals who also wore pictures around their necks of relatives who had suffered Italian cruelty or of themselves. Each told a brief story to the PM," Trupiano added. It was a day Berlusconi would not forget.[19]

Qaddafi's former Arab allies, Qatar and Saudi Arabia, bore him similar grudges. The Saudis despised him since his attempt to sponsor the assassination of Crown Prince Abdullah in 2003 through U.S.-based Muslim Brotherhood operative Abdulrahman Alamoudi. The Qatari emir nurtured a deep personal hatred for Qaddafi because he felt slighted by disparaging comments Qaddafi had made about him at an Arab summit. Through Ali al-Tarhouni, an economics professor at the University of Washington until he was named finance minister for the provisional government in Benghazi in February 2011, the Qataris arranged to finance the rebellion by agreeing to buy and market billions of dollars' worth of Libyan crude oil from ports controlled by the rebels.[20]

Qaddafi's biggest mistake may have been to alienate the British and the French by refusing their demands for lucrative oil concessions and big-ticket arms purchases. "If only Qaddafi had spent a few billion dollars buying French Rafale fighter jets," the French publisher of *Intelligence Online* told me, "President Sarkozy would have supported him all the way." As it was, Sarkozy pounded a never-ending drumbeat for war and was widely seen as leading the NATO charge to remove Qaddafi.

For President Obama, it was not about the money. It was about principle. Qaddafi the secularist, who like Mubarak had crossed swords with the jihadis, was a relic of the past, while political Islam was America's friend.

ARMING THE ENEMY

Warning signs appeared almost immediately that aiding the rebels seeking to overthrow Qaddafi would empower jihadi groups. Al Qaeda in the Islamic Magreb (AQIM), the local franchise, had issued a call to jihad against Qaddafi just one week after the fighting began, calling his overthrow "the first stage" of imposing Islamic law in the country. More warning signs soon followed.

The real story of Benghazi involved the same sort of blowback that journalists castigated Ronald Reagan for creating in the 1980s in Afghanistan, when U.S. weapons went to radical Islamist groups that ultimately spawned al Qaeda. The big difference was the response of the American media. When U.S. weapons started flowing to al Qaeda–affiliated groups on Obama's watch in Libya, and later, in Syria, few in the national media seemed to care.

As the battle for Ajdabiyah continued to rage, rebel leader Abdel-Hakim al-Hasidi boasted to the Italian newspaper *Il Sole 24 Ore* that he and his men had cut their teeth fighting Americans. After the September 11, 2001, attacks on America, al-Hasidi had gone to Afghanistan to join the resistance against the "foreign invasion," where he fought side by side with the Taliban and al Qaeda. Captured in 2002 near the Afghan border in Peshawar, Pakistan, he was handed over to the Americans, who interrogated him then rendered him to the Libyan authorities. He remained in a Libyan jail until 2008, when he benefited from the terrorist rehab program championed by Qaddafi's son, Saif al-Islam.

Al-Hasidi was one of the *Die Hard in Derna* crew. As he told the Italian reporter, after his release from jail he personally recruited "around twenty-five men" from Derna, just east of Benghazi, to fight the Americans in Iraq. Some of them, he said, are "today on the front lines in Ajdabiyah," the battle whose outcome so preoccupied President Obama that it compelled him to seek a much broader UN mandate to topple Qaddafi.

His fighters were "patriots and good Muslims, not terrorists," he insisted. He added, "Members of al-Qaeda are also good Muslims and are fighting against the invader." Al-Hasidi was a member of the Libyan Islamic Fighting Group, which formally joined forces with al Qaeda in 2007, as Chris Stevens reported to Washington at the time.[21]

Al-Hasidi was not the only al Qaeda member or former terrorist detainee who joined the Libyan rebels and received arms and assistance from the U.S.-led coalition. Abu Sufian Ibrahim Ahmed Hamuda Bin Qumu, known in Guantánamo as detainee 557, had an extensive terrorist pedigree. Another *Die Hard in Derna* former LIFG member, he cut his teeth with Osama bin Laden and his Arab Afghans fighting the Soviets in Afghanistan in the 1980s. After escaping from a Libyan jail in 1993, where he was serving a sentence on drug charges, he reconnected with bin Laden in Sudan. At one point, he drove a truck for Wadi al-Aqiq, one of bin Laden's front companies.

In 1998, Bin Qumu returned to Afghanistan and joined the Taliban. Thanks to a tip-off from informants assisting the Qaddafi Foundation to locate Libyan jihadis who were fleeing Afghanistan after the 9/11 attacks, the Pakistani police arrested him at the Plaza Hotel in Peshawar in late 2001. After an initial round of interrogations, they turned him over to the Americans, who sent him to Gitmo, where he enjoyed three square meals a day, luxury soccer fields, and a prayer mat for five and a half years, thanks to the U.S. taxpayer.[22]

In August 2007, Bin Qumu was voluntarily repatriated to a Libyan jail through the intervention of the Qaddafi Foundation, following the successful return of al-Rimi the year before. As noted previously, Chris Stevens took up the cause of the returnees with the Libyan authorities, making sure they weren't tortured and could receive family visits. He even devoted Christmas Day 2007 to meet with the detainees in a Libyan jail. In the State Department's bureaucratese, they called Stevens' visit with the man whose group later claimed responsibility for killing him, a "whereabouts and welfare visit."[23]

Former federal prosecutor turned author Andrew C. McCarthy could see what was coming. "The rebels are not rebels—they are the Libyan mujahedeen," he wrote in *National Review* at the very start of the civil war. "Like the Afghan mujahedeen, including those that became al-Qaeda and the Taliban, the Libyan mujahedeen comprise different groups. What overwhelmingly unites them, besides opposition to Qaddafi, is sharia," the supremacist doctrine of Islamic law. "The Libyan mujahedeen will exploit us but never befriend us . . . [[I]f we empower them, we will eventually rue the day."

After his release by Saif al-Islam in 2010, Bin Qumu formed the Ansar al-Sharia *katiba* (brigade) with funding and weapons from Iran and Qatar, and played a key role in radicalizing the anti-Qaddafi rebels. In a Facebook interview, he "discoursed at length about his resentment of the United States, which he accused of torturing him during his Guantánamo detention, an experience that he said will not go away."[24] So deep was his hatred of Westerners that, when a Human Rights Watch team interviewed former detainees after the fall of the Qaddafi regime to promote their cause, Bin Qumu was the only one they located who refused to meet with them.

In the end, he got revenge, but on the one person who had actually tried to help him.

ELUSIVE STINGERS

Al Qaeda in the Islamic Magreb and its affiliates were already looking beyond Libya; indeed, they felt a softer target than Qaddafi could be found in neighboring Mali, and thus they began to pillage abandoned Libyan government arms depots, carting away everything they could find into the vast Libyan desert to the south for use another day. According to Chadian President Idriss Deby, a Qaddafi ally, they had even acquired SAM-7 surface-to-air missiles. "AQIM is becoming the best equipped army in the region," Deby told the French-language weekly *Jeune Afrique*.[25]

Fears that al Qaeda would acquire surface-to-air missiles, or MAN-PADS, in Libya were very real. The country was awash with weapons. During his forty-two-year reign, Qaddafi had purchased more than 20,000 MANPADS from the Soviet Union and China, and no one really knew where they were. While most modern military aircraft had electronic countermeasures to detect and divert heat-seeking missiles, civilian airliners were ripe targets. Indeed, sheer luck foiled an al Qaeda attack on an El Al plane filled with Israeli tourists in Mombasa, Kenya, in November 2002. The pilots watched in horror as two missiles streaked past them while they were climbing after takeoff. A single hit would have brought the plane down and caused several hundred deaths.

Photographs and YouTube videos during the first weeks of the fighting showed rebel fighters posing with SAM-7 and more modern SAM-24 missiles plundered from Qaddafi's arsenals. An Australian TV reporter, Ben Knight, reported in early March 2011 that rebel fighters had shown him a U.S.-built Stinger missile, far more sophisticated and deadly than the Soviet-era SA-7. "In fact, they are claiming to have shot down another jet fighter today as well as another helicopter," Knight reported.[26]

When I asked him about this incident, Knight told me that the rebels were eager to show off their weaponry. "I have a vague memory of being shown the missile in the back of a pickup truck before they drove off. I would have been using the term *Stinger* as a generic term for missiles with that capability, after discussion with other correspondents on the road who are more experienced in military hardware than I am."[27]

But other, military, experts did see Stingers.

THE QATARI CONVOY

The French have maintained a military force in Chad, known as Operation Épervier ("Sparrowhawk"), ever since Libya occupied the uranium-rich Aouzou Strip along their common border in the Sahara in 1986. In February 2008, the French helped Chadian President Idriss Deby fend off a deadly challenge to his power after Sudanese-backed rebels seized control of the capital, N'Djamena. By this point, Deby and Qaddafi had become allies, and the French made emergency arms shipments of Milan anti-tank missiles and other equipment through Libya to Chad to bolster Deby's forces.[28]

As part of Operation Épervier, the French regularly sent armored patrols from their northern base at Faya Largeau, to listen, watch, and roam the vast reaches of the Sahara along the Libyan and Sudanese borders. So it was on April 4, 2011, that an Arabic-speaking communications specialist picked up a suspicious military transmission not far ahead of their position.

They're having problems with a truck that got stuck in the sand, he told his commanding officer. They're not locals. They have an accent from the Gulf.

The commander gave the orders to intercept the unidentified convoy and politely inquire why they were heading toward the Libyan border from Sudan in the midst of a civil war. When the French patrol caught up with them, they were stunned.

"It was a large convoy of military trucks and pickups," a senior U.S. Special Forces commander familiar with the incident told me. "The people in charge were Qatari special forces. They were carrying Stinger missiles and French-made Milan anti-tank rockets for the Libyan rebels."

The French commander radioed back to Faya, which in turn alerted Paris via Épervier headquarters at the Sergent-chef Adji Kosseï Airbase in N'Djamena. His men had surrounded the Qatari convoy. With their ERC-90 Sagaie wheeled tanks they had them outgunned, but the commanding officer gave orders to keep the atmosphere relaxed. After all, Qatar was a huge weapons client of France, and the Qatari officer in charge quickly made known that he had gone to school at the French military academy at St. Cyr. He ordered his Pakistani orderlies to set up a hexagonal white tent with a portable generator hooked up to an air-conditioning unit, and invited the French commander inside to take tea while they waited for a response from Paris.

Several hours later, after much back-and-forth by radio and the intentional use of an easy-to-intercept satellite phone, the answer came in. "They were told to stand down and to shut up," the Special Forces officer said.

A senior advisor to French President Sarkozy explained to me exactly where that order came from. "It went all the way up to the Chef d'Etat Major," the French equivalent of the chairman of the Joint Chiefs of Staff. "Since these were American weapons, and this was clearly an intelligence operation, he would have referred it to the CIA Station Chief in Paris. If the agency wasn't okay with it, we would have stopped it."[29]

The French weren't the only ones to learn about the Qatari arms convoys to the Libyan rebels. In September 2011, once Qaddafi and his lieutenants had fled Tripoli, a reporter from the *Wall Street Journal* stumbled upon documents relating to this and related incidents in the bombed-out office of Abdallah Senoussi, Qaddafi's brother-in-law and chief of his internal security apparatus. "One memo contained intercepted phone calls between military commanders in Chad who reported Qatari weapons convoys approaching Libya's southern border with Sudan, apparently intended for anti-[Qaddafi] forces," the *Journal* reported. "Another intelligence memo, dated April 4, warned that French weapons, including Stinger antiaircraft missiles and Milan antitank rockets, were making their way to Libyan rebels via Sudan."[30]

The game was on.

My sources believe that the Qataris delivered a total of four hundred Stinger missiles and fifty launchers, known as gripstocks, to jihadi groups in Libya in those early months of the civil war. They acquired some of these missiles from U.S. Army depots located at Camp Virginia and Camp Arifjan in Kuwait. My sources say the missiles were initially sold to Saudi Arabia as part of a massive $60 billion arms deal in 2010, and then retransferred quietly by the CIA to Qatar with the assistance of Saudi Intelligence Chief Prince Bandar bin Sultan, at the beginning of the insurrection. "Nobody in the administration wanted U.S. fingerprints on this deal, so the actual transfer was done between the Saudis and the Qataris," a knowledgeable source who was involved in previous covert Stinger transfers told me.

I was skeptical that the Saudis would work so closely with the Qataris, whom they despised. I pointed out that one of Osama bin Laden's front men in London boasted to me in early 1998, when I did the first-ever detailed profile of bin Laden to appear in any major U.S. publication, that a limousine bearing Qatari diplomatic plates was bringing them $50,000 in cash every month to support their propaganda efforts against the Saudi royal family, and that the Saudis were well aware of the Qatari support for their enemies.[31]

"The Saudis didn't have the special forces personnel trained to the level of the Qataris, so they held their noses and used the Qataris as proxies," a special operations officer with decades of experience told me. "This was done with the full knowledge and approval of the administration, which did not want an American military footprint in Libya. Brennan and Panetta thought they had a foolproof plan that laid the groundwork for the later gun-running to Syria," he added.

Former CIA operations officers, knowledgeable about covert Stinger transfers to the Afghan mujahideen in the 1980s, told me they did not believe the CIA had the appetite for such risky transfers today. Afghanistan continued to haunt the agency, making them reluctant to do it again. If such an operation were ever exposed, they knew they would have to lawyer up, and that it would probably bankrupt them. Beyond that, the State Department had engaged since 2002 in a worldwide effort to collect missing MANPADS, sending out diplomats and spies to buy them back from black-market vendors, militias, and governments. Some of these programs have been very successful and have taken thousands of these missiles off the streets. Others have become giant slush funds for Arab intelligence agencies, eager to get U.S. walking-around money to buy off political rivals, tribal enemies, and jihadi groups. "The CIA was taking these weapons off the street, not distributing them," a former senior CIA operations officer told me.

"There's not a single operations officer anywhere in the world who is stupid enough to do this kind of thing without a presidential finding," another former CIA operations officer who had been involved in covert weapons transfers and the arming of proxy forces told me.

The CIA declined to comment on this story.

A great deal of hype surrounds Stingers. As mentioned in the incident with the Australian journalist earlier, journalists and commentators have a tendency to call any shoulder-fired surface-to-air missile a Stinger, with images of bearded Afghan mujahideen in mind. Accounts of how many Stingers the CIA actually provided to the Afghan muj also vary wildly, from five hundred to two thousand. So do accounts of how many of these missiles went missing. The most famous incident occurred in 1987, when Iran acquired a truckload of Stingers from a muj leader named Younis Khalis—either confiscating them from him when he strayed across the Afghan border into Iran in May 1987, or buying them. They acquired an estimated twelve missiles from Khalis and used some of them to fire unsuccessfully at U.S. aircraft in the Gulf.[32]

In March 1988, Qatar put Stingers on display during a military parade in Doha that attracted the frenzied attention of U.S. diplomats and the CIA. The State Department dispatched Assistant Secretary of State Richard Murphy to visit the island kingdom that June, demanding to get the serial numbers of the missiles so that the United States could track their origin. The Qataris acknowledged buying twelve missiles on the black market, but refused to provide their serial numbers or reveal their supplier. The Reagan administration made their recalcitrance public, prompting Congress to ban U.S. arms sales to

Qatar until they returned the missiles, which they never did. Since the United States wasn't selling any weapons at the time to Qatar, the arms ban was lifted two years later without fanfare. Qatari rival Mohammed bin Zayed of Dubai later told U.S. officials that the Qataris had purchased the missiles from the "Taliban," a shortcut meaning one of the more radical Afghan jihadi groups.[33]

But, at the same time, the agency might be trying to cover its own tracks. With more than 70,000 Stingers produced to date, the United States has been selling these deadly missiles all across the Middle East, including countries such as Turkey, Saudi Arabia, and the United Arab Emirates, all of whom supported the anti-Qaddafi fight. While Raytheon Missile Systems has signed a consent agreement with the State Department after admitting to violating seventy-six of the 480 export licenses it reviewed, it does not appear that these violations involved unauthorized Stinger sales by Raytheon's foreign clients.[34] "If there was a Stinger sale to Qatar, it most likely would have been done by the intelligence community," a U.S. official knowledgeable about the licensing process told me. Going through the licensing process, with all its notifications to Congress and the public, would have been too risky.

Attorney Joseph diGenova, who represented U.S government employees who were being prevented from meeting with members of Congress because they were involved in or knowledgeable about the Benghazi attacks, told Fox News that his clients believed as many as four hundred Stingers were delivered to the Libyan rebels. For such deliveries to be legal required a presidential finding, a classified document authorizing the U.S. intelligence community to conduct a specified covert operation.

Whether Obama actually signed such a finding is one of the key questions of the Benghazi investigation. Failure to have done so could be an impeachable offense.

Just days before the French patrol stumbled upon that Qatari arms convoy in the Sahara, an Obama administration official made an authorized leak to Reuters reporter Mark Hosenball, stating that the president had indeed signed such a finding "within the last two or three weeks." The secret order "authoriz[ed] covert U.S. government support for rebel forces seeking to oust Libyan leader Muammar Qaddafi."

When asked to comment on the story, White House spokesman Jay Carney issued a nondenial denial. "As is common practice for this and all administrations, I am not going to comment on intelligence matters," he said.

Hosenball went on to explain:

> In order for specific operations to be carried out under
> the provisions of such a broad authorization—for
> example the delivery of cash or weapons to anti-Qaddafi
> forces—the White House also would have to give ad-
> ditional "permission" allowing such activities to proceed.
>
> Former officials say these follow-up authorizations
> are known in the intelligence world as "Mother may I"
> findings.[35]

President Obama and other top members of his administration made a series of statements in the weeks before this report came out, suggesting that they were contemplating shipping arms to the Libyan rebels. "It's fair to say that if we wanted to get weapons into Libya, we probably could," Obama told ABC News anchor Diane Sawyer. "We're looking at all our options at this point."

But for the chairman of the House Permanent Select Committee on Intelligence, such a venture was fraught with danger. Recalling the blowback from Saudi-sponsored arm sales to the Afghan mujahideen in the 1980s, Republican Congressman Mike Rogers said he opposed arms deliveries to the Libyan rebels "at this time. . . . We need to understand more about the opposition before I would support passing out guns and advanced weapons to them."

The leak about a supposed presidential finding came out just as Secretary of State Hillary Clinton, Defense Secretary Robert Gates, the chairman of the Joint Chiefs, and the director of national intelligence were conducting classified briefings with congressional committees on the ongoing U.S. military strikes against the Qaddafi regime. By statute, the administration is required to brief presidential findings to the top Republican and Democrat of the House and Senate intelligence committees, as well as the majority and minority leaders of both houses of Congress, before any operations take place.

Surprised by reporters in the corridor, Clinton "refused to say why the order was not discussed in the briefing." But the top Democrat on the House Intelligence Committee, Maryland Representative Dutch Ruppersberger, was categorical: "I have no knowledge of what [the president] signed."[36]

So, did Obama actually sign a presidential finding authorizing the CIA to arm the Libyan rebels? Michigan Republican Pete Hoekstra, who had been the ranking member of the House Intelligence Committee until he retired from Congress just two months before the Libyan Civil War began, thinks not. "It wouldn't necessarily have been needed, even later, to arm the Syrian rebels," he told me. "They might have been able to do it under existing authority and, knowing Obama, they probably did."

Despite the incessant, even vindictive criticism of the previous administration by President Obama and his top aides, the Obama White House never revoked the Bush-era presidential findings that authorized a broad range of intelligence operations, including covert arms deliveries. "That is absolutely astonishing," Hoekstra said. "If you go into the classified holdings room in Congress and ask to see the binder of presidential findings, you'll see that it is pretty small," he added, holding up his thumb and forefinger less than an inch apart. "When one of them comes in, you notice."

"Given what I have seen over the past four years, I am not surprised that we gave MANPADS to the Libyan rebels, and that they leaked to our enemies and eventually showed up in Afghanistan, where they were used against U.S. forces," Hoekstra said. "The real scandal in Benghazi was finding out that we had trained and armed the same guys who later attacked us." [37]

Senator James Inhofe, the ranking Republican on the Senate Armed Services Committee, told me that arming the Libyan rebels was legal: "The president has tremendous powers. I think he exercised those powers." But what happened later, when weapons from Libya ended up in the hands of rebels in Syria, was a different story. "Arming the rebels in Syria—they can do that and nobody has to know," Inhofe said. "We wouldn't know it, probably not even today, unless somebody had leaked it out of the White House. You would never know about it." [38]

Ali Aujali was the Libyan ambassador to Washington for Qaddafi, but joined many of his colleagues in defecting to the rebel cause early on during the civil war. He retired on June 30, 2013, after representing the Transitional National Council and eventually the newly elected Libyan government in Washington after the war. He is still appreciative of the help the Qataris gave the rebel cause.

"When we were desperate and had no weapons and no money to buy them, Qatar came forward to help the Libyan people," Aujali told me. "They provided all our needs, with as much as they can, and were one of the first countries to come forward. Qatar helped us in a very tough time." [39]

By having the CIA encourage the Qataris to deliver weapons to the Libyan rebels while not getting involved in the logistics of the operation, the Obama administration was trying to have it both ways: make sure that the rebels got the weapons, while assuming no responsibility, and paying no political price if things went wrong.

Obama was finally joining Hillary's war, and virtually dared Congress to stop him.

6

INTO THE DANGER ZONE

Secretary of State Hillary Clinton appointed Chris Stevens her special envoy to the Libyan rebels in early March 2011, sending him to Valetta, Malta, to join Ambassador Gene Cretz and other U.S. diplomats who had been evacuated from Tripoli when the U.S Embassy was shut down a few weeks earlier.

As the civil war turned increasingly bloody and the U.S. military involvement intensified, Hillary sent Stevens into Libya to become her eyes and ears regarding the Libyan rebel council. Such a move was unprecedented in recent history. In Iraq and Afghanistan—and in other cases before them—the CIA handled such tasks. Stevens booked passage on a Greek high-speed ferry, the *Maria Dolores*, and arrived in Benghazi on April 5, 2011, along with a retinue of heavily armed guards and several armored cars. This unusual step by the State Department later gave rise to breathless claims that Chris Stevens was a CIA officer operating under State Department cover. As I have mentioned earlier, I found no evidence to support those claims.

"WE CAME, WE SAW, HE DIED"

Now that the NATO air strikes were starting to erode Qaddafi's forces, Hillary wanted to put her stamp on the Libyan rebellion. She could smell success, and wanted to own it. "We came, we saw, he died," she joked with a reporter when she finally visited Libya, the day after Qaddafi was killed.[1]

A State Department account of these unusual days, published the year before Stevens' murder, called it "expeditionary diplomacy." Stevens said the Libyans were "genuinely grateful to the United States for supporting their aspirations for freedom, as demonstrated by the greeting the team received." As he told the department's official chronicler, when they got off the ferry in Benghazi they were warmly welcomed with "British, French, Qatari and American flags at Freedom Square, the vast open area in front of the Benghazi courthouse." They were also greeted with nearly constant automatic weapons fire. This prompted Chief of Security Keith Carter to keep Stevens on board ship that first night, as eight Diplomatic Security Service agents and a young political officer, Nathan Tek, scoured the town in armored cars in search of secure lodgings. They eventually settled on a suite of rooms at the formerly government-run Tibesti Hotel on an inlet just behind Benghazi's Mediterranean Sea coast, a luxury palace that had been taken over by international news organizations and diplomats, and moved in the next day.

With his love of the Arab world and his romantic attraction to Sufi Islam, Chris Stevens was the perfect expression of the worldview shared by Hillary and Obama, which saw political Islam as a positive force for change. He already knew many of the Libyan rebel leaders from his previous posting to Tripoli, including some of the most radical among them. From his new outpost at the Tibesti Hotel, he expanded his contacts and put a very public face on the U.S. effort to forge the Transitional National Council (TNC) into a coalition capable of ruling the country. "My mandate was to go out and meet as many members of the leadership as I could," Stevens said. Traveling in a heavily armored convoy, Stevens went out to meet the commanders of the various *katibas*, or rebel battalions, as well as "members of the emerging civil society and newly free news media," wrote Mario Montoyo, one of the Diplomatic Security Service agents guarding Stevens. He was the American viceroy, the personal representative of the most powerful leader on earth. Everyone wanted to be seen with him.[2]

The TNC leaders came from a wide variety of backgrounds: "former Qadhafi-era officials who had defected, academics, lawyers, doctors, military officers and volunteer fighters—who were united in a desire to overthrow Qadhafi." Stevens' job was to evaluate these leaders, each of whom had his personal *katiba*, and determine who should get what. The official story, of course, was that the United States was only providing "nonlethal military assistance" to the Libyan rebels "for the protection of civilians and civilian-populated areas." How you can "protect" civilians from enemy artillery and ground attacks with "nonlethal" equipment is anyone's guess. The notion that the United States left the arming of the rebels to others was a convenient fiction.

DS Agent Montoyo revealed two other aspects of Stevens' mission to Benghazi that were not well known at the time. First was Stevens' responsibility for "launch[ing] the U.S. government's cooperative program with the council to collect dangerous weapons such as shoulder-fired anti-aircraft missiles." The MANPADS program began during the Bush administration as an outgrowth of international counterterrorism cooperation. Although diplomats were in charge, the actual groundwork was done by former Special Operations troops, such as Glen Doherty, who hired on as contractors. Stevens' job was to use his skills as a diplomat to convince rebel leaders to turn over their weapons (for the right price), so U.S. military personnel then could collect them, disarm them, or destroy them. Before anyone realized that all but two dozen of the 300 or 400 Stingers shipped to the rebels by Qatar had disappeared, the State Department was touting its MANPADS collection effort as a big success story. After the attack on the Special Mission Complex, the program dropped off a cliff. Indeed, the United Nations concluded in March 2014 that "thousands of MANPADS were still available" in Libya.[3] The U.S. program appeared to have actually *aided* the proliferation of these weapons, not staunched it. And Chris Stevens was right in the middle of it.

The second little-known task handed to Stevens was to perform "the unseen and sometimes underappreciated management tasks that make an outpost run—paying bills, buying provisions, negotiating leases with the landlords," Montoyo wrote. When a car bomb exploded outside the Tibesti Hotel on June 1, 2011, those more mundane tasks took over. Security Chief Keith Carter ordered the team to move in with the CIA spooks and special operators at a walled compound in Western Fwayhat beyond the Fourth Ring Road, known as the Annex. At the time, eight intelligence officers were based in the immense four-villa compound with its sculpted lawns, high walls, and state-of-the-art security, guarded by six U.S. Marines. At its peak, the CIA Annex housed another thirty-five U.S. intelligence officers, contractors, and former Special Forces personnel running the covert side of the Obama policy to arm and train the Libyan rebels. My sources tell me that the Annex also included a sophisticated NSA listening post to keep track of the opposition, especially the Iranian intelligence officers and Quds Force operations teams who started flocking to Benghazi in growing numbers.

The agency types were not comfortable with the ambassador so close, even if he was hunkering down behind their solid steel gates. Their presence in Benghazi was supposed to be a closely guarded secret and, besides, they were going to need the extra space as they ramped up their operations. So, over the next three weeks, Stevens' team looked around for new digs, eventually negotiating to lease a walled compound that included three comfortable villas, a swimming pool, and thirteen acres of grounds, just a few minutes' drive to the northwest of the Annex that would become the Special Mission Benghazi Compound (SMC). Stevens, Tek, and their security detail moved into that compound on June 21, 2011, and stayed there until the fall of the Qaddafi regime. Stevens signed three separate leases for the three villas, two of which had their own smaller walled-in area. This greatly facilitated the attacks, since most of the security upgrades were carried out on the portion of the compound containing the ambassador's residence. The exorbitant bill sent to the U.S. taxpayer for the facility—$70,000 per month—suggests that Stevens was in a hurry to put some distance between his diplomatic mission and the spooks.[4]

The Special Mission Compound was never hardened against attack to the Inman standards—named after Admiral Bobby Inman, who led a commission to examine embassy security after the 1983 Beirut embassy bombing. Despite that shortcoming, the State Department point person in charge of security at overseas facilities, Charlene Lamb, told Congress that they had made expensive security upgrades before Stevens and his team could move in. Among them:

- extending the height of the outer perimeter wall with masonry concrete
- adding barbed wire and concertina wire on top of that, bringing the total height to 12 feet
- installing powerful external lighting around the perimeter
- installing Jersey barriers (large concrete blocks) outside the perimeter to provide anti-ram protection against truck bombs
- installing steel drop bars to control traffic inside each of the three steel entry gates

- installing guard booths and sandbag emplacements "to create defensive positions inside the compound."

Inside the perimeter, they installed explosives detection equipment and an Imminent Danger Notification System, and reinforced all wooden doors with steel plate. All ground-accessible windows got heavy steel security grilles, while escape windows leading from the safe haven to the roof were equipped with emergency releases.[5]

The problem was that many of these security features, such as the concertina wire, were never installed. Others, including the security cameras, functioned sporadically. U.S. officials told me that in places the outside walls were in such poor repair that an intruder could simply climb over them. Deputy Assistant Secretary of State Lamb, whose office within the Diplomatic Security Service was responsible for these upgrades, was honest enough to tell Congress that money was not a problem.

It was focus, and outlook.

Libya wasn't Iraq. The United States hadn't fought its way to victory; we led from behind. President Obama wasn't Bush.

Charlene Lamb took her cues from her bosses, and ultimately lost her Diplomatic Security job as a result of the Accountability Review Board report, written by Hillary Clinton supporters.

THE TIPPING POINT

The tipping point in the civil war occurred not long after Stevens' arrival. Captured Libyan documents, discovered after Qaddafi fled Tripoli, showed that Qaddafi's own commanders knew the end was near way before the rebels or their foreign backers understood how close they were to victory.

On April 17, government troops launched a major offensive to clear the rebels from the western mountains and reassert control over the border with Tunisia. For four days straight they shelled the border town of Nalut, where the rebels had set up a powerful clandestine radio transmitter to reach nearby Berber tribesmen. Then, the rebels counterattacked powerfully, breaking the offensive and scattering the government troops. Hundreds fled the battle for their lives.

In a blistering eight-page memo, penned on April 26, Lieutenant General Mohammed al-Issawi blamed a "lack of intelligence" for his defeat. Before launching his attack, General Issawi said he had no reliable information on the numbers of rebels he should expect to encounter, their weapons, and their level of military training. However, the high command had sent him the wrong type of units for a counterinsurgency ground attack. Instead of experienced ground troops, the high command sent him three hundred air force and navy troops. "Only later did we realize . . . the rebels were more numerous than we were and they had good weapons and vehicles," he wrote. "They were not weaker than us qualitatively or quantitatively, as we had been told in our orders."[6]

On May 23, Stevens' immediate boss, Assistant Secretary of State Jeff Feltman, arrived in Benghazi for two days of unannounced talks with the rebel leadership. Already France, Italy, and Qatar had formally recognized the TNC, casting their lots with the rebels over Qaddafi. The U.S. wasn't far behind, with the State Department calling the TNC "a legitimate and credible interlocutor for the Libyan people." But, for TNC president Mustapha Abdul-Jalil, that wasn't enough. "We have tried very hard to explain to [Feltman] that we need the arms, we need funding, to be able to bring this to a successful conclusion at the earliest possible time and with the fewest humanitarian costs possible," a spokesman told the Associated Press.[7]

The rebels wanted more of everything. And they were about to get it.

ARMS DEALERS AND SPIES

As weapons and money began to flow to the Libyan rebels in the late spring and early summer of 2011, Benghazi became a hotbed of intrigue. Everyone from foreign governments to oil companies and wealthy members of the TNC seemed to have their own security detail, many of them former Special Forces troops from Western countries. Among the private military contractors (PMC) on the ground in Benghazi were SECOPEX and Gallice Security of France; the British-owned Blue Mountain Group (which later subcontracted security for the Special Mission Compound to a local militia); Garda Security Group of Canada; and Control Risk, HIS, and the Olive Group, all British. "The possible involvement of other American PMCs with known Agency links—such as Blackwater/Xe (Select), Triple Canopy, and SOCMG—is something that needs further investigation," write former special operators Jack Murphy and Brandon Webb in their early account of the Benghazi attacks.[8]

The contractors were a colorful lot. They included former U.S. military officers such as James F. Smith, a onetime Blackwater executive who emailed friends back in the States that he'd been on the ground in Benghazi "since no-fly." His Virginia Beach company, SCG International, was beating the pavement for work with the TNC and anyone else they could find. Claiming to have a CIA background, he eventually became source "LY700" for Stratfor, the U.S.-based private intelligence broker whose computer systems were hacked by Anonymous and posted online. In one of his emails, he claimed to have information about "missing SAMs" his men were tracking. Jamie Smith's intelligence impressed Stratfor Vice President Fred Burton. "Good skinny. This is what is defined as a credible source. Not some windbag Paki academic belching and passing gas," Burton wrote in one evaluation.[9]

Where there were private military contractors, there was big money to be made—the type of money people were willing to kill for.

Pierre Marziali retired from the French Army in 2007, after twenty-five years of service as a parachutist with an airborne unit. Wounded in Beirut in 1983, he had deployed to Bosnia and Rwanda. His company, SECOPEX, made headlines in 2008 when Marziali revealed they had won a large contract to train the Somali coast guard to operate against pirates. He once claimed his network of contacts within the mercenary community was so vast that he could raise a private army of 2,000 men, if a client were willing to pay.

Marziali sent a team to Benghazi to scout for work shortly after the UN no-fly zone went into effect in late March, just as Jamie Smith had done. On May 10, he left his home in Carcassonne, in south-central France, to join them in Benghazi. Just three days later, he turned up dead in a local hospital. According to the official version, he had been trying to stop an altercation between armed rebels and police at a checkpoint during the night of May 12–13 and was shot in the melee. His wife wasn't buying it. "My husband was known for his nerves of steel," said Dominique Marziali. "I don't see him throwing himself in between policemen and Libyan rebels armed to the teeth."

Further adding to the intrigue was the fact that the four colleagues in the car with him escaped unharmed, and were later taken into custody by the police where they were supposedly vigorously interrogated. "I've been told that French intelligence is conducting two separate investigations in Benghazi into my husband's death," his widow said. "The confusion about the shooting contributed to a growing feeling that a shadow war is simmering in Benghazi between the many militias under the rebel umbrella and former Qaddafi loyalists or other groups with unknown allegiances," the *New York Times* reported. "No one seemed able to say who had attacked the Secopex team, and no one seemed to know, or was willing to say, exactly why the security contractors were in Libya."[10]

The Iranians were also keeping close tabs on the Libyan rebels, and in early March 2011 had dispatched to Benghazi Lebanese Hezbollah operatives who were blending in with the local scene, according to an American defense contractor I had known for some time who was on the ground in Benghazi. "They came in through Egypt, with the help of the Muslim Brotherhood. They spread money and guns all over the ground, helping anybody who would take them. They figured they'd settle scores later on," the contractor told me.

Terrified of dissent at home, the Iranian regime was trying to claim that its anti-Western example had sparked the Arab Spring revolts in Tunisia, Egypt, and Libya. Few were buying it, which is why the early arrival of the Hezbollah teams was noticed by the TNC and by their Western advisors. Farsi-speaking Iranian intelligence officers were also observed in Benghazi, and later in Tripoli. Their presence in Libya spelled trouble.[11]

Everyone seemed to know that Stevens was the go-to man if they wanted to reach the rebel council. In many ways, he became the Libyan ambassador to the murky underworld of arms dealers, mercenaries, private military contractors, journalists, and adventurers, all seeking a profit from the war.

One of these was an American named Marc Turi, a registered arms broker who divided his time between Arizona and the United Arab Emirates. Smelling opportunity, he applied to the State Department for a license in March 2011 to ship former Soviet Bloc weapons to the rebels. He also brought Stevens into the loop, emailing him his proposal. Stevens wrote back saying that he would share Turi's information with the appropriate people back at the State Department.

Despite the extra push, State turned him down. By statute, commercial military sales are notified to Congress, and approved licenses regularly wind up in public databases on the arms trade compiled by organizations such as the Stockholm International Peace Research Institute (SIPRI). Having a data trail back to the State Department on arms shipments to the Libyan rebels was not exactly a covert operation. Turi reapplied in May 2011, this time saying that he planned to sell $200 million worth of weaponry to Qatar. That license was approved. Turi later told the *New York Times* that once the weapons reached Qatar "what the U.S. government and Qatar allowed from there was between them."

Turi Defense Group, Inc., was a U.S. government–licensed arms supplier that claimed to have "the international presence, personnel networks and procurement infrastructure to deliver wide-ranging support for any mission, anywhere." That may be hype, but Turi was for real. He was also competing on contracts to supply weapons, armored SUVs, and protection services with some of the biggest-known PMCs worldwide.

Two months after the State Department approved his weapons deal with Qatar, agents from the Department of Homeland Security raided his home near Phoenix, Arizona. Turi believes his operation was shut down because the Qatari arms pipeline to the Libyan rebels was out of control. "They just handed them out like candy," he remarked.[12]

QATAR ARMS THE ISLAMISTS

There was growing alarm within the administration that the Qataris were arming the wrong people. "Within weeks of endorsing Qatar's plan to send weapons there in spring 2011, the White House began receiving reports that they were going to Islamic militant groups," the *New York Times* reported. President Obama complained to the emir in person about the arms shipments, asking for "transparency about what Qatar was doing in Libya."[13]

But the Qatari arms shipments just kept on coming. And it was the type of thing that was hard to keep quiet. Even TNC president Mustapha Abdul-Jalil, who had helped free LIFG members when he was Qaddafi's justice minister, was complaining that the jihadis were getting all the money and arms, while the TNC itself was left begging.

"The Qataris are not to be sneered at when it comes to arms smuggling," says Simon Henderson of the Washington Institute for Near East Policy. "In Benghazi, they were making sure the weapons got through to the best fighters. They were working alongside the CIA, but not necessarily with them. And the CIA often didn't know what they were up to."

From their base at the Tibesti Hotel, where Chris Stevens had initially set up shop, the Qatari Special Forces officers donned blue jeans and T-shirts and blended in with the rebels. "The only thing that made them stand out was their military haircuts," a contractor who was on the ground at the time told me.

As the Qataris ramped up the deliveries, they no longer took the risk of land convoys through the desert. "They just flew them into Benghazi on Qatari aircraft," Henderson says. To do so required that they notify NATO to allow their aircraft to land in Libya despite the UN-authorized no-fly restrictions. That was definitely not discreet.

The Qataris, of course, were also participating in the NATO no-fly zone, but that appears to have involved a bit of military overreach. "They had decided to fly four Mirage 2000 fighter jets along with a C-17 full of support gear and possibly weapons into Benghazi," Henderson recalls. "Someone forget to get landing rights or to calculate the flight path and they wound up making an emergency unplanned fuel stop at the Royal Air Force base on Cyprus. Otherwise, they would have landed in the Med."[14]

Emir Sheikh Hamad bin Khalifa al-Thani wasn't being shy about his support for the rebels in Libya. By late June 2011, rebel troops began strutting around in Qatari-supplied desert camo in Benghazi, and erected a giant billboard of the Qatari emir outside TNC head-quarters. Souvenir stands were selling Qatari white and maroon flags outside the Tibesti Hotel and on the streets. In Misrata, the Qataris had set up a field hospital to treat wounded rebels and their family members. The director of Misrata airport told Western reporters in mid-July that he was awaiting NATO clearance so the Qataris could fly in "humanitarian" aid and evacuate severely wounded rebels.[15]

On August 6, 2011, a Qatar Emiri Air Force Boeing Globe-master III C-17 made a corkscrew landing into Misrata, a military tactic the aircrew had learned from their American trainers and used just in case Qaddafi loyalists were lurking in the vicinity with Russian surface-to-air missiles. "The plane offloaded six pick-up trucks which were packed with ammunition and minutes later it flew off again," Reuters reported. As it took off, a local photographer captured the grey, maroon, and white aircraft with "Qatar" written in large letters on the underbelly.[16]

The commander of U.S. Special Operations Command, Admiral Eric Olson, had vigorously promoted the sale of these aircraft to Qatar in 2008 as a way for the U.S. to develop greater transparency into the small but secretive world of Qatari Special Forces. When the aircraft were first delivered, they were flown by mixed American and Qatari aircrews. The State Department saw the Special Forces relationship with Qatar as a convenient way to "gather intelligence . . . to learn more about Qatari intentions and actions or inactions on counterterrorism efforts, relations with and perceptions of Iran, and internal [Qatari government] decision-making."[17] The Qataris wanted U.S. protection from their powerful northern neighbor, Iran, and U.S.

expertise to expand their influence in Jihadiland. After one meeting with Admiral Olson in March 2009, the commander of the Internal Security Forces, Sheikh Abdullah bin Nasser al-Thani, "expressed his desire to see the Qatar Armed Forces engage in more special operations activity" with the United States Libya would put those desires to the test.[18]

The aircraft that landed in Misrata, with tail number A7-MAB, had rolled off the Boeing assembly line in Long Beach, California, on September 10, 2009. The commander of the Qatar Emiri Air Force airlift selection committee, Brigadier General Ahmed Al-Malki, pointed out that the unusual paint scheme—similar to that used by Qatar Airlines at the time—was "intended to build awareness of Qatar's participation in operations around the world." [19]

That turned out to be an understatement.

In the weeks after that initial August 6, 2011, flight bringing ammo into Misrata, both Qatari C-17s would fly into rebel-held airports on regular rotations, carrying severely wounded rebels in U.S.-supplied palletized hospital trauma pods to Malta for further treatment, and returning to Libya chock full of supposed humanitarian supplies. The Qatar Air Force became so fully identified with Qatar's commercial airline that two years after the civil war an angry mob stormed the tarmac at Benghazi airport to prevent non-Libyans arriving on a Qatar Air flight from Doha from disembarking. They also prevented Libyans from boarding it for the return flight. "According to a Benghazi Local Council member at the time, the militiamen accused Qatar of interfering in Libya's internal affairs," the local *Libya Herald* reported. Anti-Qatar sentiments reached such a boiling point that Qatar Air suspended all flights to Benghazi for two months. When they tried to resume them, gunmen occupied the Qatar Air office at Tripoli Airport and prevented the commercial airliners from landing.[20]

JIHADIS ON THE MARCH

The United Nations Panel of Experts investigating violations of the arms embargo on Libya and arms smuggling to the rebels traveled to Benghazi in July 2011 as the civil war continued to rage. Sources within the rebel Ministry of Defense told the panel that since the beginning of the uprising, "Twenty flights had delivered military materiel from Qatar to the revolutionaries in Libya, including French anti-tank weapons launchers (MILANs)." Additional supplies of arms and ammunition were coming in by road across the Tunisian border. In a letter sent to the Qatari authorities on August 10, 2011, the panel asked about a Qatari aircraft "that had allegedly landed in Misrata on 6 March 2011 transporting weapons and ammunition," and about the "reported presence of Qatari military personnel on the ground." The Qataris denied sending arms, saying that they were only providing humanitarian assistance.[21]

The Qataris did acknowledge sending "a limited number of military personnel" into rebel-held areas to "provide military consultation to the revolutionaries, defend Libyan civilians, [and for the] protection of aid convoys to the civilian population whether coming through the sea, air, or land." It was a pretty open admission that the Qataris were training the rebels and helping secure the rebel supply lines to the outside world, and those supply lines weren't just bringing in food, tents, and blankets. The Qatari rebuttal also contains this passage:

> The State of Qatar notes in this regard that it supplied
> those Qatari military personnel with limited arms and
> ammunition for the purpose of self-defence and to
> enable them to carry out the above-mentioned tasks,
> especially since they were directly targeted by Qadhafi's
> troops.[22]

The UN panel contacted the Swiss government after a news report by the Swiss TV channel SF1 showed the rebels using ammunition manufactured by the Swiss company RUAG Ammotec. "The box of ammunition clearly stated that the ammunition had been exported to the Qatar armed forces in 2009 by a Swiss company, FGS Frex," the UN Panel wrote. The Swiss government confirmed that they had issued an export license to Qatar for one million rounds of 7.62 x 51mm NATO Ball ammunition, used primarily in sniper rifles and light machine guns, on condition the ammo not be re-exported with prior authorization. When they asked the Qatari ambassador to Switzerland in late November 2011 what had happened, he replied that "the transfer of the aforementioned ammunition to the Libyan opposition was a misadventure in the course of his country's support of the NATO engagement in Libya," and reassured the Swiss that his government "took appropriate measures to prevent similar errors in the future." [23]

Of course, by then the war was well over.

The UN experts contacted more than three dozen governments to pursue reports of arms smuggling into Libya, and found significant shipments of weapons to the rebels coming from the United Arab Emirates, Ukraine, Serbia, Albania, and more. Someone was paying for all the weapons. With the exception of the UAE shipments, that someone was the Emir of Qatar.

Qatar's jihadi clients had their sights on bigger prey than just Qaddafi. They were actively expanding southward into Mali, Niger, and Nigeria, and all across northern Africa.

On June 12, 2011, soldiers from Niger intercepted an armed convoy crossing the border from Libya, about eighty kilometers north of the uranium mining zone of Arlit. Several people were killed in the clash, and two SUVs full of armed fighters escaped into the desert. When the Niger authorities examined the vehicle they had captured, they discovered an insurance card issued in Benghazi in 2010. They also found forty boxes of Semtex, each containing sixteen kilograms of the deadly plastic explosive: 640 kilograms in all. In theory, it was enough to bring down one thousand commercial jetliners. The car also contained 355 detonators and $90,000 in cash.

A few days later, a Nigerien citizen named Abta Hamedi turned himself in to the authorities, saying he had been in one of the two cars that had escaped during the clash. He said that the convoy was headed to Mali, where they were planning to meet with an al Qaeda cell. "In his statement to the authorities, Abta Hamedi said that the explosives were destined for Al-Qaida in the Islamic Magreb," the UN report states. The Semtex was produced by the Czechoslovak state company VCHZ Synthesia and came from two separate lots sold to Libya in 1977 and 1980. The UN experts later discovered a stock of Semtex boxes "piled up in the open air in a site in the desert outside the town of Gharyan," which they considered "a potential source of the illicit trafficking."[24]

The jihadis were on the march, and the NATO-led military operation against Qaddafi had opened a Pandora's box, giving them access to arms and explosives beyond their wildest dreams. This was an entirely foreseeable consequence of U.S. policy that went virtually unmentioned by the national media, which had jumped all over President Reagan for the blowback from arming the Afghan mujahideen in the 1980s.

DEFYING THE WAR POWERS ACT

In a June 15, 2011, report to Congress, the White House argued that U.S. military action in Libya, which by now had cost taxpayers $715 million, did not require congressional notification under the War Powers Act, even though American pilots continued to fly combat missions, Air Force and Air National Guard reservists were doing the bulk of the in-flight refueling for the coalition, and the United States was providing 70 percent of the coalition's intelligence assets.

The most astonishing claim came at the end of this thirty-two-page document, in which the White House asserted that no jihadi groups were actively engaged in the fighting—at least, not alongside the TNC. It was a carefully parsed deceit.

> We are not aware of any direct relationship between the TNC and al-Qaeda, Hezbollah, the Libya Islamic Fighting Group (LIFG) or any other terrorist organization. There are reports that former members of the LIFG, which had been initially formed as an anti-Qadhafi group, are present in Eastern Libya and that some of them were fighting with opposition forces on the front lines against the regime. . . . The TNC has consistently and publicly rejected terrorism and extremist influences and we have not observed any TNC support or endorsement of the LIFG.

The report goes on to make equally extraordinary claims about the secular nature of the TNC and to sweep aside the dangers posed by previously banned Islamist groups that were now supposedly participating in Libyan society.

"From public press reports, we understand that the Libyan Muslim Brotherhood has declared its support for moderate Islam, emphasized the important role of women in society-building, and formed a relief organization in Benghazi," the report stated. Undermining these rosy statements was a classified annex that included an assessment of extremist groups in Libya, and a separate threat assessment of MANPADS, ballistic missiles, and chemical weapons in Libya.[25]

The claim about moderate Islam would be proven wrong fewer

than six months later, once the rebels toppled Qaddafi and began to impose Sharia law on the entire country. But the denial that al Qaeda and other terrorist groups were deeply embedded in the rebel alliance was flatly contradicted by information the administration *already* possessed, starting with meetings Stevens was having with tribal chiefs and key rebels, including LIGF leader Abdelhakim Belhaj, now a top rebel commander with his own *katiba*.

If Chris Stevens had been able to look out into the crowd that stormed his villa on the night he was murdered, he would have recognized some familiar faces. He knew all the rebel leaders, all the *katibas*. And those relationships would doom him, not save him.

THE MURDER OF GENERAL ABDELFATTAH YOUNIS

General Abdelfattah Younis was a big catch for the rebel team. After his defection in Benghazi at the start of the insurrection, General Younis was appointed chief of staff of the rebel forces and injected a degree of order and professionalism into their disparate ranks that had been sorely missing. A member of the large and closely knit Obeidi tribe in Benghazi, General Younis was well liked and regularly toured rebel positions in his camouflage uniform; his charismatic presence alone seemed sufficient to boost the moral of the fighters.

At every meeting he had with Special Envoy Chris Stevens, he begged the American to deliver heavy weapons to the rebels, who were under siege in Misrata. He wanted armed helicopters, antitank rockets, and SAMs. At one of these meetings, at the end of April 2011, he claimed that Qaddafi was planning to use chemical weapons against the rebels.

"He will fight up to the final drop of his blood," Younis said. "He has been offered chances to leave and he refused them all. Most probably he will be killed or commit suicide. Unfortunately he still has about 25 per cent of his chemical weapons, which he might use as he's in a desperate situation. He always says: 'You will love me or I will kill you.' " [26]

This was coming from a man who had spent the past forty-two years at Qaddafi's side, so, even if he was exaggerating Qaddafi's chemical weapons arsenal—which Stevens knew the Americans had destroyed—his psychological evaluation of Qaddafi was sobering. Stevens urged the State Department to formally recognize the TNC as the legitimate government of Libya, a milestone they reached on July 15, thanks in good measure to the moderating presence of General Younis in the rebel camp.

On July 28, 2011, the TNC summoned Younis to Benghazi, following the inconclusive fourth battle for the control of Brega, an important oil port on the coast road leading to Benghazi. Council leader Mustapha Abdul-Jalil reportedly accused General Younis of secret contacts with the Qaddafi camp, which he claimed was the only thing that explained the lackluster performance of the troops under Younis' command at Brega.

A convoy of *technicals*, pickups armed with twin-tube anti-aircraft guns and other weapons, screeched to a halt in front of General Younis' headquarters near the front lines at two in the morning, and told the guards they had been sent by the TNC to escort the general back to Benghazi. He wasn't overly concerned, since they said that they belonged to the 17th February Martyrs Brigade, a unit that was under his command. According to family members, they showed Younis an arrest warrant signed by a judge and by Ali Essawi, the deputy head of the council.

The arrest was a well-coordinated affair, not some fly-by-night venture. Before getting into the armored SUV escorted by the gun trucks, he contacted his family to say that he had called Essawi, who confirmed that the warrant was legitimate. [27] Later, when he was named the chief suspect in Younis' murder, Essawi denied it. "I never signed any decision relating to Abdel Fattah Younis," he told a Libyan television reporter.[28] As they headed toward Benghazi, the general's escort swelled into a huge posse of men who blocked side roads and sent word ahead of them so the gates at the checkpoints stood open and they could pass without stopping.

When the TNC announced his murder on July 28, they initially claimed his convoy had been ambushed during a moment of lax security along the road. But family members say that an eyewitness saw the general enter the Garyounis Military Camp on the outskirts of Benghazi, where he was questioned by four judges and eventually released. A Special Forces officer who was at the compound said that two men from the 17th February Martyrs Brigade team escorting the general opened fire from their car with automatic weapons as they were leaving the camp. "The men's leader was shouting 'Don't do it!' but they shot Younis and his two aides, and took the bodies in their car and drove away," the officer said.[29] The murderers later burned the bodies of the three men—a sign of ultimate disrespect in Islam—and dumped them in a grassy field not far from the general's home on the outskirts of Benghazi.

Who ordered the rebel militiamen to murder their own military chief? A member of his Obeidi tribe told reporters that the general had lots of enemies within the TNC leadership. Suspicions subsequently fell on TNC president Mustapha Abdul-Jalil, who was detained for questioning after the revolution but ultimately released.

As Qaddafi's interior minister, Abdul-Jalil had been instrumental in the release of Islamist prisoners under the rehab program sponsored by Saif al-Islam al-Qaddafi. General Younis was at odds with his pro-Islamist views. Soon after General Younis was killed, Abdul-Jalil named former Libyan Islamic Fighting Group leader Abdelhakim Belhaj head of the Tripoli Military Council, essentially making him

General Younis' replacement. So much for the White House and State Department fiction, which influential members of Congress, such as Senator John McCain, lapped up, that the rebels had no relationship to the LIFG or other al Qaeda–affiliated groups.

The Islamists blamed General Younis for the massacre of some 1,200 LIFG members in Abu Selim prison in 1996, when he was Qaddafi's interior minister. The Qataris also hated General Younis because they saw him as the biggest obstacle to their plans to stack the rebel council with the Islamists that they had bought and paid for.

TNC Finance Minister Ali Tarhouni announced that the general had been murdered by renegade members of the Abu Obeida al-Jarrah Brigade, the *katiba* that the TNC had sent to Breda to bring the general in for questioning. "So secretive is the Abu Obeida al-Jarrah Brigade—said to be one of at least 30 semi-independent militias operating in the east of the country—that until yesterday few in the rebel capital had ever heard of it," the *Daily Telegraph* reported at the time of his murder.

And just what happened to this shadowy *katiba*, run by an Islamist named Ahmed Abu Khattala, now accused by a TNC minister of murdering the rebel military chief? The TNC put them in charge of internal security in Benghazi, essentially operating as a secret police force. It was a pretty extraordinary turn of events, embarrassing for the rebels, and embarrassing for their Western supporters.

"Some rebel fighters say privately admit that the Islamists are stronger than is generally admitted," the *Telegraph* reported.[30] One of the Abu Selim inmates who managed to escape the prison massacre was Abu Khattala himself. He later joined forces with former Guantánamo inmates in Ansar al-Sharia, the *katiba* that claimed responsibility for the attack on the U.S. Special Mission Compound in Benghazi. The Justice Department indicted him eleven months later for his role in the September 11, 2012, attacks.

Norman Benotman, a former LIFG member who recanted the use of violence and became a go-to analyst of the rebels from his perch in London, said the Abu Obeida brigade was one of many independent revolutionaries operating outside the command structure set up by General Younis. Benotman claimed the LIFG had disbanded in August 2009 but regrouped when the anti-Qaddafi uprising began, now calling itself *Al-Haraka Al-Islamiya Al-Libya Lit-Tahghir*, the Libyan Islamic Movement for Change. "Are there Islamists and jihadists in Libya? Yes, of course," he said. "But they use the term 'jihad' as a 'just war' for their homeland, not as a transnational crusade." They had joined the TNC and "accept the idea of a new democratic Libya," he claimed.[31]

A Library of Congress report, "Al Qaeda in Libya," identified the murder of General Younis as a seminal event that empowered the Islamists inside the TNC. "Abdelhakim Belhaj, former LIFG emir, was appointed by the TNC's president as the military commander of Tripoli, controlling the twenty-thousand-strong Tripoli Military Council (TMC). Later, with his political ally, Ali Sallabi, a prominent cleric with links to the international branch of the Muslim Brotherhood, he forced the resignation of Mahmoud Jibril, the former TNC chief executive, whom they probably perceived as too secular. These two Islamist leaders are expected to play a major role in the new Libya, partly because of the support they are believed to enjoy from Qatar," the report states.[32]

A pro-Qaddafi website claimed that protesters critical of the feckless TNC investigation into the general's assassination were pointing fingers at Qatar for keeping the truth from coming out. "What is the capital of Libya: Tripoli or Doha?" one banner read. During a protest, angry young men burned the Qatar flag flying at the Tibesti Hotel, widely viewed as the seat of Qatar in Benghazi.[33] Of course, until early

June 2011, it was also the seat of the U.S. government's special envoy, Chris Stevens.

"General Younis was the one person who could have steered the Libyan revolution in a different direction," former CIA clandestine service officer John Maguire told me. "As the father of Libyan special forces, he would have been a rallying point with credibility for the post-Qaddafi stabilization effort. He would have kept the Muslim Brotherhood at arm's length, and kept the arms system under control."

Maguire left the agency after serving as deputy chief of station in the massive Baghdad embassy in 2004, and continues to track jihadi groups as a private contractor. He believes the assassination of General Younis was planned by Abdul-Jalil and Belhaj to cement their control over the rebels.

"Belhaj picked the right guy to kill. He eliminated the one guy in the Libyan system who would have been a problem for them because of his Special Forces background and his ties to the West. He was the most serious risk they would have faced. He would have kept the Muslim Brotherhood at arms length, and kept Qaddafi's massive arms stockpiles under control," Maguire told me.[34]

With Younis out of the way, the Islamists and their Qatari backers were in charge. And, now, they had the Americans in tow.

7

HOUSE OF WAR

The Hillary Clinton/Obama doctrine of separating the jihadis from the proponents of political Islam was getting people killed, and not just in Libya. In Afghanistan, the United States was supporting President Hamid Karzai (good), while fighting the Taliban (bad). But at the same time we were fighting the Taliban of the Haqqani faction (bad), we were negotiating with the Taliban of Mullah Omar (now good), apparently forgetting that Mullah Omar was the leader of the Islamic Emirate of Afghanistan who had sheltered Osama bin Laden. As the White House increased the use of armed drones to kill suspected terrorists, ordinary citizens throughout the Muslim world felt targeted and joined the jihad. American soldiers were dying as a result of this intellectual muck.

Hillary Clinton told Congress there was no point trying to "distinguish between good terrorists and bad terrorists." But what exactly did that mean? Did the United States plan to fight or to talk with the Haqqani faction, asked Representative Ileana Ros-Lehtinen (R-FL)? "Both," Clinton answered. The U.S. policy was to "fight-talk-build," she explained.[1]

The inability of the administration to tell the difference between friends and foes was as profound as it was dangerous. Friends, you help; foes, you defeat. That's a concept as old as warfare itself, but this administration seemed to stand it on its head by threatening traditional U.S. friends and befriending our new foes. Obama's terrorism advisor, John Brennan, further muddied the waters by claiming that jihad was not a problem for the United States because it was a holy struggle.

> The President's strategy is absolutely clear about the threat we face. Our enemy is not "terrorism" because terrorism is but a tactic. Our enemy is not "terror" because terror is a state of mind, and as Americans we refuse to live in fear. Nor do we describe our enemy as "jihadists" or "Islamists" because jihad is a holy struggle, a legitimate tenet of Islam, meaning to purify oneself or one's community, and there is nothing holy or legitimate or Islamic about murdering innocent men, women, and children.[2]

I first learned about jihad and the doctrines of Islamic law from dissident Iranian ayatollah Dr. Mehdi Rouhani in Paris in the 1980s. He explained that jihad was indeed about "purifying" one's community, as Brennan explained, but by purging it of corruption, deviancy, and non-Muslims.

Islamic doctrine, or Sharia law, divided the world into two competing spheres: the *Dar al Islam*, and the *Dar al Harb*, Rouhani taught. The House of Islam was the land and territories controlled by Muslims, which needed to be purified of corruption and deviancy. This was the task al Qaeda arrogated to itself by waging jihad against the ruling family in Saudi Arabia. The *Dar al Harb* (House of War) was the rest of the world, the infidels. The duty of the good Muslim, Rouhani explained, was to help make non-Muslim countries submit to Islamic rule. For the most part, this was to be done without violence, just as Muslims in Brussels have established Sharia zones in the administrative capital of Europe and claim that by having large families they will become the majority in all of Belgium by 2030 and submit that nation to Sharia law.[3] However, it was also the duty of Muslims to wage violent jihad when called upon by a legitimate ruler. "If you want to find a moderate mullah in Iran," Rouhani joked, "look no further than Ayatollah Khomeini. He's the most moderate of them all."

The administration's attempt to parse between violent and non-violent Islamist groups was a distinction without a difference. The only distinction that really mattered was the line dividing the Islamists from truly secular groups, who rejected the idea of Islamic law governing the political and social institutions of society. As we have seen, from Tunisia to Egypt to Libya, the better-organized and better-funded Muslim Brotherhood and its allies quickly defeated their secular competitors. This is the group whose motto since its creation in 1928 by Egyptian Hassan al-Banna has been: "Allah is our objective; the Quran is our law, the Prophet is our leader; Jihad is our way; and dying in the way of Allah is the highest of our aspirations."

The Muslim Brotherhood credo had the merit of being much clearer than Hillary Clinton's description of administration policy.

When asked about the Muslim Brotherhood at a House Intelligence Committee hearing as the Brotherhood was poised to take over Egypt, Director of National Intelligence Lieutenant General James Clapper called it "largely secular." Clapper was just repeating what he was being fed by U.S. government intelligence analysts, who had been steeped in the politically correct doctrines of the Obama administration. To NBC News Chief Foreign Correspondent Richard Engel, who had worked in the Middle East for many years and seen how the Brotherhood operates, Clapper's comment showed a "wild misreading of this organization."[4]

THE AGENCY ANALYST

One of those intelligence analysts, Quintan Wiktorowicz, was hired by the White House in January 2011 to implement the president's outreach policies toward the Muslim world. It is no coincidence that he took up his new position right at the start of the Arab Spring. He lent an academic patina to the White House policy of shunning moderates and empowering Muslim supremacists from Dearborn, Michigan, to Benghazi. Obama instinctively knew that he preferred Muslim Brotherhood spiritual guide Sheikh Yusuf al-Qaradawi to Google executive Wael Ghonim. Quintan Wiktorowicz explained why he was right.

Wiktorowicz had just returned from two years at the U.S. Embassy in London, from 2009 to 2011, where he'd been sent by the CIA's Counterterrorism Center to study Britain's Prevent program. Set in motion after homegrown Muslims killed fifty-two commuters during attacks on London subways and buses on July 7, 2005, Prevent was touted as a deradicalization program, which would prevent future terrorist attacks by marginalizing violent groups and steering young at-risk Muslims away from radical Muslim clerics. It proved to be an abysmal failure, says Dr. Sebastian Gorka, a specialist in Islamic ideology and irregular warfare.

"Just as Britain was waking up to the fact that all their counter-radicalization policies were bankrupt and were actually empowering the bad guys, [Wiktorowicz] was sitting there taking notes," Dr. Gorka told a conference of U.S. policymakers in September 2013. Shortly after Wiktorowicz left London, the British ditched the Prevent program and revamped their approach to the Islamist threat. "So he's got the wrong recipe, and this bad recipe is what we're now living in America," Dr. Gorka said.[5]

Wiktorowicz separated the Islamists into two main groups: those he called *purists*, or nonviolent Islamists, and the jihadis. The purists included preachers such as al-Qaradawi, who championed *dawa*, proselytizing the faith. They also included political Islamist groups such as the Muslim Brotherhood. Wiktorowicz argued that the United States should invest in these allegedly nonviolent Islamists, in the hope that they could deter or marginalize their violent brethren, al Qaeda and its affiliates.

But there was a problem. "Every single al Qaeda leader was first a member of the Muslim Brotherhood," Dr. Gorka pointed out. "It's a completely overlapping diagram. There is no hermetic separation between purists, politicians, and violent terrorists."

Beyond that, the two groups of Islamists were bound by the same ideology. "Despite the fact that the nonviolent jihadists, the stealth jihadis, outnumber the violent jihadis by factors of thousands, they both have the same theocratic goals: undemocratic societies from Washington to Cairo. They argue amongst themselves, but the arguments are almost exclusively about timing and tactics," Gorka explained.

Wiktorowicz was instrumental in getting the FBI and the U.S. counterintelligence community to abandon all teaching on Islam that highlighted the Islamic ideology of supremacy and jihad. He also helped craft new guidance for federal and local law enforcement authorities that pushed political correctness to silly and dangerous extremes. He argued successfully that the United States was not at war with jihadi Islam, but with "violent extremism," an amorphous, undefined "ism" that had nothing to do with Islam. He succeeded in getting the White House to issue that view as an official policy statement in August 2011.[6] Muddying the waters even further, Wiktorowicz argued that al Qaeda and obscure sovereign citizen groups in the United States posed equal threats to the United States, even referring to them in the same sentence as "violent extremist groups" the U.S. government had to combat.[7]

The tragic irony is that Wiktorowicz himself seemed to acknowledge the fatal flaw of his own approach. "For reformist Salafis (i.e., the purists, or political Islamists he championed), there is great concern that the Muslim community is not ready to engage in jihad, either against incumbent Arab regimes or the United States," he wrote. "It is not that jihad is *rejected* as a tactic of religious transformation, rather, reformists believe that several prior phases are necessary before a jihad is permissible."[8] In other words, as Dr. Gorka put it, it's just a matter of timing and tactics.

Dr. Patrick Sookhdeo, a renowned scholar of Islam who advised the British General Staff on Islamic law, warned that the Obama administration's approach of cozying up to the Muslim Brotherhood was foolhardy.

"The [Obama administration] view that Islam and the United States are compatible is extraordinarily short-sighted. Islamists all share the same aim, the creation of an Islamic state ruled by Sharia. This is an end result completely incompatible with freedom, equality or democracy. The threat from al Qaeda is an ideology which is rooted in a classical interpretation of Islam," Dr. Sookhdeo explained. "Trying to win the war of ideas without ideology is foolhardy."

From Egypt to Libya to Syria, the United States tried to separate violent from nonviolent Islamists, supporting the political Islamists in the hope they would neutralize the jihadis. "I believe this is a very dangerous policy," Sookhdeo declared. "We are engaging the wrong kind of Muslims. Instead of dealing with those who share our values, who share our ethos, our fundamental principles, we engage and back those who seek our destruction."[9]

EXTORTION 17

Among the Americans who fell victim to the administration's confused policy were some of our most undisputed heroes, the best of the best of the Special Forces community: SEAL Team 6, also known as Naval Special Warfare Development Group, or DEVGRU. These are the men who took out Osama bin Laden in a well-planned raid in May 2011.

You and I were never supposed to know about the involvement of SEAL Team 6 in the bin Laden raid. But, because of leaks at the very top of the Obama administration, we know an awful lot about Operation Neptune Spear that once was highly classified. Those leaks ultimately may have cost the lives of seventeen members of SEAL Team 6 as well as thirteen other U.S. warriors in Afghanistan when their Chinook CH47-D helicopter was shot down by the Taliban on August 6, 2011. It was the single biggest loss of life in the history of the Navy SEALs.

Karen and Billy Vaughn were the parents of one of those men, thirty-year-old Aaron C. Vaughn. "The day after the bin Laden takedown, we got a frantic call from Aaron in Afghanistan, telling us to clean our social media of any reference to him or his buddies," Karen told me. " 'Mom,' he said: 'we've been outed. There's chatter that they're going after our families. You're in danger.' "

Aaron Vaughn was referring to braggadocio comments made by Vice President Joe Biden on May 3, 2011, when he identified "our brave Navy SEALs" for the successful bin Laden raid. Secretary of

Defense Robert Gates was disgusted at Biden's loose lips, and chastised him publicly for putting the family members of the Navy SEALs in danger.

And Biden wasn't alone. At an awards ceremony at CIA Headquarters on June 24 to honor the men and women who had participated in the thirteen-year manhunt of bin Laden, then CIA director Leon Panetta disclosed the name of the specific unit that had carried out the raid and its commander, even though a producer of the Hollywood film *Zero Dark Thirty* was present. The event was classified top secret, as were Panetta's remarks. Needless to say, filmmaker Mark Boal did not have security clearance.[10] The White House was pushing for full cooperation with the filmmakers in the hope of using the movie as a prop for the president's reelection campaign the following year. Security be damned.

Just one week after Biden's initial disclosure, U.S. military intelligence reported that a force of more than one hundred Taliban fighters were making their way into the Tangi Valley in Afghanistan, where they knew U.S. Special Forces teams were operating. "The Taliban's goal was to shoot down a NATO helicopter in retaliation for the bin Laden raid," Karen Vaughn told me, summarizing what she had learned from a 1,300-page after-action report by Army Brigadier General Jeffrey Colt.

When Aaron Vaughn came back for home leave that summer, "something was different," Karen remembers. "He had a haunted look, as if he knew something was about to happen."

Twenty-five-year-old Navy cryptologist Michael Strange also came home that summer on leave. "Usually it was, see you in three months, make sure the beer is cold," said his father, Charles Strange. "This time, my son grabbed me by the biceps and said, 'Dad, I got to make out a will.' "

The call sign of the helicopter that flew them to their death on the night of August 6, 2011, was *Extortion 17*. The after-action investigation, stamped SECRET, was declassified for the families, who made it available to me. Despite an extraordinary level of detail, it left many key questions unanswered.

For starters, the mission itself made little sense. Earlier in the day, a team of U.S. Army Rangers had choppered into the Tangi Valley seeking out "Objective LEFTY GROVE," the code name for Taliban commander Qari Tahir. During the fighting, a force of nine or ten suspected Taliban escaped and moved deeper into the valley. The Rangers initially requested that an AC-130 Spectre gunship take them out, but the Task Force commander "was unable to determine whether the group was armed and therefore could not authorize the strike," wrote Brigadier General Jeffrey Colt.[11] Why the Task Force commander was unable to determine that the men were armed when the Rangers reported that they were walking in single file, carrying AK-47s and RPG-7 tubes, went unexplained in the report.

At half-past midnight, the Task Force commander issued a warning order for SEALs to deploy an IRF (Immediate Reaction Force) in support of the Rangers, to capture or kill Qari Tahir and his men. As the situation on the ground evolved, he decided to expand the mission from seventeen to thirty-two men, figuring there could be more Taliban, and because of the probability the SEALs would have to fight their way out of the valley during daylight hours. When *Extortion 17* took off, thirty-eight men were on board: seventeen Navy SEALs, five U.S. Navy special operations support personnel, three U.S. Air Force special tactics airmen, five U.S. Army Reserve airmen (who flew the aircraft), seven Afghan commandos, an Afghan civilian interpreter, and a working military dog. None of them would survive.

Instead of dropping the thirty-eight-man IRF at a known landing zone a mile or so from the fighting, the mission commander chose a hot site with a fifty-foot *qalat* (tower) at one end, and groups of Taliban on two sides. The shooters were waiting in the *qalat*. According to other pilots on the mission that night, they fired two, possibly three projectiles at the lumbering chopper when it was two hundred meters away. The second one hit *Extortion 17* in the aft rotor, just above the engine, bringing it crashing to the ground in five seconds from an altitude of approximately 150 feet. According to the report, the chopper burst into flames as soon as it hit the ground.

"They told us the bodies were so badly burned that they had to be cremated," recalled Charlie Strange. "But in the pictures I was given later, the only visible sign of damage to my son was his left ankle. He was holding his sidearm as if he was just about to fire. He went down fighting—and no one told me."

Aaron Vaughn was also found intact and unburned on the ground, although every bone in his body was crushed from the fall. And yet the military told the Vaughns the same story about having to cremate the bodies because they were burned beyond recognition.

"There were so many things that just weren't right about that mission," Billy Vaughn said. "We want answers."

I accompanied Billy and Karen Vaughn; Charlie and Mary Ann Strange; attorney Larry Klayman and his assistant, Dina James, in June and July 2013 to take their story to Representative Darrell Issa, chairman of the House Committee on Oversight and Government Reform, and Representative Jason Chaffetz, his National Security Subcommittee chairman. They pledged to help the families get answers to some of the disturbing anomalies of the mission that ended in the catastrophic shoot-down of *Extortion 17*.

The congressmen wanted to know why were the SEALs put on a CH-47D, a chopper with technology dating from the 1960s, flown by U.S. Army Reservists, and not the MH-47G Special Forces Chinook they normally used. The SpecOps version came with a Thunder package of twin fuselage-mounted Gatling guns, side-door M240 machine guns, integrated night vision (FLIR), and upgraded engines so it could fly in low, hot, and hard into a hostile landing zone.

Normally, Army Special Operations Aviation (ARSOA) flight crews drove the SEALs and other special operators into battle. But *Extortion 17* was flown by U.S. Army Reservists who didn't train with the SEALs, were unfamiliar with the terrain, and who loitered over the landing zone for over six minutes as the Taliban lurked below. So lumbering was their approach that the AH64 Apache pilots flying backup behind them started to make fun of their flying skills just seconds before they were hit.

It wasn't as if the Task Force commanders weren't aware of the vulnerability of the older Chinooks. (Representative Issa called them "Sh-thooks," a flashback to his time in the U.S. Army at the end of the Vietnam era.) Just forty-five days before the downing of *Extortion 17*, two CH-47Ds were forced to abort an insertion mission in the Tangi Valley when they were engaged by Taliban with RPGs. They were replaced by an ARSOA MH-47G Chinook a few hours later. Although the Taliban also fired RPGs at the MH-47G from multiple locations, the ARSOA pilots successfully inserted a Special Forces team. "[N]o damage was reported to the aircraft," General Colt noted in his report.[12]

"We were scared to death every time we had to go out in those choppers," an Army Ranger who was present at the crash site that night told the Vaughns. "Everyone knows they're not safe. They can't avert attack like the MH-47Gs. We knew our lives were in danger every time we stepped into one.

"The MH-47 flies low and fast, like a roller-coaster ride, due to its quick, agile abilities in air," he explained. "The conventional CH-47Ds fly really high and slow with no evasive maneuvers. They're a huge target up there; like a train coming in for landing. They do six to eight pushups before landing, while the MH-47 burns straight in."

Even more disturbing: As *Extortion 17* was preparing to take off, the seven Afghan commandos were removed from the aircraft and replaced with seven other Afghans, without any explanation. So abruptly were they replaced that no one changed the flight manifest, so that when the Afghan commander sent out condolences after the crash, he notified the wrong families and had to call them a second time to tell them their sons were still alive. "A SEAL commander told me later this was a 'big, big deal,'" Billy Vaughn said.

A special operations officer who had been on the base that night told the Vaughns that the first seven Afghans had refused to board the aircraft, as if they had known something was about to happen. Why were unvetted Afghan commandos even allowed on the same aircraft as our Navy SEALs? With the rise of green-on-blue attacks carried out by Afghans against NATO troops, the SEALs were always on guard around them. As Aaron once told his parents, the Afghans were "loyal only to the highest bidder."

Brigadier General Colt and his team never interviewed any of the Afghan commanders when compiling their 1,300-page report. While no reason for this appeared in the report, the families were told by General Colt and others it was to protect the "sensitivities" of the Afghan government.

Few Americans realize that, since 2009, the Afghan high command has had veto power over coalition military operations, to the point of being able to choose targets and cancel specific operations ahead of time. Before any combat mission could get approved, the International Security Assistance Force (ISAF) briefed the Operational Coordination Group (OCG), which was made up of the leaders of the Afghan National Army, the Afghan National Police Force, and the

Afghan National Director for Security. Under a special agreement between President Obama and President Hamid Karzai, no member of the OCG could be vetted by the United States. "Our intel men fear this group more than any other," Karen Vaughn said. "Specifically, they fear Pakistani infiltration."

Was it really a lucky hit with an RPG that brought down *Extortion 17*? The threat of shoulder-fired missiles was on every commander's mind in Afghanistan. The WikiLeaks release of the Afghan war logs showed that NATO spokesmen did their best to bury field reports of MANPADS use, claiming falsely that NATO aircraft were being downed by RPGs and automatic weapons fire, not missiles.

Brigadier General Colt spent a considerable amount of time interviewing intelligence officers (J2, or S2, depending on the unit) who were tracking Taliban arms-smuggling routes from Iran and elsewhere, to determine whether a Stinger or other MANPADS had brought down *Extortion 17*. The S2 and his assistant at Bagram Airfield, just north of Kabul, had the lead for conducting threat assessments for NATO aircraft in Afghanistan. They told Colt there was little likelihood of a shoulder-fired missile being present in the Tangi Valley, because the weapons were "very expensive" and difficult to move because of the terrain. "[F]or an insurgent leader to have a MANPADS, usually there's a lot of [redacted] reporting that goes along with it, because, for lack of a better analogy, a forty-year-old guy in America buys a Viper, his neighbors are going to know about it. That's kind of what a MANPADS is in [redacted]. It is a status symbol. It is significant. It is expensive, and it means that you are the leader of that valley." [13]

Of course, the high-value target the Navy SEALs were supposed to engage that night, Qari Tahir, *was* the leader of the Tangi Valley Taliban. If anyone was going to have a MANPADS—the Afghan equivalent of the Dodge Viper—it would be him. However, that went unmentioned by the S2 or by Colt in the interview.

Heat-seeking missiles have a different signature than unguided RPGs. With MANPADS "you see a different kind of sparkle, smoke trails, corkscrews," the S2 said. Because of this, "we believe the event for EXTORTION was not a MANPADS. It just—it doesn't fit the MANPADS profile," he said.

Perhaps. But the pilots of the chase chopper and the gunship both reported seeing a very different type of signature, a blinding flash, which was consistent with a heat-seeking missile, not an RPG. It was possible that because of the very close range at which the missile was fired and the slow speed of *Extortion 17* (58 mph), the missile had no time or need to change direction to hit its target. It could have missed the engine intake by inches and hit the aft rotor blade.

I asked a senior Special Forces officer who has decades of combat experience to review the after-action report with me. He came away convinced that *Extortion 17* was hit by a Stinger or possibly a Russian-made SA-24.

He also was made suspicious by the cremation of Aaron Vaughn, Michael Strange, and the other SEALs, who apparently survived the initial impact and were found dead in defensive positions. "Now, perhaps I am behind the times, but we have *never* cremated the remains of the dead," this officer said. "We always put them in a coffin with a seal that said 'Remains NOT viewable' and sent an officer along to make sure they did not open the coffin. When there were not enough remains we added lead bars to the coffin. I know this because I participated in several burials like this." [14]

Billy Vaughn still burned with anger when he recalled the visit he and his wife received from Admiral William McRaven, the recently appointed commander of U.S. Special Forces Command. "He spent three hours with us, and repeatedly stood up for Obama. At one point, when we were saying something he didn't like, he pounded his fist on the table. 'President Obama is not just a good commander-in-chief.

I know this man. He's a good *man*. He got us out of Iraq and is getting us out of Afghanistan, whereas Bush got us into both wars.'

"He also tried to tell us that the CH-47D was 'just as capable' as the MH-47. We already knew that was bullsh-t," Billy told me. "I told him in the end to get out of my house."

The families didn't learn of the ultimate insult from their government until later, and that's what convinced some of them to retain attorney Larry Klayman to bring a lawsuit against Vice President Biden and CIA Director Panetta for the illegal disclosures of classified information that they believe cost the lives of their sons.

"When they greeted us at Dover, they gave us a DVD of the ramp ceremony back at Bagram," Karen recalls. "It took me a while before I was able to watch it. But when I did, I couldn't believe what I saw."

In the video, a Muslim cleric is invited to speak over the dead bodies, American and Afghan, whose remains were commingled. "I wanted to know why there was a Muslim cleric at my son's funeral. I wanted to know what he said."

Karen told me she sent the DVD out to a professional translator, then had the translation checked by seven other Arabic-language experts, including people in government, because she couldn't believe what the cleric had said. "He damned my son to eternal hellfire in the name of Allah," Karen said.

Every time the Vaughns visited with members of Congress, they insisted that they watch the two-minute video of the cleric's remarks, which included standard Koranic verses used at Muslim funerals.

> Amen, I shelter in Allah from the devil who has been
> cast with stones. [During the Haj in Mecca, one of the
> rituals is to cast stones against a pillar representing
> the devil.] In the name of Allah, the merciful forgiver.

The companions of the fire [the sinners and the infidels who are fodder for hell fire] are not equal with the companions of heaven. The companions of heaven [Muslims] are the winners. Had we sent this Koran to a mountain, you would have seen the mountain prostrated in fear of Allah. [Mocking the God of Moses as being weak.] Such examples are what we present to the people, to the people so that they would think. [Repent and convert to Islam.] Blessing are to Allah, the God of glory. And peace be upon the messengers [prophets] and thanks be to Allah the Lord of both universes [mankind and Jinn].[15]

There it was, right out in the open. The U.S. military invited a Muslim preacher to damn the souls of U.S. troops killed in battle, and to mock them. That has got to be a first.

"When I met Barack Obama at the ramp ceremony at Dover Air Force Base, he came up to me and put his hands on my shoulders," Strange said. "And he said, 'Michael changed the way America lived.' So I grabbed the president of the United States by the shoulders and said, 'I don't need to know about my son. I need to know what *happened* to my son.'"

There is often tension between the parents of SpecOps warriors and the widows of their sons, who are worried about survivor benefits. Despite that, Karen Vaughn and Michael Strange said they had received phone calls from both active-duty and retired members of SEAL Team 6, "thanking us for what we were doing."

The downing of *Extortion 17* showed just how dangerously misguided was the Obama policy of giving the Afghan government advance notice of U.S. Special Forces operations, without vetting the recipients of that intelligence or restricting it to a need-to-know standard. More U.S. soldiers died in Afghanistan during the first two years of the Obama administration than during the entire eight years of George W. Bush. The rules of engagement in Afghanistan—changed twice under Obama to make it more difficult for U.S. troops to defend themselves—were getting American soldiers killed.

"My son went to Afghanistan to fight the enemy," Billy Vaughn said. "Now the brass are saying, we're there to win hearts and minds." As the funeral service at Bagram showed, that wasn't working so well.

In Benghazi, our Muslim allies would turn against us directly.

8

WEAPONS, WEAPONS, EVERYWHERE

For the Libyan rebels, the war was going their way. As the NATO bombing campaign intensified, Qaddafi's forces steadily weakened. On August 23, 2011, the rebels stormed Qaddafi's Bab al-Azizia compound in Tripoli, only to find that Qaddafi himself had already fled. They consolidated their hold over the capital the next day.

The United Nations recognized the TNC as the official government of Libya in mid-September, and the United States sent Ambassador Gene Cretz back to Tripoli to reopen the embassy shortly afterward. Chris Stevens remained in Benghazi as the special envoy to the TNC for another two months. Abdelhakim Belhaj, the former LIFG chief whom Stevens had befriended in Qaddafi's jail, moved to Tripoli. As head of the Tripoli Military Council, Belhaj now was in charge of TNC military operations as the rebels mopped up the remaining pockets of resistance from Qaddafi loyalists.

With the rebel victory, the emphasis of the CIA outpost in Benghazi quickly shifted. Instead of distributing weapons they now tried to gather them, especially the four hundred Stingers and thousands of Russian-made shoulder-fired portable air defense systems (MAN-PADS) that were floating around.

But there was a problem: The missiles were nowhere to be found. At least not in Libya.

LEAKING MANPADS

One place the missing MANPADS turned up almost immediately was the Sinai, where terrorist cells, supported by the Muslim Brotherhood in Egypt and working closely with their Hamas brethren in Gaza, opened fire on Israeli Cobra helicopters in August 2011 during a series of running gun battles that killed eight Israelis. In September, authorities in Egypt intercepted eight SA-24s being brought by smugglers, most likely from Libya, into the Sinai. Libyan SAMs turned up in Algeria as well, where the authorities near the vast natural gas facility at Amenas seized fifteen SA-24 and twenty-eight older SA-7 missiles in mid-February.[1] This is the same facility that was subsequently attacked by an al Qaeda affiliate led by Mokhtar Belmokhtar, who boasted of mobilizing Algerian al Qaeda members to fight Qaddafi.

Hillary Clinton made an unannounced trip to Libya on October 18, 2011, to congratulate the Libyan people on their victory. She smelled a photo op for a future presidential campaign.

ABC News called it a "dangerous diplomatic mission," noting that her security team had required her to leave the comfort of her specially outfitted Boeing 757 in Malta and switch to a U.S. military C-17 with canvas seats for the remainder of the journey, since the military jet was "equipped with defenses against surface-to-air missiles." It showed that the State Department was capable of laying on security when it really mattered to the boss. It also showed that the Diplomatic Security people soberly viewed the threat of the leaking MANPADS.

While in Libya, Secretary Clinton announced that the State Department planned to spend $40 million to help the new Libyan government to collect weapons left over from the civil war and would hire additional personnel to complement the fourteen State Department contractors already working the project.[2]

Who were those contractors? The ARB report mentions two South Africans. My sources tell me they also hired local fixers, former militiamen, and Special Forces troops from Arab North Africa and Africa as well. In some countries, the State Department Bureau of Political Military Affairs hired companies such as the Danish Demining Group and the HALO Trust to carry out Explosive Ordnance Disposal (EOD) work. They also hired DynCorps, which considered its work on MANPADS collection so sensitive that it referred my questions on the subject to the State Department.[3] Some of these contractors worked out of the CIA Annex in Benghazi. Others, such as Glen Doherty, worked out of another CIA Annex in Tripoli. As this book goes to press, the State Department still has not answered how many of these former Special Forces operators had been training the Libya rebels in military tactics and weapons handling on the U.S. taxpayer's dime.

There was certainly cause for concern. There had been widespread looting all across the country of Qaddafi's armories since the start of the insurrection, and the CIA and the State Department contractors had quietly been trying to track the missing MANPADS and, where possible, gather them up for use another day. Heightening those concerns was the discovery by Western reporters just two weeks after Qaddafi abandoned Tripoli of a gigantic weapons storehouse in Tripoli.

The sign outside said "Schoolbook Printing and Storage Warehouse." But behind the double steel doors, which had been ripped off their hinges by looters, the only books to be found were weapons manuals. It was the Arms R Us of how-to guides: how to operate Soviet Bloc multiple rocket launchers, how to assemble and operate surface-to-air missiles, how to use French-built wire-guided antitank missiles. The warehouse was split into several buildings, filled with mortar shells, artillery rounds, antitank missiles: thousands of pieces of military ordnance that were completely unguarded more than two weeks after the fall of the capital.

Reporters found empty crates that once had contained SA-7b Grail MANPADS, a somewhat modernized version of the original Soviet shoulder-fired heat-seeking missile, the type that could be used by terrorists to shoot down civilian airliners.

Human Rights Watch researcher Peter Bouckaert was there and stumbled upon nine empty crates labeled "9M342," the Russian designation of the more modern SA-24 Strella. "These are some of the most advanced weaponry the Russians made," Bouckaert said. "They need to get somebody here to secure some of this." Weapons researchers subsequently found that Russia had exported only the twin-tube vehicle-mounted version of the SA-24, and not the gripstocks needed to fire them from the shoulder. The Strella is the Russian equivalent of the U.S.-made Stinger.

Each crate once held two missiles. Paperwork found at the site showed that they had come from nine different consignments, which in all totaled "2,445 crates containing 4,890 missiles."[4]

A military official in the new government gave almost direct corroboration of what the reporters and the Human Rights Watch weapons analyst believed they had found, saying that *five thousand* SA-7s from Qaddafi's stockpiles had gone missing. "Qaddafi's Libya bought about 20,000 SAM-7 missiles, Soviet- or Bulgarian-manufactured," said General Mohammed Adia, who was in charge of armaments at the TNC's newly formed defense ministry. He was speaking at a Qaddafi-era arms depot in Benghazi, where the rebels were ceremoniously destroying missiles the folks at the CIA Annex had found.

"More than fourteen thousand of these missiles were used, destroyed or are now out of commission. Most of them were stockpiled in Zintan," General Adia said, shortly before Clinton arrived in Tripoli. (Zintan is in southwest Libya.) "About 5,000 of the SAM-7s are still missing. . . . Unfortunately, some of these missiles could have fallen into the wrong hands . . . abroad."[5]

A Western security official based in Benghazi chimed in. "Libya has become an arms bazaar for all the world's leading terrorists. Everyone is here trying to do a deal and get their hands on Qaddafi's weapons. We are in a desperate race against time to prevent them falling into the wrong hands."

Al Qaeda supporters in Tunisia and Algeria were among those who were actively seeking to acquire the missiles, another intelligence official told British reporter Con Coughlin. Some of the missiles had been smuggled into Gaza, where the Israelis would track them down relentlessly for over a year. Not to be left out, the Iranians had sent a team of Revolutionary Guards intelligence officers to Benghazi "to buy dozens of missiles which can then be passed on to the numerous terrorist groups it supports throughout the region, including the Taliban in Afghanistan," Coughlin wrote.[6] We will hear more about the Iranian agents in subsequent chapters.

In Washington, John Brennan told a security conference that the news of the missing MANPADS raised "lots of concerns," and that the United States had pressed the rebel government to secure the weapons. "Obviously, there are a lot of parts of that country right now that are ungoverned," Brennan observed.

A rebel spokesman said that the military police were aware of the schoolbook warehouse in Tripoli, since it was only a quarter mile from a well-known barracks once used by the Khamis Brigade, the crack commando unit led by Qaddafi's son of the same name. "The military police were aware of this and they took charge of it; they're the ones who secured it," said Abdulrahman Busin.[7]

In a supreme irony, I can reveal that the man who controlled the looted missiles and what became of them was Abdelhakim Belhaj, the former head of the al Qaeda–affiliated Libyan Islamic Fighting Group. Belhaj was now the top military official in the new regime. And he was a dedicated Islamist who hated America and was devoted to the global jihad.

Senator John McCain met Belhaj in Benghazi at the start of the revolution along with Special Envoy Chris Stevens and naïvely hailed him and his *katiba* as freedom fighters. "They are my heroes," he told National Public Radio. "I have met with these brave fighters, and they are not al Qaeda," McCain said. He dismissed questions about their background, absurdly claiming that none of them had any record of supporting radical Islam. "To the contrary: They are Libyan patriots who want to liberate their nation. We should help them do it," he said.[8]

The United States did help them. And Belhaj repaid the favor by handing out these deadly missiles to every jihadi group in the region, many of them controlling territory well within range of major civilian airports.

ISLAMIST VICTORY

Two days after Hillary Clinton's surprise visit to Libya, NATO repeatedly bombed a large convoy of military vehicles in Sirte that they suspected was harboring top Qaddafi loyalists and perhaps Qaddafi himself. YouTube video soon surfaced of a dazed and bloodied Qaddafi being pulled off the back of a pickup truck by rebels and dragged along a dusty road, where they beat him repeatedly in the head. Some analysts analyzing the footage believe a rebel sodomized him with a knife. The dictator was dead.[9]

Moderates in the Arab world were already bemoaning the marriage of convenience between NATO and radical Islamist groups, and accused the West of hijacking the Arab Spring. As a result, Libya "has become a hub of extremism and lawlessness, with a plethora of military aid being collected by an assortment of armed Islamists who aim to exclude others from power." While the United Arab Emirates, another key arms supplier to the rebels, was attempting to play a moderating role by helping to arm and train the national police force, Qatar continued to support the Islamists, "which undermines the ability of non-Islamists to compete for power." [10]

The Obama administration crowed that its low-profile approach of leading from behind had won the day. National Security Council deputy for strategic communications Ben Rhodes made sure journalists compared Obama's recipe for regime change in Libya with that of George W. Bush, and noted that under Obama, not a single U.S. soldier had died (at least, not in Libya). One administration official slipped a quote into the media slipstream that would get replayed repeatedly in the 2012 election campaign. "Reagan targeted Qaddafi; Bush targeted bin Laden. Obama has done both." Of course, with such obvious spiking the football, he (or she) spoke anonymously. [11]

It was too much even for London's liberal *Guardian* newspaper, which ripped into the sycophantic U.S. media, chickenhawk Republicans, and the "narcissistic myopia of the American political establishment."

The administration's " 'strategy' in Libya—if indeed it merits the term—has been incoherent and contradictory," wrote commentator Michael Boyle. "If this is a victory, it is one produced by circumstance and a fair amount of luck, rather than any ingenious plan from Washington."

The new government made its intentions clear from the start. TNC chairman Mustapha Abdul-Jalil announced that the government's first official decree authorized polygamy, banned interest, and proclaimed that Islamic Sharia law was now the "basic source" of all Libyan law.

The Islamization of Libya began in earnest. Abdul-Jalil began appearing in public with a *Zabibah* mark on his forehead, a dark bruise ostensibly caused by repeated prostration during the five-times-daily Muslim prayers that was worn as a mark of piety by devout Muslims. In Benghazi, the black flag of al Qaeda was now flying over the courthouse, and gangs of Islamist militiamen in brand-new SUVs drove through the streets at night, shouting "Islamiya, Islamiya! No East, nor West," a clear intimidation to secular Libyans. There is no record that Obama, Hillary Clinton, or any other senior administration official uttered a word in protest.[12]

David Gerbi, a Libyan Jew whose family had fled to Italy during the wave of anti-Jewish riots in 1967, felt the sting of Sharia firsthand. Buoyed by reports in the Western media about the secular nature of the insurgents, he returned to Libya at the peak of the civil war in the spring of 2011 and joined the rebel ranks to liberate his country. Once Qaddafi fled and the TNC took charge, he went to Tripoli in hopes of reopening the main synagogue in Tripoli, a city that had been 25 percent Jewish as recently as 1941.

After weeks of administrative labors, he got a permit from the TNC and set to work. He hired a crew of neighborhood residents to help him clear the garbage piled around the synagogue complex in the Hara Kabira, a sand-choked slum that was once the heart of Tripoli's thriving Jewish quarter. Gerbi was photographed with a sledgehammer, ecstatically breaking through the cinder-block barrier the rebels had erected to block the magnificent main doors of the old synagogue. But, the very next day, a massive anti-Jewish protest in nearby Martyrs Square (known as Green Square during the Qaddafi era) forced him to abandon his dreams. He spoke to reporters, tears in his eyes, as he learned that the TNC had rescinded his permit and was now planning to deport him, bowing to the Islamist mob. The TNC dismissed his case as a "matter of no importance."[13]

Just one month after Secretary Clinton's victory lap in Tripoli, Belhaj went to southern Turkey at the behest of Mustapha Abdul-Jalil, now interim Libyan president, to meet with radical elements among the Syrian rebels. Belhaj shared their Salafist worldview: First defeat the renegades and apostates among the Muslims, then crush the infidels. He traveled with them clandestinely to tour their bases along the border with Turkey, and pledged to deliver weapons to enable them to shoot down Syrian government aircraft. He also discussed sending Libyan fighters to train the Syrians. "Having ousted one dictator, triumphant young men, still filled with revolutionary fervor, are keen to topple the next," the *Daily Telegraph* reported.

It was supposed to be a secret mission, but word of his trip leaked out when he was detained by a rival militia at the Tripoli airport. They claimed Belhaj was carrying a passport and a suitcase of cash after traveling to Qatar. Only President Abdul-Jalil's personal intervention got him released.[14]

Did Belhaj tell Chris Stevens of his intention to meet with the Syrian rebels? Did Stevens ask him to scout out the rebels to determine which ones would be worthy of U.S. support? For now, the record of their conversations, though reported by Stevens to Washington, remains under the wraps of classification.

But one thing was sure: Everyone wanted a piece of Qaddafi's arsenal, and Abdelhakim Belhaj held the keys to the warehouse.

9

THE THREAT MATRIX

When Special Envoy Christopher Stevens returned home from Benghazi on November 17, 2011, he could legitimately claim, "Mission accomplished." Qaddafi's forces had been defeated, and the rebels that he and Secretary of State Hillary Clinton had so ardently supported were now in control of the government. Aid from the United States and its allies was shifting from arms and security-related assistance to civilian projects and democracy training. As the U.S. diplomatic missions in Tripoli and Benghazi reduced their security footprint, Washington placed a bet that the central Libyan government would tame the post-revolutionary chaos and that the hundreds of armed militias who had fought Qaddafi would melt away.

Documents that have come to light as a result of congressional investigations led by Representatives Darrell Issa, Jason Chaffetz, and Ed Royce reveal an almost constant tug-of-war between State Department managers in Washington, D.C., who were trying to reduce the number of Diplomatic Security and other support personnel in Libya, and the diplomats on the ground who were experiencing the threats firsthand. "The administration made a policy decision to place Libya into a 'normalized' country status as quickly as possible," Representatives Issa and Chaffetz wrote to President Obama as they began their investigation. The policy makers were intent on "conveying the impression that the situation in Libya was getting better, not worse."[1]

They were wrong.

THE FELTMAN MEMO

On December 27, 2011, Assistant Secretary of State Jeffrey Feltman, whom I knew from Beirut as a straight-shooter, submitted a detailed request to Undersecretary of State for Management Patrick Kennedy, outlining future operations in a downsized U.S. mission in Benghazi in the same villa compound Stevens and his team first rented earlier that year.

With government operations back in Tripoli, they didn't need to have a full-scale operation in Benghazi with seventeen diplomats and security personnel, as before. Eight was plenty, along with two additional slots for temporary duty officers. Given the deteriorating security situation in Benghazi, Feltman insisted that half of them be permanent Diplomatic Service security officers, one for each diplomat assigned to the post.[2] That never happened.

Why maintain a presence in Benghazi in the first place? "Many Libyans have said the U.S. presence in Benghazi has a salutary, calming effect on easterners who are fearful that the new focus on Tripoli could once again lead to their neglect and exclusion from reconstruction and wealth distribution," Feltman argued. Beyond that, the TNC was telling U.S. diplomats that it planned to shift the headquarters of the National Oil Company to Benghazi, making the eastern city once again the commercial capital of Libya.

Feltman also recommended that State maintain its presence in Benghazi separate from the CIA-run Annex, where an additional twenty-four U.S. personnel were based. "Although all the groups have come to the conclusion that co-location is the best and most economical option for continued presence, the State presence cannot be accommodated at the Annex, and the current State facility is not large enough to permit co-location," he wrote.

But co-location was the only legal option under the provisions of the Omnibus Diplomatic Security and Antiterrorism Act of 1986 and the Secure Embassy Construction and Counter Terrorism Act of 1999 (SECCA). Both laws were written in response to catastrophic terrorist attacks on U.S. embassies: Beirut, in April 1983, and the simultaneous attacks in Kenya and Tanzania in July 1998. The law mandated rigorous security standards that had to be incorporated in all new U.S. diplomatic facilities overseas (the Inman standards), and a set of slightly lower standards for existing buildings, known as the Overseas Security Policy Board (OSPB) exceptions. The SECCA explicitly stated that the secretary of state "may not delegate the waiver authority with respect to the collocation" requirement, and must notify Congress whenever he or she exercises such a waiver. In writing this requirement into law, Congress expected that "waivers used by the Secretary would be infrequent, and therefore, considered more seriously in the instances such a waiver is exercised."[3]

And yet the State Department never submitted a waiver to Congress for Benghazi. Admiral Michael Mullen, the co-chairman of the State Department's Accountability Review Board, later claimed that "temporary facilities [such as Benghazi] fell through the cracks" and did not require waivers or congressional notification.[4] That is simply not true. The ARB report tried to create an ingenious loophole: The Special Mission Compound "was never a consulate and never formally notified to the Libyan government," and was therefore exempt from the SECCA and OSPD requirements. That also was not true.

This was just one of many aspects of the Benghazi debacle that directly engaged the responsibility of Secretary of State Hillary Clinton. But the ARB never interviewed Clinton, and Mullen later told congressional investigators that the board found no evidence that would have given them reason to interview her.[5]

It was a stunning whitewash.

STAY-AT-HOME DIPLOMATS

Meanwhile, on the ground in Libya the collapse of the old regime and the constant jockeying for power among rival militias, many of them backed by foreign powers, had created a "security vacuum," according to multiple emails and cables sent to Washington by U.S. diplomats and security personnel. Major cities such as Tripoli, Benghazi, and Derna had become free-fire zones, rife with carjackings, phony checkpoints, kidnappings, and firebombings. The release of sixteen thousand convicted criminals in the waning days of the Qaddafi regime further complicated matters, as they formed roving criminal gangs that preyed on the weak or unprotected.

The chief regional security officer at the U.S. Embassy in Tripoli, Eric A. Nordstrom, penned a series of alarming reports on the security situation for his superiors in Washington, in the hope of getting Washington to send the full complement of security officers they had promised. The lack of Diplomatic Security (DS) agents in Benghazi frequently meant the diplomats who were supposed to be meeting with locals and reporting on political and economic developments were confined to quarters. "It is my understanding from my UK colleagues that they have a five person team assigned to just their head of mission," Nordstrom wrote in one February 2012 email. The U.S. team in Benghazi was lucky to have "two DS agents . . . guarding a compound with two other [Department of State] personnel."[6]

Shawn Crowley was on the ground in Benghazi and was frustrated at the lack of response from Washington. "Apologies for being a broken record," he wrote, "but beginning tomorrow Benghazi will be down to two [DS] agents. . . . Since one agent needs to remain on compound to protect the other USG employees, this leaves just one DS agent to travel off compound. RSO procedure at present calls for two agents to do so. We have no drivers and new local guard contract employees have no experience driving armored vehicles. What this means is that we will be all but restricted to compound for the vital February 12–18 timeframe," when the Libyans planned to celebrate the first anniversary of their revolution.[7]

The problem was aggravated by the anticipated loss of two of the three Mobile Security Detachments (MSD), who had been ordered home by the State Department. (The first left in March, the second in early May.) These were special units of six Diplomatic Security officers, composed mostly of former Special Forces personnel, who provided heavily armed close protection for U.S. diplomats and visiting VIPs such as senators and congressional delegations in high-threat environments. They moved about at high speed in armored SUVs with tinted windows, wore body armor and ballistic shades, and carried assault rifles. Most of them also had Heckler & Koch machine pistols visibly strapped to their thighs. These were the men who were normally sent to secure diplomats and visitors in a war zone. State didn't want them staying in Libya because they made Libya look like another Blackwater operation, just like Iraq. And that didn't fit the narrative the politicians had decided to push.

In a follow-up email to colleagues in Washington, Nordstrom explained that he had been forced to confine diplomats to quarters in Benghazi for "upwards of ten days" because of the lack of security. "I've been placed in a very difficult spot when the Ambassador tells me that I need to support Benghazi but can't direct [DS agents] there and been advised that DS isn't going to provide more than 3 DS agents over the long term."[8] When it came to ordinary Americans, however, the State Department showed far greater caution, maintaining its advisory against all but essential travel to Libya. "The Embassy's ability to assist U.S. citizens in the event of a crisis remains extremely limited," a note from the embassy warned.[9]

Back in Washington, Chris Stevens backed up the requests for more security from his temporary desk at the State Department. So did the deputy chief of mission in Tripoli, Joan Polaschik, who flew to Washington to make the case directly with the person directly in charge of allocating Diplomatic Security assets at overseas posts, Deputy Assistant Secretary of State for International Programs Charlene Lamb.

Polaschik knew from their email exchanges prior to her visit that Lamb was reluctant to send more DS agents, whose salaries came out of her budget. So she came with a workaround: Why not reauthorize the sixteen-member Special Forces team that was already on the ground in Tripoli on loan from the Pentagon? Called a Site Security Team, or SST, this heavily armed group of special operations warfighters had been assigned to the embassy in Tripoli once it was reopened in late 2011 by the Pentagon's African Command (AFRICOM), to provide a visible deterrent against attacks. Since the Pentagon paid their salaries, they were essentially cost-free. They were scheduled to wind down operations by April 5. Why not reauthorize them so they could protect the Benghazi operation as well as Tripoli? Polaschik asked.

Lamb was adamant. Libya was going to "normal" status, and the "building," that is, the State Department political leadership, didn't want to make it appear as a war zone by having heavily armed troops guarding U.S. diplomats.

But Libya *was* a war zone, Polaschik insisted. The sound of gunfire was constant, with "armed militias beyond control of the central government" shooting up the towns in broad daylight. "Until the militias were off the streets and a strong national police force is established, we will not have a reliable, host government partner that is capable of responding to the embassy's security needs. It is likely that we will need to maintain a heightened security posture for the foreseeable future," she added.[10]

Ambassador Cretz was "taken aback by Lamb's position," and emailed Elizabeth L. Dibble, the principal deputy assistant secretary of state for Near Eastern affairs, to complain. "I fail to see the logic as to why DS would not support an extension of SST, unless DoD is against it which we have no inkling of." (The SST commander, Lieutenant Colonel Andrew Wood of the Utah National Guard, testified that his Defense Department superior, Lieutenant General Carter Ham, repeatedly told him *and* the State Department that the SST could remain as long as it was needed.) Cretz's email went on, "The bottom line is we will be severely impacted without them and no one here is arguing that there has been any improvement in the security situation which would argue that they are no longer needed."[11]

There is nothing ambiguous about that exchange. Our diplomats on the ground were begging for more security, but Washington turned them down.

When she was hauled over the coals during a congressional hearing just one month after the attacks, Charlene Lamb claimed the SST was no longer needed because the embassy had gone on a lower threat posture and was using locally trained militia to provide security, a solution that had worked in Yemen. That was too much for Nordstrom, who corrected her in public. "There was an emergency action cable dated in March [2012] that specifically references" why they went to a lower threat posture, he reminded her. "[T]he tone of that was that *since we had no choice* [emphasis mine], because we did not have the assets, we had no other option but to move to a model, not unlike in basketball, moving from man to man defense to a zone defense."[12]

In other words, the State Department wanted no reminder of the need for an armed military presence to protect our diplomats. As one of Lamb's deputies told Nordstrom at the time, "this is a political game," a *policy* decision that came from the very top. Hillary's war had been a shining success. That was the message they wanted to convey. And the secretary made clear to all her underlings that inconvenient facts would not be allowed to get in the way.

In the end, Lamb allowed the SST to remain until early August 2012, then pulled the plug—with fatal results.

After learning of Lamb's stonewalling in Washington, Nordstrom wrote ominously: "I hope that nobody is injured as a result of an incident in Benghazi, since it would be particularly embarrassing to both DS and [Lamb] if it was the result of some sort of game they are playing."[13]

DESCENT INTO CHAOS

By February 2012, the jihadi groups who controlled Benghazi began harassment and terrorist attacks on Western diplomats, NGOs, and international organizations, determined to drive Westerners out of the country. Royal Dutch Shell and other international oil companies took notice and started pulling out their expats.

On February 19, 2012, armed militiamen intercepted two female CIA analysts as they were returning home from Benghazi's Benina airport. After a chase through the dark streets at one in the morning, the militiamen forced them to pull over and demanded to see their passports and inspect the vehicle and its cargo. Within minutes, the car was surrounded, with no hope of escape, as additional militiamen flocked to the scene. One of the women managed to make a few phone calls, and eventually the chief of base at the CIA Annex got the 17th February Martyrs Brigade, the militia provided by the Libyan government to serve as a "quick reaction force" in case of trouble, to send a team to the checkpoint to negotiate the women's release. In reporting back to the embassy, the women noted there had been twelve separate checkpoints on the road back from the airport. They told Nordstrom the checkpoints were "aggressive, with armed personnel who attempted to open the doors of their vehicles." But they insisted that they did not allow the militiamen to see their cargo. Ordinary diplomats would not have been driving themselves back from the airport, but would have had a Diplomatic Security officer with them. And CIA operations officers would have had a protection detail from the heavily armed Global Response Staff (GRS) guarding the Annex. The analysts fell through the cracks.[14]

On March 8, neighborhood residents sent everyone at the Special Mission Compound into high alert when they set off a series of explosions just outside the gates. A subsequent investigation determined that the explosions were "fish bombs" made of gelatin plastic explosive detonated at a wedding ceremony.

The next day, one hundred carloads of armed Salafis stormed a famous Sufi Muslim shrine in Zliten, where they clashed with large crowds of local residents and gun-waving militiamen. Had they succeeded in destroying the shrine, which contained the tomb of revered Sufi scholar Sidi Abul-Salam al-Asmar al-Fituri, they would have ignited a sectarian war.

On March 12, an American and a Ukrainian journalist were detained by the Kufra military council and held for several hours of questioning.

On March 16, armed demonstrators in Benghazi favoring autonomy from Tripoli clashed with supporters of the TNC. Gunfights erupted after Friday night prayers on Freedom Square and spilled over into the surrounding streets. At least one person was killed and five others injured.

On March 18, armed intruders dressed in military fatigues and wearing balaclavas broke into the British school in Benghazi and robbed the staff of valuables.

On March 22, a small convoy of Toyota Hilux vehicles screeched to a halt at the rear gate of the U.S. mission compound in Benghazi, firing AK47s and demanding that they be let in. The local guard on duty at the gate fled at the sound of gunfire, but managed to activate the internal defense alarm, which alerted the two DS agents at the compound and the 17th February Martyrs Brigade unit responsible for security in the neighborhood. The Martyrs Brigade eventually determined that the would-be intruders were from the al-Awfea *katiba*, under the control of the Libyan Ministry of Defense, and were patrolling the area after a firefight in search of a missing suspect.

Four separate incidents occurred the next day, March 23. Members of the Zintan Militia smashed through the front door of the five-star Rixos al-Nasr hotel in Tripoli and opened fire with automatic weapons in the lobby. They abducted the hotel's general manager, a Turkish citizen named Sukru Kocak, and took him to the Fallah neighborhood of Tripoli, where he was beaten in the head until his right eardrum burst. He was eventually released after the Turkish embassy and other officials intervened with the TNC. The militiamen were upset because Kocak had asked one of their men to pay up or leave the hotel after staying there for six months without paying.

Former rebels upset with the growing chaos stormed the Imsa'ed border crossing with Egypt, closing it until further notice to protest the corruption of the Ministry of Interior officials who had been put in charge of the border. The former rebels accused the officials of smuggling. The border post was reopened two days later.

Meanwhile, soldiers and police officers held a protest march through the streets of Benghazi, calling for the full reactivation of the army and state security services as the only means of restoring law and order. In a separate incident, armed militiamen blocked the Sidi Khribish coastal road, demanding that the Tripoli government pay them their promised government salaries.

Celebratory gunfire was so frequent that the embassy now referred to it as "CSAF"—celebratory small arms fire. The RSO and his team reported several incidents each week when CSAF impacted in various diplomatic compounds in Tripoli and Benghazi. In one such incident, a senior enlisted man from Colonel Wood's SST was hit in the forearm by a falling bullet while taking out the garbage. "It was just raining bullets," Wood told me.

On March 26, a gunfight erupted at high noon in front of the Dutch embassy in downtown Tripoli involving automatic weapons and heavy weapons fire. That evening, a separate gun battle erupted outside the Polish embassy, involving unknown individuals and Libyan Military Police.

Western countries had launched a number of programs to educate Libyan women, and these were also targeted. On March 28, for example, someone tossed a hand grenade into the courtyard of the Women's Higher Vocational Center in Derna.

All of this was just a warm-up act. On April 2, 2012, a mob attacked two armored cars carrying British diplomats in Benghazi, a serious wake-up call to the British government of the deteriorating security situation. The British convoy had apparently strayed into a demonstration pitting rival militias, one of which served as the local Traffic Police. When the shooting got hot, the Interior Ministry deployed a third security force known as "Al-Nayda" or the "Al-Shorta Police" to restore order. The demonstrators may have mistaken the British armored cars for rival militia vehicles.

When Lieutenant Colonel Andy Wood arrived in Libya that February to assume command of the SST, the scene on the ground was absolute chaos. "Think of it like waking up one day and all forms of government just stopped. You woke up in the morning and there was no federal government, no state government, and no county government. In fact, there were no policemen at any intersections. The lights are all off. So it's just, do whatever you want. It doesn't take long for the best and worst of people to come out, especially in a post-Qaddafi era society, where there was a lot of revenge killing going on and local militias," he told me.[15]

Wood and his men never knew, when they ventured beyond the embassy compound to escort the ambassador, what they would encounter. "The same checkpoint you'd go through in the morning would be manned by somebody the exact opposite in the afternoon," he said. This led to frequent stand-offs, with militiamen at the checkpoint leveling their guns at the U.S. diplomatic motorcade. "Some militias didn't understand the whole diplomatic immunity thing," Wood said with a wry laugh.

Libyans who worked for the embassy were also harassed. One local-hire employee was abducted at a checkpoint on his way home from work. Luckily, he had a tracker on him, so Wood and his men were able to follow his abductors to a warehouse they were using as a base for their *katiba*. Wood got his contacts at the TNC security apparatus to hire another militia to attack the warehouse. "It was a matter of playing one militia off another," he said. "And then, the very next day that same militia you were fighting, now you are supporting them to do something else."

Benghazi was even worse than Tripoli. Wood remembers his first trip to Benghazi, along with the RSO, Eric Nordstrom that spring. "The guys in Benghazi had been telling us, 'It's crazy here, it's crazy here.' When I got there I took a look round and said, holy sh-t, this place *is* bad," Wood recalled. They had just one ordinary DS agent at the compound and no one from the MSD teams. "The compound was huge and open, with walls you could jump over," he told me. Some of the security cameras weren't working, and there were spots where the new security fencing that had been installed suddenly stopped. It was the type of place that would be impossible to defend if attacked, Wood thought.

That's the picture as Colonel Wood, Eric Nordstrom, and all the security people on the ground could see it. But that's not what the State Department was saying—then, or now.

"SIR, ANSAR AL-SHARIA GOT THEIR FUNDING"

In the midst of all the chaos, Ansar al-Sharia was moving in and ramping up their operation. They were the big dog, and they had a plan. They were using their association with the Libyan government-backed 17th February Martyrs Brigade, which provided the armed guards who patrolled the perimeter fence of the Special Mission Compound in Benghazi, to gather intelligence on U.S. personnel movements and operations.

The more Colonel Wood watched them expand and train and plot their attacks, the more worried he became. Every night, as he went to bed, he tried to picture his counterpart in Ansar al-Sharia. What was *he* planning to do tomorrow? How was *he* training his forces? What weak points had *he* identified in the Americans' security? Where would *he* attack next? When?

Over that fateful spring, he watched the ranks of Ansar al-Sharia swell as battle-hardened foreign jihadis and trainers began flowing in from places like Iran to mentor their Libyan recruits. Wood felt that he and his men were being out paced. Ansar al-Sharia was nimble. *They* were quick. Whereas the Americans and their Libyan counterparts were just hunkering down, waiting for disaster to strike.

Lieutenant Colonel Wood had spent his entire military career on the white side of Special Operations. At heart, he was a teacher, a trainer. In previous deployments to places like the Philippines, Yemen, and Kosovo, he had helped the local military train elite squadrons of Special Operations fighters who knew how to hunt terrorists, and how to do it without terrorizing the local population. Joint Special Operations Command (JSOC) had programs and the expertise to do this. It was one reason they sent senior commanders around the world to schmooze with their military brethren in countries like Pakistan, Qatar, Saudi Arabia, the Philippines, Yemen, or Nigeria. Wood knew he wouldn't be staying in Libya indefinitely and, that sooner or later, the Libyans would have to get control of the situation on the ground by themselves if the embassy was ever to be secure. The solution was to stand up a counterterrorism proxy force in Libya that could rout out groups like Ansar al-Sharia on their own. But he was having problems getting the paperwork approved.

The vehicle for such training was Section 1208 of the Defense Authorization Act. Since 2005, the Pentagon has used Section 1208 "to support proxy forces," according to a November 30, 2011, West Point study. "Until recently, such proxy forces would principally have been paid for with CIA funds under the authority of a presidential finding for covert action . . . [and] will seldom be acknowledged openly." Separate from the host nation's armed forces—and, at any rate, Libya didn't have any central government-controlled security force at that point— the proxy force "works principally (though perhaps not exclusively) for the United States." These proxy forces can be "irregular forces, groups, or individuals who work with [U.S.] Special Operations Forces for counter-terrorism purposes."[16]

"That's my program," Wood told me.

Wood had already identified Libyan candidates for the new counterterrorism force, and had begun basic training, which didn't require formal approval. They had a training camp up on the coast outside of Tripoli where they put them through the equivalent of the Rangers' Special Operations Preparations Course, four weeks of intensive physical training and basic soldiering skills, such as tying knots and learning how to use a GPS for land navigation. But now they needed to move on to lethal things, and for that he needed a piece of paper.

At one point, Wood learned through intelligence channels that the Iranians who had deployed to Libya to train Ansar al-Sharia had gotten authorization to bring their dependents to Benghazi. When Wood saw that, it all clicked.

At the time, Rear Admiral Brian Losey was in charge of all Special Operations troops in Africa. (His official title was commander, Combined Joint Task Force—Horn of Africa.) He was getting hundreds of emails from officers spread across dozens of countries every day. How was Wood's email going to stand out?

"Sir," Wood began. "Ansar al-Sharia has had their funding approved. Their training is approved. They now have approval to bring dependents in country. So we need to be doing the same thing here, or they're going to be ahead of us, and they'll win."

It worked. Admiral Losey blasted off an email back to Special Operations Command at MacDill Air Force Base in Florida, asking them to green light the funding. Colonel Wood and his JSOC troops could now teach their Libyan protégés how to kill al Qaeda. The Libyans were eager. They were fast learners. And they were really proud, Wood saw. They understood that their enemy was all outsiders, not Libyans. "They wanted to become the premier CT unit in Libya," he told me.[17]

Lieutenant Colonel Wood had another motive, he confided. He felt genuine affection for Ambassador Stevens, and wanted to do everything he could to keep his team in place until the security situation sorted itself out. He understood that the State Department higher-ups wanted the military out of the way, so by getting the SST transferred back to DoD authority (Title 10), he was hoping to get around the political objections to having them remain in Tripoli.

But Washington stood firm. They wanted Libya back to normal, and so Lieutenant Colonel Wood and his special warriors were sent home by mid-August, a month before the Benghazi attacks.

With their protection stripped away, Ambassador Stevens and Sean Smith never had a chance.

THE MISSILE POSSE

Secretary of State Hillary Clinton and her top political cadres eagerly sought opportunities to tout their "success" in Libya. They dispatched an assistant secretary to a Washington think tank to peddle the story that quick thinking and smart action had nipped in the bud the threat of proliferation from Qaddafi's vast arms stockpiles.

In hindsight, the presentation by Assistant Secretary of State for Political-Military Affairs Andrew J. Shapiro at the Stimson Center was remarkable for its candor. "Currently in Libya, we are engaged in the most extensive effort to combat the proliferation of MANPADS in U.S. history," he said. Shapiro explained that Qaddafi's Libya had "accumulated the largest stockpile of MANPADS of any non-MANPADS producing country in the world." The United States was determined to prevent the shoulder-fired missiles from getting into the hands of terrorist groups.

Shapiro revealed that he had sent to Benghazi one of his deputies, Colonel Mark Adams, in August 2011, three months *before* Secretary of State Hillary Clinton announced the missile collection effort. Working out of the CIA Annex, Adams and his men teamed with British intelligence and Special Forces officers and scoured the country inspecting ammo dumps and bunkers. "Thus far these teams have helped to identify, recover, and secure approximately 5,000 MAN-PADS and components," Shapiro said. "We the United States acted quite quickly because of the threat posed by MANPADS to get people on the ground as quickly as we could. And these teams have been partnered with TNC personnel at every step of the way. . . . We have used our technical expertise to assist their efforts to secure weapons. And we plan to continue to train the Libyans to take care of their stockpiles on their own." [18]

That sounds like a tremendous success story; and for the most part, it was—except for the fact that the CIA estimated that Qaddafi had upwards of 20,000 MANPADS. In other words, the United States could only account for one out of four. Asked if he had changed his assessment of the security of that vast arsenal since December 2011, when news reports began to surface of weapons being shifted to Syria, Shapiro was sanguine. "We believe the majority of these weapons are still in Libya, and do not have evidence to change that impression."

That was simply not true.

In addition to the incidents I mentioned in the previous chapter, there was a steady stream of intelligence reporting of Russian, French, and, possibly, American surface-to-air missiles moving from Libya to neighboring countries. Shapiro hinted that leakage across Libya's porous borders was a problem that State understood and had under control. His team had "prepared little pamphlets to let a border security guard identify what a MANPADS is," he said. "Thus far, we have had some success in working with the neighboring countries in the region. . . ."

The director of national intelligence, Lieutenant General James

Clapper, contradicted Shapiro's happy talk in congressional testimony just two weeks later. Noting the State Department's "active and aggressive program" to round up the missing missiles, Clapper acknowledged that State only had "recovered about a quarter of them, about 5,000 MANPADS."

Some of the missing 15,000 missiles had probably been destroyed in the course of the NATO bombing campaign. "But the truth is that the MANPADS and other weapons are distributed all over the place, in homes, in factories, in schoolhouses," Clapper said. "[I]t's all over. So there is—there is a concern, obviously about recovery of these—of these weapons."[19]

And while the Obama White House was still claiming they had eliminated the al Qaeda threat by having SEAL Team 6 take out Osama bin Laden—an operation vetoed three times by Obama advisor Valerie Jarrett as too risky—nonpolitical national security professionals provided a more sober assessment of al Qaeda's resilience. Defense Intelligence Agency (DIA) Director Lieutenant General Ron Burgess told Congress that the United States "remain[ed] in a race against their ability to evolve, regenerate leadership and launch attacks." While al Qaeda losses in 2011 had forced the group and its affiliates to focus on "self-preservation and reconstitution," they still possessed the ability to carry out transnational attacks in Europe and against the United States. "Al Qaida in the lands of the Maghreb, or AQIM, acquired weapons from Libya this year, kidnapped Westerners and continues its support to Nigeria-based Boko Haram," Burgess said.[20]

There were also reports that MANPADS were turning up in battlefields as far afield as the Horn of Africa and Afghanistan, where they were being used to kill Americans.

On February 18, 2012, an American U-28A spy plane disappeared on approach to Camp Lemonnier in Djibouti, six miles out from the Djibouti international airport, killing all four Special Operations airmen on board. They had been on an ISR mission (Intelligence, Surveillance, and Reconnaissance] against al Qaeda targets in the region when they went down. The four men—Captain Ryan Hall, 30; Captain Nicholas Whitlock, 29; First Lieutenant Justin Wilkins, 26; and senior airman Julian Scholten, 26, were assigned to special operations and intelligence squadrons based at Hurlburt Field, Florida, under the command of Colonel Jim Slife, commander of the 1st Special Operations Wing. The Pentagon reported they had observed "no hostile fire" in the vicinity of the downed aircraft.[21]

But the classified military intelligence SIPRNet (Secure Internet Protocol Router Network) was jammed with reports of a claim by al-Shabaab, the al Qaida affiliate in neighboring Somalia, that they had downed the modified Pilatus-12 turboprop with a surface-to-air missile.[22] Lieutenant Colonel Andy Wood heard about it as far away as his base in Tripoli. "They were hyperventilating" about the Djibouti incident, he told me. The intelligence analysts "thought it was one of these SA-7s" that went missing in Libya. Was this like the unexplained helo crashes in Afghanistan that the Pentagon was covering up to keep the public from learning about the missing MANPADS?

Sources within the Air Force Special Operations community tell me that the crash investigation eventually attributed the crash to pilot error caused by spatial disorientation. Both the shape of the debris field and the black box recordings showed that the pilot had flown the aircraft into the ground, apparently unaware of his position. The lack of moonlight that night, added to lights on the ground, "produced what aircrews call a 'black hole' where it is difficult to gauge the plane's attitude and position," according to the crash investigation. "All of us who have flown Special Operations missions know that night missions, especially those flown when using night vision goggles, are inherently risky," former Air Force Special Operations Master navigator

Dick Brauer, Colonel, USAF, told me.[23] Brauer is a founder of Special Operations Speaks, a group that has relentlessly pushed Congress and the administration to reveal the truth about what happened in Benghazi.

A few days after the U-28A went down, al-Shabaab fighters fired what appeared to be surface-to-air missiles at two Somali helicopters outside of Mogadishu. According to eyewitnesses, the missiles were "chasing" the helicopters, a clear sign that they were heat-seeking MANPADS, not unguided rocket-propelled grenades (RPGs). "It really gave us a surprise, because what they shot was chasing the helicopter with every angle it turned, but the helicopter only just managed to escape," one witness told a regional newsletter.[24]

Israel had also begun to see more advanced SA-24 Igla-S missiles turn up in Gaza. They believed as many as 480 of these missiles, purchased by Iran from looted Libyan government stockpiles, wound up in the hands of Hamas.[25] While Israel's Apache AH-64 attack helicopters and most of their combat aircraft were equipped with flares aimed at spoofing the infrared heat-seeking sensor that spotted aircraft engines, commercial airliners had no such protection.

Everyone dealing with the missing MANPADS knew how deadly they were. "We realized we were walking on eggs," a former U.S. Air Force pilot detailed to the State Department on the MANPADS mission told me. "All it took was one of these fired at a civilian airliner and hundreds of people could die."[26]

It was a zero-sum game. And our side was down by five thousand.

THE NATIONAL SECURITY STAFF

The tragic irony is that counterterrorism professionals knew removing Qaddafi had opened a Pandora's box that the United States could not close with the wave of a magic wand.

The fall of Qaddafi had created a "new and worrisome set of openings for terrorist organizations," the State Department's coordinator for counterterrorism, Daniel Benjamin, told Congress. "Loose Libyan weapons and the return of refugees and mercenaries to their countries of origin across the Sahel has greatly increased the internal pressures faced by these countries."[27]

Al Qaeda in the Islamic Magreb (AQIM), historically the "weakest of Al Qaeda affiliates," probably had benefited the most, Ambassador Benjamin added. The intelligence community subsequently concluded that AQIM participated in the attack on U.S. facilities in Benghazi on September 11, 2012, in conjunction with the Iranian-backed Ansar al-Sharia group.[28]

A top Defense Department official who specialized in Africa seconded Ambassador Benjamin's comments: "Based on proliferation concerns following the regime change in Libya, DoD has also incorporated Man Portable Air Defense System, or MANPADS awareness and mitigation training in our mil-to-mil engagements" throughout Africa.[29]

A third official, who provided written answers to questions for the record submitted at the same congressional hearing, acknowledged that the increased flow of arms to terrorist groups caused by the Libyan revolution had "profoundly affected the region."[30]

Even the Obama White House was worried that its support for the Libyan rebels and the ouster of Qaddafi had benefited al Qaeda and was creating new threats across Africa and beyond. Under counterterrorism advisor John Brennan's authority, the White House established a special task force, led by National Security Staff, to conduct "assessments and training, and [offer] assistance throughout the region." This previously undisclosed White House unit "has provided funds for fuel and spare parts to border security forces" to countries neighboring Libya, to help them stanch the flow of weapons from Qaddafi's stockpiles.[31]

A source who was posted to a U.S. government facility in Libya in 2012 told me that two "arms negotiators who reported directly to the National Security Staff at the White House" came to Tripoli and Benghazi while he was there to "manage contractors" engaged in the hunt for the missing MANPADS.

"We watched missiles in small numbers leave through Algeria for Mali," this source told me. "We could see it happen. We had drone feeds. And I kept wondering, can't we put a bomb on that thing to stop this? We know it was sold here, it crossed the border over there. What the hell! But we were told to just let it go." The military chain of command, which ran up through AFRICOM headquarters in Stuttgart, Germany, refused to arm the drones without approval from the president, and the White House appeared content to sit back and watch the missiles disappear into the hands of al Qaeda.

What was this mysterious National Security Staff, and what was its legal authority? The Obama administration quietly announced its creation in May 2009 as a means of "ending the artificial divide between White House staff who have been dealing with national security and homeland security issues." [32]

Every president has sought to put his stamp on the National Security Council, the only agency in government tasked with coordinating operations across multiple cabinet departments. Created by Congress under the National Security Act of 1947, for more than four decades its activities were restricted to coordinating U.S. responses to foreign security threats to the nation. President Clinton expanded its responsibilities by creating the National Economic Council. After the September 11, 2001, attacks on America, President George W. Bush created the Homeland Security Advisory Council in an effort to facilitate communication between overseas intelligence-gathering agencies and domestic law enforcement.

But nowhere is the NSC authorized to carry out *operations* under its own name or to obligate money or conduct military-style training, as State Department official Donald Yamamoto revealed that the National Security Staff was doing. Indeed, that's what made the Iran-Contra affair under President Ronald Reagan a scandal and, for some, an impeachable offense. Under White House authority, National Security Council employees engaged in the actual conduct of U.S. foreign policy by raising money for the Contras, meeting with intelligence operatives overseas, and selling U.S. weapons to Iran, all without congressional oversight. It was a clear violation of the National Security Act.

So, what were negotiators from the National Security Staff doing in Libya? Why would they be conducting training and committing U.S. taxpayer dollars? And why would they be managing contractors who ostensibly were working under State Department authority to buy up MANPADS?

My sources say that *all* the contractors engaged in the MANPADS buyback were foreign nationals, "mainly North Africans and Africans, not white guys who would stand out in Libya." One advantage to hiring non-U.S. citizens was the lack of oversight and accountability, especially if they were hired directly by the White House, not by a statutory federal agency. That put them beyond the reach of Congress. The whole operation was off the books.

They also weren't that effective; at least, that's what the military was told. "The guys in the East, close to Egypt, were having no luck," a source who was posted to a U.S. government facility in Libya in 2012, who was following these activities, told me. "They were getting nothing, so they closed down the contract due to lack of performance." That happened not long after the South African street operators were kidnapped and released by the jihadis in late April.

A second group of contractors was more successful. With the assistance of U.S. military advisors from Tripoli, they found a few smaller depots in the Tripoli area, and one in the southern desert, which they blew up in the spring of 2012. "The Qaddafi people who knew where that stuff was at were not very open. They didn't know how they would be treated, so they used [information on MANPADS] to barter their position," my source said.

The State Department tried to disguise the role of the contractors in its public statements. In written replies to questions from Senator Lindsey Graham in February 2012 on the MANPADS collection effort, it said that the $40 million spent in Libya had "underwritten surveys of more than 1,500 bunkers at 134 Ammunition Storage Areas (ASAs) by Libyan-led inspection teams. Thus far these teams have helped to identify, recover, and secure approximately 5,000 MANPADS and components."[33]

However, not all of the weapons were being destroyed. Both the Saudi government and the Qatari government had leased enormous warehouses down by the port of Benghazi, where they had stockpiled weapons during the civil war to distribute to rebel groups. Those same warehouses were now being used to store weapons for the Syrian rebels. Since it wasn't a U.S. government operation, it was never reported to Congress, even though the White House, the CIA, and the State Department all knew what was going on.

The MANPADS collection effort screeched to a halt in Benghazi in late April or early May 2012 for a simple reason: The White House wanted to give the Qataris time to purchase the weapons and send them to the Syrian rebels. United Nations arms inspectors subsequently identified no fewer than twenty-eight rotations of Qatari Air Force C-17 military transport aircraft between Qatar and Turkey in a single four-month period, which they believed carried weapons for the Syrian rebels the aircraft had picked up in Libya.[34]

It was a neat operation, completely off the books, and there were no U.S. fingerprints. It also appeared to be completely illegal.

HILLARY'S SLUSH FUND

In her annual budget request for Congress, presented just seven months before the Benghazi attacks, Secretary of State Hillary Clinton saw no additional threats to U.S. diplomats and State Department facilities in the region and asked for no additional funds to beef up security, beyond ongoing construction projects, none of which were in Libya. "As President Obama has said, 'The tide of war is receding,' " Clinton reminded members.

That was the narrative the administration wanted Congress and the voters to hear. Obama had ended the war in Iraq and was winding down the war in Afghanistan. The world was a safer place because Obama had supposedly eliminated the causes of terrorism, which were U.S. aggression in Iraq and Afghanistan. Of course, that conveniently ignored the fact that the Iraq conflict had been going on since 1990, and that the eight years of it fought on President Bill Clinton's watch gave us the infamous United Nations Oil-for-Food scandal, which ultimately led to the 2003 war.

But Secretary Clinton *did* ask Congress to authorize a $770 million slush fund that could be used at her discretion anywhere in North Africa or the Middle East. She called it the "Middle East and North Africa Incentive Fund." She likened it to money that was spent after the collapse of the Soviet Union for democracy building. She insisted that the fund be "flexible and easily employed," but offered no specifics for how it should be used except to say that "this budget request also will allow us to help the Syrian people survive a brutal assault and plan for a future without Assad." [35]

How was that money spent? How much of it went to Islamist groups with ties to al Qaeda? Was any of it spent on illegal arms purchases? So far, the State Department has not answered these questions.

10

THE ARMS PIPELINE TO SYRIA

The Obama White House style of government was to move from crisis to crisis, rather than employing a sustained, coherent policy. As the flames of the Arab Spring leapt across the region, they shifted their attention from Egypt to Libya, and now they were ready to move on to Syria, without much care as to the mess they left behind. So it was that U.S. Ambassador Gene Cretz told reporters during one of many events to commemorate the first anniversary of the uprising against Qaddafi that Washington was looking at a new target.

"This past year, Libyans demonstrated courage, bravery, and heroism, overcoming forty years of enforced silence with a deafening roar," Cretz said. "Libya's freedom stands in stark contrast to the oppression ongoing in Syria; we join Libyans in standing with the Syrian people in their protest against tyranny."[1]

Whether Cretz's statement was a slip or an intentional warning, it signaled a noticeable shift in U.S. policy. And it gave all the appearances of a green light to arms smuggling operations that official Washington continued to deny.

THE *LETFALLAH II*

Islamist charities were springing up all across Libya, responding to Ambassador Cretz's call to aid the Syrian rebellion. They began collecting weapons and supplies to send to jihadi groups battling secular dictator Bashar al-Assad. One of these, calling itself Knight Suleiman Israh in Misrata, cobbled together several shipping containers full of weapons, and contracted to load them onto a Syrian-owned cargo ship, *Letfallah II*, in early April 2012.

Registered in Sierra Leone, the *Letfallah II* was owned by Mohammad Khafaji, a Syrian citizen, whose Khafaji Shipping Company was registered in Tegucigalpa, Honduras, but operated out of Tartus, Syria, as Khafaji Maritime Co. Khafaji later told United Nations arms inspectors that he had been contacted by a shipping agent in Libya, who put him in touch with a Lebanese national who wanted to ship twelve containers from Misrata to Tripoli, Lebanon.[2]

That's how the arms smuggling business worked. It was layers upon layers of cutouts and front companies, created in the expectation that the apparent complexity would deter simpler spirits from investigation and provide the participants with plausible deniability.

The *Letfallah II* docked in Khoms, Libya, on March 28 and stayed in port for six days before moving fifty-five nautical miles down the coast to Misrata on April 4. Most shipping companies spend a great deal of time and effort to keep their ships busy, since every day they idle in port is another day they have to pay crew, insurance, and port fees. But apparently Khafaji—or whoever was paying him—didn't care. The *Letfallah II* spent six more days in Misrata with its Automatic Identification System (AIS) disabled. Since 2002, the International Maritime Organization has required all cargo vessels of three hundred tons or more to carry AIS equipment to reduce the risks of maritime accidents. Turning off the system—and thus, disappearing from the live charts kept by the IMO and its affiliates—is a trick used by smugglers and others (such as Iran) seeking to disguise the movements of their vessels.

Why the additional six-day wait? Perhaps whoever had organized the weapons collection was having problems. When the ship was intercepted in Lebanon, only three containers were seized, not twelve as initially planned. One of the seized containers bore the distinctive green, white, and red logo and two-foot letters of the IRISL, the Islamic Republic of Iran Shipping Lines.[3]

Once the ship was loaded in Misrata, it began a deceptive zigzag course across the Mediterranean. The first stop was the commercial port of Gulluk, Turkey, a picturesque fishing and tourist haven in the Aegean Sea on Turkey's Turquoise Coast, where it stopped from April 14 to 16. Then, it disappeared for five days, with its AIS equipment switched off, surfacing again on April 21 in Alexandria, Egypt, where, according to the UN report, it offloaded construction materials it had picked up in Gulluk. After leaving Alexandria on April 24, it disappeared yet again for several days until a Lebanese navy vessel intercepted it off the coast of the northern Lebanese port of Tripoli on April 27. Official reports suggested that the Lebanese navy was suspicious because the *Letfallah II* was riding high in the water and reasoned it would be easier to contain a potential gunfight and prevent civilian casualties if they diverted the ship to Selaata, a military port just down the coast.

The ship's manifest showed that it was carrying engine oil. Lebanese customs officials offloaded three containers of smuggled weapons, which had been sealed in Misrata, and loaded them onto trucks with a heavy military escort and drove them down to Beirut. The Lebanese authorities arrested Ahmad Hussein Khafaji, the captain of the ship (and brother of the owner), as well as a local Lebanese agent. They kept Khafaji in jail for over a year. He was pending trial as this book went to press.[4]

Many questions about the *Letfallah II* shipment remain unanswered, despite an extensive investigation by the UN Panel of Experts appointed by the secretary general to report on the implementation of UN Security Council resolutions imposing an arms embargo and an assets freeze on Libya.

The Turkish government initially rebuffed UN inquiries about the ship. Turkey's permanent representative to the UN, Ertuğrul Apakan, sent an indignant letter to the secretary general and to the president of the Security Council, rejecting "unfounded allegations" that Turkey was conspiring with Libya to send weapons to the Syrian rebels. The *Letfallah II* "neither docked at any Turkish port nor has any affiliation with Turkey in terms of its registration or operator," Ambassador Apakan wrote on May 15, 2012. That turned out to be untrue.[5]

The Turkish government eventually admitted that the *Letfallah II* had docked in Turkey, and declared it was carrying three containers of "combustible engines" (*sic*) as its cargo, the UN inspectors reported in March 2014.[6]

An official inventory provided by the Lebanese government showed that the Libyans had scoured the local arms market for whatever they could find. The impounded containers included SA-7 and SA-24 MANPADS, French-made MILAN antitank missiles, AK-47s, Belgian-made FAL assault rifles, Dragunov sniper rifles, Soviet machine guns, RPG-7 launchers, antitank recoilless rifles, mortars, 40 kg of Semtex plastic explosive, and a wide variety of ammunition, grenades, rockets, and mortar rounds. (See appendix II for a full inventory.)

The UN experts report insisted that the containers had been "sealed in Misrata and were still sealed when they were seized by the

Lebanese authorities." However, photographs released by the Lebanese army showed half-empty containers stacked haphazardly with crates of weapons, as if they had been looted.

"I would say certain items were removed by the Lebanese at the request of the U.S. government so we were not more embarrassed than we already were," a senior U.S. official with access to intelligence reporting on the *Letfallah* seizure told me.

What items might have been removed? According to initial reports that surfaced in Arabic language newspapers in Beirut, the *Letfallah II* was carrying 100 Stinger missiles, and may have been intending to offload them at night onto dinghies on a deserted beach north of Tripoli in the rugged Akkar region of northern Lebanon bordering Syria. The Arabic-language reports estimated the street value of the weapons on board the *Letfallah II* at $60 million.[7]

The Akkar region was a well-known source of illegal weapons for smugglers supplying the Free Syrian Army, my sources say. Among the best known of these was a Lebanese man named Yehya Jassem al-Dandashi, aka Abu Maasam, who worked out of the Tripoli neighborhood of Bab al-Tabbaneh.

Abu Maasam was arrested well after this incident and charged with supplying arms to the Free Syrian Army. His arrest made news because Sunni supporters blocked traffic in Tripoli demanding his release. And they won.[8]

When the UN experts finally traveled to Lebanon in late December 2012—a full seven months after the *Letfallah II* seizure—they found some of the weapons were "damaged or missing components," making them inoperable. They found SA-7s without batteries, open weapons crates, broken weapons, and ammunition for weapons that were apparently missing. "The Panel concludes that this materiel was not prepared and shipped by experienced or qualified personnel, or that it was done in haste," their March 2013 report states.[9]

"The Qataris paid for that shipment," a U.S. official who was on the ground in Libya at the time the *Letfallah II* set sail for Lebanon and had access to the intelligence, told me. "The Qataris were there with a lot of money. They were throwing money around like crazy. You'd see signs up everywhere in Tripoli seeking help for 'our brothers' in Syria. They were doing weapons collection to send over there, separate from the U.S. effort." [10]

The coordinator of the UN Panel of Experts insisted that the Lebanese authorities showed them no Stingers as part of the confiscated *Letfallah II* shipment, only ten SA-7b rounds with one gripstock and six batteries, and two SA-24 rounds, the most advanced MANPADS currently made by Russia. (The gripstock is the guts of a shoulder-fired missile. It attaches beneath the missile container and includes the targeting electronics package, the trigger, and the thermal battery.)

"I am sure we saw everything," said Salim Raad, describing the panel's inspection of the cargo in December 2012. "There were no Stingers on board the *Letfallah*," he told me. Raad left the panel not long after we spoke. [11]

Was this just one more story where nonexpert journalists and excitable locals called every shoulder-fired surface-to-air missile a Stinger?

My sources say that Egyptian customs broke the seal on "one of the *twelve* containers containing arms" during the ship's stopover in Alexandria, and tipped off colleagues in Lebanon who called on the navy for assistance.

So was it twelve containers, as the ship's owner, Mohammad Khafaji, told the UN panel? Or was it just the three partially filled containers the Lebanese authorities said they confiscated? And what happened to the Stingers and other weapons, if they indeed had been on board and subsequently looted? "It is fifty-fifty as to whether the Stingers were removed by the Lebanese military or they found their way into the hands of Hezbollah," a U.S. military source who followed the case told me.

The *Letfallah II* was just the beginning. The arms pipeline to the Syrian rebels from Libya was just getting started.

INTO THE SINAI

Weapons from Qaddafi's arsenals were also moving via smuggling routes in the Sudan to Muslim Brotherhood networks into the Sinai Peninsula, and from there to Hamas into Gaza. Hamas was the Palestinian branch of the Muslim Brotherhood, as U.S. court documents in the epic 2007–2008 Department of Justice prosecution of Muslim Brotherhood front organizations in the United States showed.

One consequence of Obama's successful effort to remove pro–U.S. Egyptian strongman Hosni Mubarak was the collapse of Egyptian government control over the vast Sinai Peninsula. In principle, a United Nations Observer Force, headed by U.S. diplomat David Satterfield, was responsible for monitoring the security situation in the Sinai to ensure that the Egyptian government lived up to its commitment to keep the area calm. Ever since the Camp David accords in 1979, the Sinai had been Israel's most secure border. But, with Mubarak gone, the UN observers were overwhelmed. Terrorist groups allied with al Qaeda, the Muslim Brotherhood, and Hamas had reopened long-dormant smuggling routes and desert hideaways, and were using them to bring vast quantities of weapons looted from Qaddafi's to the front lines with Israel.

The Egyptian security services were worried that the goal of the terrorist groups was to provoke an Israeli incursion into the Sinai, which would exacerbate Egyptian-Israeli tensions and play into the hands of the Muslim Brotherhood, who were poised to take over the country. So, they negotiated a quiet deal with Israel so the Egyptian Air Force and army aviation could fly its own combat patrols into the Sinai, which was technically a violation of the Camp David accords.

In October 2011, as Qaddafi's regime was collapsing, Egyptian security officers seized two separate shipments of weapons from Libya in trucks crossing the Sinai bound for the Gaza Strip. One of the shipments included a small number of surface-to-air missiles.

Despite the stepped-up interdiction effort, the shipments just got larger. Six months later, in May 2012, the Egyptian border patrol intercepted a convoy of three trucks arriving from Libya carrying fifty Russian-built MANPADS, assorted ammunition, and hashish. The Egyptians believed the weapons were bound for Gaza or for al Qaeda training camps in the Sinai. Subsequent reports said the shipment had included 120 MANPADS.[12]

The Egyptians were also worried by reports that Iran was involved in the arms smuggling from Benghazi. An Egyptian security source told the Saudi daily *Asharq al-Awsat* they had "observed Iranian businessmen and nationals, carrying Western passports, entering the country following the collapse of the Mubarak regime."[13]

The Egyptians had known for years that the Iranian regime had no compunctions about dealing with radical Sunni terrorist groups, as security officials in Cairo told me when I first investigated Osama bin Laden's networks in February 1998 for *Reader's Digest*, well before he became a household name. So it was no surprise for them to discover the Iranians working with the Muslim Brotherhood now.

JOHN BRENNAN'S IRON CLAW

Qaddafi's missing missiles were also moving south through the desert into Niger and Algeria, along smuggling routes established by Qaddafi years ago to fuel revolts up and down the African continent. With the victory of the revolution, control over the lucrative arms trade routes was up for grabs.

"We'd been watching those smuggling routes for decades," says John Maguire, who retired from the CIA after serving as deputy chief of station in the sprawling CIA base in Baghdad in 2004, at the peak of the insurrection against the U.S. occupation.

Once Abdelhakim Belhaj was appointed head of the Tripoli Military Council, the former emir of the al Qaeda–affiliated Libyan Islamic Fighting Group seized control of the southern smuggling routes. "We began to see those routes fill up with weapons that Belhaj controlled, especially IGLA missiles," Maguire told me. This was a family of improved Russian MANPADS that included the SA-16, the SA-18, and the Stinger-equivalent SA-24, also known as the IGLA-S.

So-called ant traders (individual smugglers) and Belhaj lieutenants brought the weapons through the desert of southern Libya and handed them off to Abdulwahhab Hassan Qayed, the new head of Border Control and Strategic Institutions in the Interior Ministry. Qayed, who co-authored a book with Belhaj and other senior Libyan Islamic Fighting Group members, was the brother of Abu Yayha al-Libi, al Qaeda's second in command. Al-Libi was also a former colleague of Belhaj's from the early days when they cofounded the LIFG. Captured in Afghanistan and held by U.S. forces at the Bagram interim detention facility, he escaped along with several high-profile al Qaeda terrorists on July 10, 2005, and rejoined the senior al Qaeda leadership. "So, Belhaj was essentially delivering these missiles into the hands of al Qaeda central," Maguire told me.[14]

From southern Libya, Maguire said Qayed's network took them to Agadez, a well-known smuggler's paradise in the north of Niger, which boasted an open-air arms market in the desert on the outskirts of town, where you could regularly hear sellers demonstrating their wares to suspicious buyers. "We used to call it the Peshawar of Africa,"

Maguire said. It was a wild town that for generations had linked north to south, where traffickers in all forms of illicit merchandise—human slaves, refugees, weapons, and drugs—made and lost fortunes in a day.

In late 2011, a source who worked as a security advisor to President Mahamadou Issoufou of Niger came to him with troubling information: They had spotted SA-7s from Qaddafi's arsenal in Mali. At first, Maguire kind of ho-hummed the tip-off. After all, the missiles were old, and they had been roughly stored in un-air-conditioned bunkers in the desert, so the batteries were all dead. Then the source got his attention: He described a new type of battery that Maguire instantly recognized, because of its distinct features as a CIA-designed replacement first used in Afghanistan in the 1980s.

In the summer of 2012, the source provided him with a photo and told him that Belhaj had moved 800 IGLA missiles to Agadez and set up a repair shop where they could be reconditioned using the CIA batteries. Four hundred of the missiles had already been refurbished and were gone. Gone. Another 400 were still in the pipeline. The work was being done by Egyptian military technicians.

"We knew they were Egyptians because they have a unique military ordnance color that nobody else uses," Maguire said. "The Libyan missiles are green. The Egyptian ordnance has this unique shade of brown. Brown Egyptian gripstocks, installed by the Egyptian military technicians, were being fitted onto the green Libyan missiles. Then they were reboxed and moved out onto the market." Maguire provided me with a photo, available in the Online Reader's Guide, which clearly shows the Egyptian retrofit. While glaringly obvious to anyone in the know, none of the people involved in the refurbishing operation paid any attention to the telltale military color schemes.[15]

Maguire's big worry was that al Qaeda would use the missiles in a coordinated assault on U.S. civilian airliners, in a replay of the Bojinka plot al Qaeda had developed in 1995 but was never able to carry out.

"All the airports coming into North Africa are located near the coast, so it would be relatively easy to have a dozen guys on beaches in different countries attack a dozen airliners all at once. We know that al Qaeda revisits operational concepts they think are viable. The U.S. airline industry cannot sustain another major event like that."

The stakes were high, so Maguire and his colleague, another former CIA operations officer, decided to take the information to the agency. Maguire had gotten to know CIA Director General David Petraeus in Iraq, so he emailed him a general outline of what he had uncovered. Petraeus thanked him and said he would get his guys on it immediately. They'd probably want to send someone to meet with the source in Niger, he wrote. Maguire gave him the information they would need to contact him. Petraeus said they should get together ten days later to go over the case in more detail.

And then all hell broke loose.

Maguire's source called him not long afterward in a panic. President Issoufou was about to throw him out of the country, even though he had a diplomatic passport. Someone in the agency had obviously gotten to the Niger president. "They told him, 'you're not going to get the drones or anything else we've promised you until you toss this guy,'" Maguire's colleague told me. "And then they went to the French, and had the source taken off an airplane in Madrid and told him that if tried to go back to Niger he was going to get whacked."

The source then tried twice to come to the United States, but he was turned back at European airports because he had been put on the U.S. no-fly list. "Someone was ripping into him with an iron claw," Maguire said.

Maguire's colleague had working relationships with senior officers at Joint Special Operations Command (JSOC) and at the FBI, so he thought he would appeal to them for help. He proposed bringing Issoufou's intelligence advisor to the United States to brief JSOC and the FBI on weapons leaking out of Libya. "The agency came down on both of them, with the president's approval, with an iron claw. They said to bugger off; they were in control. They were handling it," he told me.

Except that they weren't handling it; or at least, they weren't stopping the weapons from moving back into the black market arms pipeline, where they were snatched up by al Qaeda.

What was really going on? Neither Maguire nor his colleague ever learned what the White House had in mind when they shut down the operation to take those missiles out of the hands of al Qaeda. But they did learn one thing: The man who gave the orders to go after their source and to make life difficult for them with their other U.S. government contracts was President Obama's counterterrorism advisor, John Brennan.

"Something really sensitive was going on," Maguire's colleague said. "John Brennan fancies himself as an operator. When you look at all the JSOC operations in Libya, in Yemen, in Afghanistan and Pakistan, it's Brennan. My guess is, it was an operation that went out of control that they were desperately trying to put back in the box."

As Ty Woods' father, Charles Woods—who is also a lawyer— pointed out to me, it is illegal under U.S. Code 18, Section 2339, to provide material assistance to a terrorist group, whether you are a private citizen or an official of the U.S. government.

SECRET WAR?

Former JSOC operators Jack Murphy and Brandon Webb believe that Brennan was running a "secret war across North Africa" out of his White House office, with little or no coordination with the CIA.

> With JSOC, Brennan waged his own unilateral operations in North Africa outside of the traditional com-

mand structure. These Direct Action operations . . . were "off the books" in the sense that they were not co-ordinated through the Pentagon or other governmental agencies, including the CIA. With Obama more than likely providing a rubber stamp, the chain of command went from Brennan to (JSOC commander Admiral William) McRaven, who would then mobilize the men of ISA (Intelligence Support Activity), SEAL Team Six, or Delta Force to conduct these missions.

Sometime prior to Sept 2012, this JSOC element was directed by John Brennan to conduct combat operations in Libya. These operations targeted a high-level al Qaeda operative. . . .[16]

The target most frequently mentioned was Ezedin Abdel Aziz Khalil, aka Yasin al-Suri. He was one of the last senior al Qaeda leaders still on the loose. The Treasury Department identified him in July 2011 as the man now responsible for the al Qaeda fund-raising and logistics network, putting him at the heart of every major terrorist operation conducted by the core al Qaeda organization.

Most significant was the Treasury Department's assertion that al-Suri was operating from a base in Iran, under an agreement with the Iranian government. "Iran is the leading state sponsor of terrorism in the world today. By exposing Iran's secret deal with al-Qa'ida allowing it to funnel funds and operatives through its territory, we are illuminating yet another aspect of Iran's unmatched support for terrorism," Undersecretary for Terrorism and Financial Intelligence David S. Cohen announced.[17]

Al-Suri's importance to al Qaeda was further highlighted when the State Department offered a $10 million bounty for information leading to locating him. This put him on a par with the Taliban leader

Mullah Omar and the head of al Qaeda in Iraq. Only al Qaeda leader Ayman al-Zawahiri had a greater price on his head, $25 million. In announcing the inclusion of Yasin al-Suri in the Rewards for Justice Program on December 22, 2011, a Treasury Department official pointed out that it was "the first such reward offered for a terrorist financier."[18]

Murphy and Webb believe the Benghazi attacks came in retaliation for "late summer [2012] JSOC operations that were threatening the al Qaeda–aligned militant groups (including Ansar al-Sharia) in Libya and North Africa," operations that most likely included the targeted assassination of "a known associate of al-Suri."

John Brennan, JSOC. Yasin al-Suri, al Qaeda, Iran. They were lethal combinations.

"The global network of espionage is a dark underworld, full of ruthless individuals, a moral vacuum where ego and self-gratification generally rule," wrote former CIA polygrapher Kevin Shipp in his memoir, *Company of Shadows*. "Billions of dollars flow invisibly across continents to fund operations pitting countries against each other, and fund undercover individuals working against their own government. Even on the good side, there are individuals who are career-hungry for the next catch, even a small one, to further their reputation. Fortunately, there are a few who do it purely out of patriotism, especially in America."

But those white knights didn't always survive the dark forces working against them.

11

OBAMA'S AMBASSADOR

On May 26, 2012, Chris Stevens returned to Tripoli in glory. After helping to mold the disparate rebel factions into an effective coalition that unseated Qaddafi, he was coming back as President Obama's ambassador to the new Libya. It was not just a promotion, but a vindication. Like Lawrence of Arabia a century earlier, Stevens showed that an intellectual who had never served in the military could broker his passion for Arabs and Islamic culture into a military victory on the battlefield through the force of his will. Now, as ambassador, he could continue to guide his Libyan protégés as they sought to rebuild their country.

The only problem was that Libya was crumbling; the country was disintegrating in front of the international news cameras. Instead of a major success story for Obama's reelection campaign and for Secretary of State Hillary Clinton's future, Libya was spinning wildly out of control. There were no Jeffersonian democrats waiting in the wings to steer the new Libya into a new era of peace and freedom. Worse, the central government didn't even control security in Tripoli. Bands of heavily armed street thugs staked claims to neighborhoods, intersections, former Qaddafi barracks, and highways. They bombed prisons and police stations, and kidnapped businessmen and foreigners for sport and sometimes ransom. It was "a volatile situation in which militias previously united in opposition to Qaddafi were now jockeying for position in the new Libya," the State Department's Accountability Review Board found. "Frequent clashes, including assassinations, took place between contesting militias. Fundamentalist influence with Salafi and al Qaeda connections was also growing."

And yet, as security got worse, the State Department in Washington *decreased* the protection of U.S. diplomats and facilities in Libya, rather than increase it to meet the needs and the requests of the personnel on the ground. As a result, the United States wound up hiring groups affiliated with the very same militia that later would attack the Special Mission Compound in Benghazi.

PROBING ATTACKS

In the spring and summer of 2012, the jihadi militias launched a series of probing attacks against U.S. and foreign missions in Libya, trying to learn more about the secret activities of the United States and its partners and to find points of weakness in their defenses.

Just one month before Stevens returned to Libya, two South African contractors were kidnapped in Benghazi as they were walking down the street for early morning coffee in a residential area of the city. The State Department's Regional Security Officer Eric Nordstrom dutifully reported the incident back to Washington. Why? Because the two South Africans were working out of the supposedly secret CIA Annex in Benghazi under light State Department cover.

In Nordstrom's bureaucratese, they were staffing a "U.S. funded weapons abatement, UXO (unexploded ordnance) removal, and demining project." But I can reveal that they were street operators working on the MANPADS collection effort.

Their kidnappers initially passed by them, did a double take, then rammed their Ford sedan into reverse and bundled them into the car at gunpoint. The kidnappers drove at high speed to a walled villa, so excited by their catch that they crashed the gate open with their vehicle. A militia member wearing a balaclava interrogated the two South Africans to find out what they were doing in Benghazi. Once he learned that they were working for the Americans at the Annex, he ordered them to be blindfolded, then transferred to another militia headquarters for more intensive interrogation. In the end, they were taken back to the street where they had been abducted, and released.[1]

Media accounts in the wake of the Benghazi attacks have tried to paint the CIA Annex as a supersecret facility that successfully eluded the curiosity of its Libyan neighbors. With input from the CIA, the *Wall Street Journal* wrote that the very existence of the Annex was a "well-kept secret" in Benghazi. "A neighbor said that he never saw Libyan security guards at the Annex compound and that the street never had any extra police presence or security cordon. 'If the CIA was living there, we never knew it,' the neighbor said."[2]

That was patently false. The CIA Annex had high-tech security barriers, surveillance cameras, and a higher perimeter wall than the diplomatic compound. Armored SUVs of the type used by diplomats and spies regularly entered and exited the compound at high speed, using arranged security codes to open the solid metal gates. The Annex was so well known among spooks and diplomats that French espionage writer Gerard de Villiers used it as the backdrop for his eerily prescient thriller, *Les Fous de Benghazi* (The Wild Men of Benghazi), which was published in France in December 2011. Some of the neighbors might not have known (or wanted to know) that the Annex was the CIA's base of operations, but the jihadis certainly did.

Just one week before Stevens returned as ambassador, Lieutenant General Carter Ham, commander in chief of United States Africa Command, made his second trip to Libya since the fall of Qaddafi. General Ham met with Deputy Chief of Mission Joan Polaschik and the top security official at the embassy, Lieutenant Colonel Andy Wood.

Also in town for the meetings was Ben Fishman, director for Libya on the National Security Staff, who was visiting Libya to conduct a gut check for the White House as Libya geared up for its first post-revolution elections. He wanted to get ground truth of the security situation, and a feel for whether the losers would accept the outcome of the elections or behave as they did in many African countries and resort to arms. For their most sensitive discussions, they traveled outside the embassy to another CIA Annex—this one in Tripoli—where they could discuss frankly without any fear of enemy eavesdropping in a specially constructed bubble known as a Sensitive Compartmented Information Facility, or SCIF.

"General Ham's purpose in visiting Libya was to advance the bilateral relationship between the U.S. and Libyan militaries," the embassy claimed. But Lieutenant Colonel Wood told a very different story.

"General Ham came primarily to meet with the Americans," he told me. "He came under the radar for the Libyans, because if a general officer shows up, that's a big deal."[3]

What was so important for a combatant commander, who reports directly to the president, to make his second trip to Libya in less than two months?

According to Wood, General Ham was worried about the spiraling insecurity in Libya, and wanted to make sure that U.S. Africa Command (AFRICOM) was doing everything it could to keep the diplomats safe.

Lieutenant Colonel Wood had sixteen Special Operations troops under his command. They had been detailed by General Ham to the State Department as a Site Security Team (SST) in February 2012 to beef up security at the embassy, an arrangement that Fishman at the White House had helped set up. Their job was to provide additional firepower to the Diplomatic Security and Mobile Security Deployment teams already in Tripoli.* As a secondary mission, they were training an elite counterterrorism force to hunt al Qaeda. The embassy protection mission put Lieutenant Colonel Wood and his men under State Department control. In bureaucratese, that meant they were under Title 22 of U.S. Code, which governs foreign relations, and reported to the ambassador. He could task them, and, more importantly, he could wrap them in the mantle of diplomatic immunity in case bad things happened when they were out on the streets.

If they were to continue training Libyan proxy forces, however, that would return them to Pentagon control, under Title 10. "I was explaining to General Ham that we were starting to do more of one, and less of the other, when he stopped me cold. He said, 'you are here [for the embassy] for as long as they need you.' He made that very clear to me and to the deputy chief of mission at the time," Wood told me.[4]

But back in Washington, the building didn't want Special Operations troops guarding the ambassador since it gave the impression that the embassy was under siege in a country at war. "We were there really incognito," Wood told me. "We were not allowed to wear uniforms. We were told specifically not even to wear [military] boots."

That pretense of normality from the Washington politicos frustrated Wood and Regional Security Officer Eric Nordstrom. They

---

*When it reopened in late 2011, the embassy had three Mobile Security Deployment teams, each with six combat-trained security officers, to supplement the roughly seventeen Diplomatic Security officers that were posted to Libya. These numbers were gradually reduced by the time the SST arrived.

knew the militias were probing. It was only a matter of time before they found a weak spot.

JOGGING IN THE STREETS

Ambassador Stevens initially thought he could contribute to the atmosphere of normality by jogging outside the compound. Wood says he was okay with that. In fact, he welcomed the challenge.

"Stevens was more of a distance runner, and a really good runner. I had no problem keeping up with [Ambassador Gene] Cretz. But Stevens, damn it; that guy could *run*," Wood told me.

Cretz used to run downtown at the soccer stadium or at the horse track. "But Stevens liked to run the streets, run the neighborhoods. He'd get up early in the morning, six or seven in the morning, when [the locals] are still sleeping. The streets are *dead*. So that's the time Stevens liked to run."

Wood says he and his team would plot out different jogging routes for Stevens ahead of time, scout them out, and make sure that they understood potential security threats and had an exit plan if things went bad.

"Things went well for a while, until one of the groups posted a picture of Stevens on their Facebook page, saying this is the American ambassador and he likes to run in this and this neighborhood, so we're going to get him." That put the kibosh on Stevens' early morning jogs for about two weeks.

The June 6, 2012, posting, on the page of the al-Moqawama (Libyan Resistance) movement, showed a smiling Stevens strolling down a Tripoli street full of shuttered shops, as two Libyan men looked on. Wood said it sent a chilling message.

The threat to Stevens in Tripoli came as the tempo of anti-Western attacks accelerated in Benghazi.

During the early morning hours of May 22, 2012, a group calling itself the Imprisoned Sheikh Omar Abdul-Rahman Brigade fired two RPG rounds at the Benghazi office of the International Committee of the Red Cross. The ICRC headquarters was located on Shara al-Andalus at the Third Ring Road traffic circle, just one kilometer from the U.S. mission compound. In a Facebook post claiming responsibility for the attack, the group accused the ICRC of trying to convert Muslims to Christianity. They also declared Libya an Islamic state. "Now we are preparing a message for the Americans for disturbing the skies over Derna," they warned.

Jihadi groups in Libya were worried that the United States was flying drones over Benghazi, Derna, and points east, and would arm them just as they were doing in Afghanistan. On June 4, their suspicions hit the roof when the deputy commander of al Qaeda, Abu Yahya al-Libi, was killed in a drone strike in northern Waziristan in the rugged badlands along Pakistan's border with Afghanistan. This was the same Mohammad Hassan Qayed whose brother now controlled Libya's southern border. Not only was he a founding member of the Libyan Islamic Fighting Group, well-known in Libya because of his origins. He also was a well-known propagandist, who featured in numerous al Qaeda videos. Many of his former comrades-in-arms were now in senior positions in the post-Qaddafi government, starting with Abdelhakim Belhaj.

At 3:45 in the morning on June 6, the local security guard at the Special Mission Compound in Benghazi awakened with a fright as his video monitor came alive with movement. A pickup had stopped along the perimeter wall and an individual wearing "Islamic dress" got out and was placing a suspicious package along the perimeter wall. The guard sounded the "emergency, imminent danger" alarm, which automatically alerted the Quick Reaction Force located at the CIA Annex. Six minutes later the bomb went off, blowing a thirty-foot hole in the compound wall. The CIA security team, along with backup from the 17th February Martyrs Brigade, "responded quickly" to the alarm, according to the official after-action report. Later that day, the Imprisoned Sheikh Omar Abdul-Rahman Brigade claimed credit for the attack, saying it was in retaliation for the drone strike against al-Libi, the al Qaeda leader in Pakistan. They even released a video of the attack and left leaflets behind promising more attacks on Americans.[5]

Despite this, the State Department's Accountability Review Board tried to dismiss the Imprisoned Sheikh Omar Abdul-Rahman Brigade as a "previously unknown group," and called their claim of responsibility for the attack against the Special Mission Compound "unsubstantiated." Those are the only references the ARB made to the Blind Sheikh and his supporters in Egypt and Libya, despite the very public involvement of the group in the events of September 11, 2012.

Regional Security officer David Oliveira later told congressional investigators that the IED attack on the Mission Compound in Benghazi made him realize that a larger attack loomed. The local guards told him they were afraid to work. "[T]hey felt that the U.S. was a target and they felt that they didn't want to work overnight. What we heard from the [Blue Mountain] representative on the ground was that some of their families might have put some pressure on them to not want to work for the U.S. mission."[6]

The State Department contract with the British-registered Blue Mountain group, who provided the local guards for the Benghazi compound, specified that they be unarmed, since Libyan law prevented private security guards from carrying weapons. A similar arrangement with local guards at the Tripoli embassy prompted several State Department diplomats to ask the SST men to train them to shoot handguns and to issue them firearms, so they could at least defend themselves. "They told us that if something like the Benghazi attacks had happened in Tripoli, it would have been much worse," Representative James Lankford, a member of the House Oversight investigating panel, told me. "When the ambassador traveled outside the embassy, he took *all* the security with him, so you had something like fifty people left behind at the embassy with no security at all."[7]

Ambassador Stevens stepped into the breach and sent multiple emails to John Moretti at the State Department's Bureau of Diplomatic Security, requesting additional security. With elections coming up in July and August, the Mission "would feel much safer if we could keep two MSD teams with us through this period [to support] our staff and [personal security detail] for me and the DCM and our VIP visitors." The State Department replied that due to other commitments and limited resources, "unfortunately, MSD cannot support the request."[8]

"After we departed Libya in early August, they were left stripped naked," Lieutenant Colonel Wood said. "We didn't feel good about leaving them like that. If I were to meet Greg Hicks today, I'd be ashamed. I'm sorry, you got let down."

THE BLIND SHEIKH

So who was the Blind Sheikh, and why was an al Qaeda–affiliated militia in Libya invoking his name?

The Egyptian imam was the leader of Gama'a al-Islamiya, which has been classified as a terrorist group by the Egyptian and U.S. governments since the 1980s. He issued the fatwa, or religious edict, that supposedly justified the assassination of Egyptian president Anwar Sadat in 1981, because Sadat had made peace with Israel. Jailed for several years in Egypt in the 1980s, he was granted a tourist visa to enter the United States in 1986. After that error was discovered, he was put on a terrorism watch list. Despite that, CIA officers working under light cover at U.S. consular offices in Cairo and Khartoum gave him six more visas to enter the United States between that time and 1990, according to a classified State Department inspector general report.[9]

Once in the United States, he began recruiting terrorist candidates to join Osama bin Laden in Afghanistan and Sudan. He was arrested in June 1993 after an undercover investigation following the first World Trade Center attack discovered that he was plotting new terrorist attacks against the Holland Tunnel and other New York landmarks. In 1996, he was sentenced to life in prison. At the time of the Libya attacks in the summer of 2012 he was seventy-four years old and still confined to the high-security federal lockup in Butner, North Carolina.

Sheikh Omar Abdul-Rahman had become a cult figure to a whole generation of jihadis. He was the man who punished Sadat for making peace with the Jews. He was the visionary who helped recruit young Muslims in the United States and send them to fight the Soviet Union in Afghanistan. And he was the tireless advocate of Sharia and global Islamic jihad who fearlessly turned his sights on the Americans.

His followers wanted him set free. And, now, with Muslim Brotherhood leader Mohamed Morsi favored to win election as president of Egypt, they had the weight of the most powerful Muslim state behind them.

Morsi sent a delegation of Muslim Brotherhood officials to Washington just days before his election, to demand that the United States release the Blind Sheikh. Among them was a man named Hani Nour Eldin, a self-avowed member of Gama'a al-Islamiya, a group that still figured on the State Department's list of international terrorist organizations. By law, members of terrorist organizations are denied entry to the United States. That is why we have a terrorist watch list: to prevent people like Hani Nour Eldin from entering the country, where they could meet up with clandestine terror support or operations cells.

Reached by a reporter for the *Daily Beast*, Nour Eldin confirmed that not only had he been granted a U.S. visa, but that the Obama administration had invited him to the White House, where he met with

Deputy National Security Advisor Denis McDonough. He also met with Deputy Secretary of State William Burns and others at the State Department.

Party spokesman Tarek el-Zomor told CNN that Nour Eldin "pressed American officials for a transfer into Egyptian custody of Sheikh Omar Abdel-Rahman," and that the request "mirrors the demands of Gamaa Islamiya members in Cairo who have protested in Tahrir Square, seeking the sheikh's release." State Department spokesman Victoria Nuland squirmed when asked how someone who is a self-avowed member of a terrorist organization could gain entry to the United States, let alone visit the White House. "We are reviewing the case of the visa issuance," she said.

It wasn't as if Nour Eldin was hiding his identity or his role in the Egyptian terrorist group. All a consular officer in Cairo had to do was visit his personal Facebook page and a bio popped up "where he very clearly says he is a member of Gamaa Islamiya, and that he was arrested in Egypt, and spent eleven years of his life in prison," said Samuel Tadros, a Egyptian researcher at the Hudson Institute.[10]

The answer is simple: The State Department, on orders from Hillary Clinton or her office director, Huma Abedin, gave instructions to the U.S. Embassy in Cairo to issue a special waiver to Nour Eldin so he could get a visa. Abedin's parents and her brother were all top officials in the Egyptian Muslim Brotherhood or related organizations.

Denis McDonough, a former Obama campaign communications director who had never held a security clearance in his life, was now sitting in on meetings with David Axelrod, Valerie Jarrett, and other political hacks, at which the most sensitive and highly classified national security information was discussed.

It was a dangerous mix. It's one thing for political leaders to rely on sound and very sensitive intelligence to make political judgments. That's the reason that the United States spends more than $50 billion per year to collect secret information on our enemies, potential adversaries, and even our friends. It's something entirely different for the politicians and their hacks to drive the intelligence, as has been the rule in the Obama White House from the get-go. If a politician such as Hillary Clinton or Barack Obama simply determines that the Muslim Brotherhood is our friend and jihad is not a threat, how can security professionals at the State Department take action to defend our embassies in the Muslim world? No wonder all the warnings from Eric Nordstrom and other RSO in Tripoli and Benghazi were simply swept under the rug. They were warning about a threat the politicians had convinced themselves did not exist.

The meeting with the Muslim Brotherhood envoy from Egypt raised the same issue as sending National Security Staff members to Tripoli and Benghazi to allegedly negotiate arms purchases on behalf of the White House. Was McDonough now running a secret political operation, totally beyond congressional oversight, to free the Blind Sheikh?

JIHAD ON PARADE

After a spate of shadowy attacks against Western interests in Benghazi, the Islamists came out in the open on June 7, 2012, putting on a show of force that immobilized the streets of Benghazi for hours.

Down near the sun-swept port, thousands of jihadis staged a sit-in on the broad esplanade in front of the local Council building, spreading a half acre of prayer rugs on the pavement.

They were a motley crew. Some were dressed in the long white *thawb* of the Arabian Peninsula and wore their beards in the style of the Saudis. Many wore white skullcaps designating them as *hajjis*, venerated for having made the pilgrimage to Mecca. Others were dressed in expensive jeans and Western T-shirts. Still others wore

Afghan-style tunics and baggy pants and the *pakol*, the pancake hat made famous by the mujahideen. It was an all-male crowd, and they were demanding that the Transitional National Council adopt Islamic Sharia law in the upcoming national elections.

"It is time to clean the homelands of Islam from filth and positivist laws that have been inherited from the West and have ruined the Islamic nation for ages," said Hani al-Mansouri, a spokesman for Ansar al-Sharia, the main sponsor of the parade. A representative of the brigade's propaganda oufit, Al Raya Media Productions Foundation, told Iran's state-run PressTV they would "continue the struggle until the country is totally liberated from non-Islamic values." [11]

Local press reports identified fifteen jihadi groups who participated in the event and vied with each other to put their military hardware on parade that day. They brought out an assortment of SUVs and "technicals," mostly Toyota Hilux pickups, with various sorts of weaponry mounted on the bed: single or twin-tubed 23mm anti-aircraft guns, heavy machine guns, 50mm turret-mounted guns. Just as on the Fourth of July here in America, they lined up ahead of time and posed for photographers before joining the parade. Each brigade, or *katiba*, had its own flag. Many of them sported the black flag of al Qaeda.

The fighters themselves were not just Libyans but came from Sudan, Mali, Egypt, Tunisia, Saudi Arabia, and Afghanistan. As they drove along the corniche, honking their horns and shouting, many of them swiveled the turrets of their anti-aircraft guns in mimicry of bravura. Eight- and ten-year-old children brandishing Kalashnikovs leaned out of car windows, pretending to let off bursts of celebratory gunfire.

Initially, they had planned to occupy Benghazi for three days, but pushback from crowds of secular Benghazi residents forced them to call it quits after a few hours. Still, given the recent attacks on the ICRC building, the UN, and the U.S. Special Mission Compound, the presence of so many jihadi fighters and so much military gear "generated concern among U.S. officials in Libya," as Representative Darrell Issa reminded President Obama.

Where was Ansar al-Sharia getting its funding and training? Clearly they were getting support from the Muslim Brotherhood in Egypt. But in one PowerPoint on the group that was shared with Ambassador Stevens and the U.S. Embassy security team shortly after the parade, slides showed arrows moving people and money from Iran to Libya via Syria and Egypt.*

It was this combination that ultimately fueled the September 11, 2012, attack on Benghazi.

PRETENDING NORMAL

As part of the return to normal operations, the State Department changed the status of the Tripoli embassy and Benghazi outpost in June 2012, putting an end to temporary duty postings (TDY) and bringing in a permanent embassy staff. They brought in new political officers, new reporting officers, a new deputy chief of mission (Greg Hicks); even a new CIA chief of station. John Martinec had just arrived as the full-time regional security officer, and was immediately brought up to speed on the disastrous state of security by Eric Nordstrom.

Hannah Draper was a junior Arabic-speaking officer who arrived in Tripoli on June 9, 2012. Although she was just twenty-seven, Tripoli was her third overseas posting in just five years of joining the State Department. She began as a consular officer in Dhahran, Saudi Arabia, and more recently was a desk officer at the U.S. Consulate in Istanbul, Turkey. An avid blogger, she was ecstatic at arriving with

---

*The CIA has denied my Freedom of Information Act request to get that PowerPoint released.

the "first wave" of permanent staff at the newly normal U.S. Embassy in Tripoli, and toured the town briefly on her way to her first assignment the day after she arrived. She drove past Martyrs Square, Bab al-Aziziyah (Qaddafi's former compound), and the Rixos Hotel, where journalists were held hostage during the early days of the insurrection. "It's incredibly moving to see these sites and to realize how recent the scars are from the Revolution," she wrote on her blog, which she called "the slow move east."

Unable to meet many Libyans because of security restrictions, she sought indices to political reality through graffiti glimpsed hastily as the embassy car sped by. "The one phrase you can see over and over is 'February 17th, Free Libya.' That's the day the uprising started, and it's the date by which this revolution is hashtagged on Twitter," she wrote. If the insight wasn't exactly brilliant, at least she was looking.[12]

The following weeks passed much in the same closeted fashion, inside embassy vehicles, traveling in convoys to meetings with TNC officials, pining for the new husband she had left behind in Istanbul, complaining about the boring embassy food. A colleague showed her some of the local currency, which still bore Qaddafi's image. "Apparently, in many shops the cashiers will methodically take a Sharpie to Gadhafi's face on every single bill before handing back a customer's change," she wrote.

The main perk to being permanently posted to Tripoli as opposed to a TDY assignment might seem minor, but to Hannah and many others it was significant. "Permanent staff have bedrooms to themselves [in the embassy residential compound] and share bathrooms with their neighbors; temporary staff have roommates. Our movements off compound are limited both because of logistics and security."

Although the State Department was calling Tripoli a "normal" diplomatic outpost, Hannah Draper and other junior diplomats were essentially assigned to quarters, only allowed to venture outside the compound accompanied by diplomatic security guards and other diplomats on official visits.

It was a make-believe world, which she filled by surfing the Internet for "Hardship Homemaking" recipes and Skyping her husband. And yet, even in this fairy tale of normality, Hannah Draper saw her own instincts change. After a while, she wrote, "I no longer flinch when I hear gunfire in the distance."

Would Hillary Clinton, Patrick Kennedy, and Charlene Lamb have changed their determination to paint Libya as a normal diplomatic outpost and assigned more security to the ambassador and his top officers if they had been hearing the daily sound of gunfire in the distance?

One thing Hannah Draper did discover, without apparently realizing the significance of it, was that the White House was tracking events in Libya—not from Washington, D.C., but from the ground. "I've not worked so closely with other branches of the US government, such as the Commercial service or the National Security Council," she wrote. "It's an interesting lesson in how the whole government works together (not always harmoniously) to inform, to determine, and to implement policy in a given country."[13]

The National Security Council? (Actually, the National Security *Staff*?) What were *they* doing involved in the day-to-day activities of the embassy in Tripoli?

NEAR MISS ON THE BRITISH AMBASSADOR

It was June 11, 2012, another sun-drenched day in Benghazi, where the clean blue sky belied the dangers lurking in the shadow just around the corner. Sir Dominic Asquith, the British ambassador to Libya, was visiting town and had gone for a leisurely lunch at the Venezia Café, one of a handful of Western eateries in Benghazi.

The Venezia was a local landmark; so much so, that taxi drivers referred to that section of the Fourth Ring Road in the posh West Fwayhat district of Benghazi as Venice Street. The avenue was studded with the occasional palm tree, towering like a lone sentinel far above the pockmarked asphalt and the low-slung residential compounds. Sometimes you could see the silvery leaves of an olive tree catching a welcome breeze in the distance, promising watered dirt and cool shrubbery beyond a dust-covered wall. The Venezia also happened to be just across the street from the rear gate of the U.S. Special Mission Compound. Asquith and other diners could watch the Libyan guards making the rounds along the perimeter wall. The record does not show who the ambassador lunched with that day, but the Venezia was well-known as a haven for militia leaders, spies, diplomats, and arms dealers, all seeking to ply their trades.

The British ambassador traveled with a security detail of five heavily armed men, many of them veterans of British Special Forces. When he was ready to leave, the drivers pulled up in a convoy of three armored SUVs, and his guards bustled Sir Dominic into their vehicle for the short drive back to the British consulate. They hadn't gone more than a thousand meters when the lead vehicle developed engine trouble and had to fall back, changing the position of the ambassador's vehicle in the small convoy. They were just five hundred meters away from the consulate when Sir Dominic's bodyguards heard the hollow sucking sound of a rocket in flight. They had no time to turn away when an RPG antitank projectile slammed into the rear of the SUV behind them, the one where the ambassador would have been.

The car swerved from the impact and the scene quickly turned chaotic. Luckily the armor plate held, but the rear window exploded and large pieces of white-hot metal ripped into the car and hit the two passengers in the rear seat, where the ambassador was supposed to be.

The driver in the front car, where the ambassador now was, floored his vehicle and sped off to the relatively safety of the consulate just down the road, leaving the other two to fend for themselves.

Lieutenant Colonel Andy Wood happened to be in Benghazi that day, attending a conference on military law in place of the defense attaché back in Tripoli. When the call came that the British ambassador's convoy had been hit, he rushed back to the Special Mission Compound, donned body armor, strapped on his 9mm Glock 22 and grabbed an M4 carbine and extra magazines. "Up until the moment I made it back to the SMC, I had been a civilian protectee," he told me, cocooned by a DS agent and members of his own SST. "Once I strapped on my gear, I became a security guy once again." He and a DS guard jumped into one of the seven Toyota Land Cruiser FAVs (Fully Armored Vehicles) parked on the compound, and roared up to the 17th February Martyrs Brigade barracks near the front gate to pick up an extra man. They were like firemen, Wood thought, suiting up whenever the alarm went off.

After what seemed like an interminable wait, a backup vehicle from the Annex finally showed up with two GRS shooters and an 18 Delta Force combat medic, and together they raced the two miles through traffic to the British compound, careful to enter through the rear gate so not to attract attention. Wood groaned when he saw the scene. "One of the British guards had been hit in the shoulder with a large fragment of steel and was bleeding out," he told me. The former Delta Force medic immediately went to work while the rest of them pulled security until the Brits could figure out what to do.

The Libyan police found leaflets at the scene from the Imprisoned Sheikh Omar Abdul-Rahman Brigade, claiming responsibility for the attack. They wanted to drive the Westerners out of Libya. They were beginning to show some success.

The near miss on the British ambassador caused the Brits to close their consulate for good. There was no way they could protect their guys from organized criminal gangs, let alone the jihadis. They were out of there.

A few days went by and Lieutenant Colonel Wood began to wonder why no one from the Annex had filed an after-action report. He began to raise Cain with the chief of base that it was a big deal, but got only shrugs in response. Eventually he called the Brits and went back to the scene of the attack so he could reconstitute exactly how it happened. He put the whole thing back together, like a crime scene, to figure out where the shooter must have stood, who must have helped them, how many they had to be, and more. And then he realized just how sophisticated the attack had been.

"They had to have had two teams in place. The guys doing the shooting had to have communications with the spotters up the road," he told me. The more he looked at it, the more he knew that his nemesis—the man he went to bed dreaming about every night—was getting outside talent. With the speed of the convoy and the confines of the street, the shooter had just a split second to take his shot. He needed to have nerves of steel, which was not something you learned your first night out. They also had a great getaway plan, Wood learned, hopping on the main ring road and blending into traffic.

Wood wrote an after-action report and shared it with the chief of base at the Annex, along with some of the pictures he had taken. He was shocked that the CIA hadn't investigated on their own, and were still under the misconception that the shooter had hit the vehicle from the front, not the rear. Later, back in Tripoli, he saw that the chief of base essentially had taken his report, put his name on it, and filed it as his own, along with a cockamamie theory about small arms being used when Wood clearly saw that there were none—no holes in the wall, nothing. (The small arms story was picked up by numerous Western press reports in the beginning, showing where they got that information.)

The chief of base's laziness made Wood angry, so he filed a more detailed version, explaining exactly how it had happened, and included the same photos he had left with the agency guys. That got people talking. They saw the photos, then compared the sparse and inaccurate CIA account against Wood's far more detailed one, and realized that the agency boys had cribbed his report.

"It was like they didn't really take it seriously, like they didn't care," Wood said. "They didn't even bother to get out and look at the site."

The CIA initially set up the Annex in Benghazi as an intelligence-and-training center during the Libyan revolution, so it included a mix of CIA case officers, analysts, NSA signals intelligence specialists, and Special Operations officers. Now that the revolution was over, the SpecOps guys had no more mission beyond providing protection to the case officers when they went beyond the wire to meet a source. So they sat around playing video games, cleaning their weapons, drinking alcohol-free beer, throwing knives with the Libyan guards. They were just waiting for the next mission, because the one they had was over.

By the end of June 2012, all that had changed.

THE SHIFT TO SYRIA

Not long after the near miss on the British ambassador, the word came from Washington that the president, urged on by Secretary of State Clinton and CIA Director David Petraeus, had made a determination that the United States was going to provide covert aid to the Syrian rebels.

The news spread like wildfire and within days the adrenaline junkies—the soldiers of fortune types, the arms dealers, and wannabes who hung around the Annex—packed their bags and headed off for southern Turkey, where the staging grounds for the war against Syrian dictator Bashar al-Assad had been established. The battlefield had moved on and they were moving with it, in search of fame, fortune, and kicks.

"The president caved on something, because all of a sudden, you had all these guys in Benghazi saying, yahoo, I'm out of here. We're going to Syria," a U.S. official who was in Benghazi at the time told me.

They went to join a small number of CIA operations officers who had been sent into southern Turkey earlier as part of a "clandestine intelligence gathering" operation, to assess the various factions of the Syrian opposition. A secondary goal was to "help keep weapons out of the hands of fighters allied with Al Qaeda or other terrorist groups," a senior U.S. official said.[14]

The State Department had already authorized $15 million in nonlethal aid for the Syrian rebels, which was being distributed through conduits in southern Turkey, but sophisticated weapons were reaching the most radical rebel groups without any apparent help from the United States. "The rebels are starting to crack the code on how to take out tanks," said Joseph Halliday, a former U.S. Army intelligence officer in Afghanistan who now worked for the Institute for the Study of War, a Washington, D.C., think tank.[15]

The CIA operators came from the agency's top-secret Special Activities Division, which was responsible for covert operations, political influence operations, and paramilitary activities. If the setup at the Benghazi Annex was any indicator, they most likely rented and fortified a compound in one of the towns along the Syrian border, where they could vet rebel leaders and dispense cash, communications gear, and other goodies.

There is a fine line between secret intelligence activities and covert operations that gets drummed into every CIA operator by the agency lawyers. "Secret intelligence activities" covers a broad spectrum of tasks that the CIA and other U.S. intelligence agencies perform every day, from communications intercepts, to monitoring the secret activities of foreign governments, to liaising with foreign intelligence organizations. By statute, the president (through the director of national intelligence or the various agency chiefs) must keep the congressional intelligence committees "fully and currently informed" on all secret intelligence activities. These briefings flow through the Cardinals of Capitol Hill, sometimes known as the Gang of Eight: the Senate majority leader (Harry Reid), minority leader (Mitch McConnell), the Speaker of the House of Representatives (John Boehner), the House minority leader (Nancy Pelosi), and the chairmen and ranking member of the Senate and House intelligence committees. They in turn make the decision whether to brief the other members of the intelligence committees.

In very rare circumstances, such as the classified evidence presented by the Bush administration in October 2002 in support of going to war against Saddam, the Cardinals agreed to allow the full membership of the House or the Senate to access the evidence by personally signing into special high secure document rooms in the U.S. Capitol complex. But, in no case, are individual members allowed to take notes of what they read in the classified holdings.

Sometimes these briefings occur so frequently that Gang of Eight members stop paying attention. That was the case when the CIA briefed Nancy Pelosi in September 2002 on the enhanced interrogation techniques—including waterboarding—that the agency was using on high-value al Qaeda detainees. At the time, Pelosi was the ranking member of the House Permanent Select Committee on Intelligence and a member of the Gang of Eight, as she is today. She later claimed she had never been told about waterboarding—or else she would have objected. So the CIA produced a ten-page log with the dates, times, and subjects of these briefings, as well as the members who were present. Pelosi had to eat her words.[16]

But, even here, the government is allowed a great deal of leeway, and can delay reporting on some ongoing intelligence activities if it deems the limitation "necessary to protect intelligence sources and methods."[17]

That's a loophole big enough to establish a CIA intelligence-gathering post in southern Turkey without more than passing notice to Congress.

Covert action (CA)—including the supply of lethal military equipment—is a different beast entirely and requires a formal presidential finding. "Nobody at the agency is stupid enough to engage in CA without a presidential finding," a former senior CIA covert operations officer told me. "It gets drummed into you. You wouldn't spend two minutes on it. So, if they were actually moving weapons from Libya to the Syrian rebels, the agency would demand a finding."[18]

Presidential findings are briefed in elaborate ceremonies that are hard to forget. A senior presidential emissary meets the Cardinals of each chamber separately in their own Sensitive Compartmented Information Facility, or SCIF. (The House Intelligence Committee had its bubble specially built inside the Capitol Building; the Senate committee had its SCIF incorporated into the design specifications of the Hart Senate Office Building, which opened in 1982.) After the principals are seated, without staff, the president's emissary opens a folder with a thick cover embossed with bright red diagonal stripes, indicating the classification or code word of the operation involved, and reads the words of the finding. During the Bush administration, Vice President Dick Cheney sometimes conducted these briefings in person. The members present are allowed to look at the document and ask questions about the operation. Then they are reminded of their obligations as keepers of the nation's secrets and warned of the damage revealing this information could do to the national interest—and to their own career.

If the Gang of Eight was briefed on a finding that authorized covert arms shipments to the Syrian rebels and they didn't oppose it at the time—which members would have made known one way or another—then they have as much interest in covering up such a finding as the president and his advisors do because they signed off on it.

A Republican member of Congress tells me he confronted Speaker Boehner about this at a closed-door meeting of the Republican conference in March 2013. This member and others were pushing Boehner to approve the creation of a select committee to investigate Benghazi, which they believed would be more aggressive in using subpoena power to elicit testimony from survivors and other sources. Boehner tried to brush off the allegation that he had been briefed on the covert arms shipments, in a manner the member found suspicious. "It was the biggest nondenial denial I have ever heard," the member who confronted Boehner told me.

Lieutenant General Jerry Boykin, a former commander of U.S. Special Forces Command in Fort Bragg, North Carolina, told me that he believes Boehner and intelligence committee chairman Mike Rogers were briefed on the covert operations being run out of Benghazi. "They don't want to admit they knew what was going on," he said.

After the Benghazi attacks, the White House tried to reel back any suggestion that it had approved a covert arms pipeline to the Syrian rebels. Citing unnamed White House sources, *New York Times* correspondent Michael Gordon wrote that Clinton and Petraeus had pitched the idea to the president in the summer of 2012, with backing from Defense Secretary Leon Panetta. "But with the White House worried about the risks, and with President Obama in the midst of a reelection bid, they were rebuffed," Gordon concluded.[19]

I have found no hard evidence that the Obama administration issued a presidential finding to arm the Syrian rebels in the summer of 2012. But I have found extensive evidence that the State Department, the CIA, and other U.S. intelligence agencies had direct knowledge of arms shipments to the Syrian rebels by Libyan government-related militias and organizations, and helped to facilitate them. "To get around the reporting requirements, they tried to call it policy," one of my sources said.

There was only one man who had the necessary cover and who was knowledgeable enough, courageous enough, and dedicated enough to the cause of the Syrian rebels to coordinate these activities on the ground.

That was U.S. Ambassador to Libya J. Christopher Stevens. After all, he was the president's personal representative, with the protocol rank of a four-star general. "Ambassador Stevens was the U.S. commander-in-chief in Libya," said Lieutenant Colonel Andy Wood. The CIA station chief and the National Security Staff arms negotiators kept the ambassador up to speed on their activities. "The ambassador was fully briefed on everything," Wood told me.

Stevens knew that the agency was moving assets to southern Turkey to aid the Syrian rebels. He knew about the missing MANPADS. He knew that Belhaj and his allies were shipping weapons to Turkey.

And he knew that he was a target.

EMERGENCY SECURITY MEETING

Two weeks after the assassination attempt on the Sir Dominic Asquith in Benghazi, Ambassador Stevens called an emergency security meeting at the CIA Annex in Tripoli. He was alarmed at the dramatic increase in anti-Western attacks. The International Committee of the Red Cross had just pulled out of Benghazi. The Brits had also left. Stevens asked his security team if the United States should do the same.

Except for Lieutenant Colonel Andy Wood, the Army Green Beret who led the Site Security Team, most of the embassy's security team had just arrived in Libya. They had a new CIA chief of station, who was just settling in. The new regional security officer, John Martinec, had been given the briefing before leaving Washington about Libya returning to normal. Lieutenant Colonel Wood felt like the odd man out, the proverbial turd in the punch bowl. But because he had witnessed many of the attacks personally and had been privy to the intelligence on the buildup of Ansar al-Sharia and the brazenness of their Iranian sponsors, he felt compelled to speak up.

"We're going to be next," he said, making a point to look not at Stevens, but directly at the new CIA chief. "Ansar al-Sharia is going through the drill. First they warn you, then they attack. They warned the Red Cross, the Brits, and us. So, we're next. We've got to change our posture in Benghazi. Staff it up. Change the profile. Do something different. Unless we throw them off, the attack is going to come and they will be the ones to pick the day and the time."

He went through the warnings Ansar al-Sharia had been posting on its Facebook page against the Red Cross, the Brits, and the United States. He detailed the recruitment tactics they used with the local population, pledging to respect Islamic law by giving their adversaries an opportunity to leave or join their cause before they attacked. He knew he was leaning forward by getting into the realm of predictive intelligence. That's what the agency was supposed to be doing, not the military. But perhaps because the new station chief had been in Tripoli for such a short time, he hadn't yet picked up on all the signs. "We've got to abandon Benghazi, or seriously beef it up," Wood said. "We are going to be hit."

Wood half expected Ambassador Stevens to hit the roof and repeat the mantra he'd been hearing from Main State about Libya getting back to normal. Instead, he pulled Wood aside afterward and asked him to lay it all out.

That briefing fed a series of cables from Stevens and the State Department regional security officers back to Main State that warned of the increasing boldness of the jihadis and begged for more security at the Tripoli embassy and the outpost in Benghazi.

In a June 22, 2012, classified cable, Ambassador Stevens wrote that "the consensus of the [emergency action committee] is a continuing presence of extremist groups and individuals in Libya, which warrant ongoing monitoring by the [emergency action committee]." Stevens met with interim Libyan Prime Minister Abdurrahim El Keib two days later, and came away convinced of the government's fecklessness. On June 25, he sent a detailed account on the precarious security situation penned by Diplomatic Security Officer Patrick Tillou, titled, "Libya's Fragile Security Deteriorates as Tribal Rivalries, Power Plays and Extremism Intensify." It stated pointedly that El Keib's "government remains reluctant to confront extremists, preferring to co-opt them instead."

That cable was a stark warning of things to come. Attacks on Western interests were on the rise, especially in the east, where the Imprisoned Sheikh Omar Abdul-Rahman Brigade claimed responsibility for the bomb attacks against the International Committee of the Red Cross and the U.S Special Mission compound in Benghazi. The group "accused the [Red Cross] of proselytizing Christianity to Libyans and described the June 6 bombing of the Special Mission Compound in Benghazi as 'target[ing] the Christians supervising the management of the consulate,' " Stevens wrote. These weren't the signs of garden-variety Islamist activism, he argued. "[T]he Al Qaeda flag has been spotted several times flying over government buildings and training facilities in Derna."[20]

Al Qaeda was on the march. And they were getting help from America's archnemesis, the Islamic Republic of Iran.

But, in Washington, no one seemed to care.

12

PRELUDE TO MURDER

Hillary Clinton was riding well above the fray. As she testified before Congress in January 2013, she didn't bother to read the cables from Ambassador Stevens and his subordinates that spelled out the security problems in Libya. Nor did she consult with Defense Secretary Leon Panetta to continue the loan to the embassy in Tripoli of the U.S. Special Forces Site Security Team led by Lieutenant Colonel Andy Wood. All she cared about was delivering a success story to the media—and to a select few influential members of Congress—in time for the U.S. presidential election in November. After that, she was out of there, to prepare her own coronation as the first female president of the United States in 2016.

"Benghazi is all about Hillary," one of my sources, who has reviewed much of the classified intelligence on the September 11 attack, said. "This was to be her great success story, the new way America fights wars. We go in from behind, stay for a few months, then get out. No boots on the ground." If a few Americans do get killed, chances are they would be former SpecOps guys working as hired guns for a private military contractor, not a group that elicited a groundswell of public support.

At least, that's what Hillary thought.

SENATOR McCAIN COMES TO TRIPOLI

Primary among Hillary Clinton's congressional allies was Senator John McCain (R-AZ), a big supporter of the administration's policy of reaching out to the Muslim Brotherhood in the belief it would spawn Arab democracies.

A big test of that policy was coming up with the Libyan elections. They had already been postponed once, because the interim government was incapable of taming rivalries among competing tribes and militias to make sure the voting would be held without a massive outbreak of violence. In Benghazi and Derna, Islamist militias were threatening a boycott and the violence to enforce it. In the oasis towns of the south, frequent clashes between warring tribes had left hundreds dead. Along the western borders with Tunisia and Algeria, a mishmash of tribes and militias battled almost daily to control lucrative smuggling routes, where Qaddafi's arsenals were seeping out of the country into the hands of al Qaeda–affiliated groups.

None of that seemed to concern the secretary of state. Instead, in a constant flow of messages, Hillary Clinton pressed Ambassador Stevens to make sure that Prime Minster Abdurrahim El Keib and President Mohammed Mugariaf understood they were not to miss the next deadline for the ballot, which was July 7. She knew Senator McCain was planning to travel to Libya to monitor the voting. She wanted the embassy to put on a good show.

Embassy newbie Hannah Draper was psyched. She was a true believer. "I will be out and about on Saturday, visiting polling stations as an accredited elections observer," she wrote in her blog. "This is the reason I wanted an assignment in Libya over all others—to be here for this historic event, to witness the rebirth of Libya as a democratic state, and to cheer for the brave Libyan people, who have waited so long for this day."[1]

She got the assignment to accompany Ambassador Stevens and Senator McCain as they toured polling places in Tripoli. "People recognized him wherever we went. When we'd arrive at a polling station, I could hear people shouting, is that McCain? Is that John McCain?"

The senator from Arizona, of course, lapped it up. "How excited are Libyans to vote? I just saw a man stop traffic and breakdance in street [sic] while his friends waved flags and cheered him on!" McCain (or an aide) tweeted from Tripoli. In a more sober statement posted to his Senate website, McCain called the vote "a historic day for the people of Libya." While the elections were "not flawless," he judged them to be "free, fair and successful."[2]

Just how "democratic," "free," or "successful" were the Libyan elections in reality? The 130 parties contending to win seats in the National Conference put forth a dazzling array of candidates: 1,207 for just the first eighty seats in the new assembly. An additional 2,501 candidates were running as individuals for the remaining 120 seats. "Rarely has anyone been confronted with what appears to be so much choice, and yet at the same time so little—and with such paucity of information," wrote a commentator in the new English-language daily, the *Libya Herald*.

Four major political parties were vying to lead the government that would emerge from the vote count, but all of them had one thing in common: They wanted Islamic Sharia to become the basis of law in the new Libya. "In fact, it is very difficult to find a Libyan, either within the parties or on the street, who would describe himself as secularist, with an overwhelming majority insisting that Islam must play an important role in political life," the *Libya Herald* analyst wrote. "Certainly, there are radicals on either end of the spectrum, but for the most part it is a matter of degrees."

The Justice and Construction Party (Hizb al-Adala wa al-Bina) was the Libyan branch of the Muslim Brotherhood. The best organized and the best funded of the four main parties, it had received open backing from the United States. Next in prominence was the Nation Party (Hizb al-Watan), led by the head of the Tripoli Military Council, Abdelhakim Belhaj, the former al Qaeda-affiliated terrorist the United States first rendered to Qaddafi, then helped to get released, and now called a "freedom fighter." Critics joked that al-Watan's purple and white banners, which seemed to be on every street corner, resembled the national flag of their purported sponsor, Qatar. They accused the party's ideological patron, Islamist cleric Ali Salabi, of being a Qatari stooge.[3]

Eric Nordstrom and his replacement, John Martinec, the new chief Regional Security officer at the embassy, put the kibosh on sending McCain to Benghazi to watch the voting. Security in Tripoli was already dicey, but Benghazi was boiling over and there was no way they could guarantee anyone's security there with the few assets they had in place. One week before the elections, demonstrators had ransacked the office of the High National Election Commission in Benghazi and burned the ballots. On July 4, a border security officer was assassinated in a drive-by shooting. On the fifth, the main polling station in nearby Ajdabiya was torched and all the ballots destroyed. On the sixth, a Libyan Air Force helicopter bringing replacement ballots to Benghazi was struck by ground fire near the airport, killing an election official on board and wounding an aide. It looked like the al-Qaeda groups were making good on their promise to shut down the elections in the east.[4]

In the end, voters gave the Muslim Brotherhood seventeen seats and al-Wattan just one, despite heavy spending by both groups during the campaign. Belhaj, the former Taliban ally Chris Stevens had visited in Qaddfafi's prison on Christmas Day, failed to win a seat in the new assembly. Defeat left him free to pursue politics by other means.

Later, McCain would reflect on that day and the ambassador he had gotten to know personally, visiting him initially in Benghazi in the heat of the revolution, and, most recently, on July 7, 2012, on the day of the Libyan elections. That morning, Stevens made cappuccino for the senator before they set out for the polls, "a task that he carried out with as much pride and proficiency as his diplomatic mission," McCain recalled.

"What we saw together on that day was the real Libya—the peaceful desire of millions of people to live in freedom and democracy, the immense gratitude they felt for America's support for them, and their strong desire to build a new partnership between our nations," McCain said.[5]

Even then—the day after Stevens' murder—McCain was incapable of seeing anything but the triumph of freedom in the Libyan elections, after the United States backed Islamist ideologues tied to al Qaeda in their successful bid to overthrow the unpalatable Colonel Qaddafi. Just like Hillary Clinton, Barack Obama, and George W. Bush before them, McCain had drunk the Muslim democracy Kool-Aid. And it blinded him to the reality of a failed state ruled by Sharia law where al Qaeda–affiliated militias owned the streets.

POSTELECTION VIOLENCE

At the next country team meeting at the embassy, the spiraling violence in Benghazi and the rise of the al Qaeda–affiliated Ansar al-Sharia prompted Ambassador Stevens to launch a formal review of their security posture. By this point, they had lost two of the State Department's three specialized Mobile Security Detachments and were relying for point protection on the Delta Force operators and Army Green Berets under Andy Wood's command. Everyone knew that the State Department had refused to extend the SST beyond August 5, and that the last MSD was also on its way out. Stevens asked the outgoing RSO, Eric Nordstrom, to draft an action request cable for the secretary of state. "According to one participant, this 'fell like a dead fish on the table,' because everyone, including Ambassador Stevens, knew that the Embassy lacked support from Washington, D.C., and could do little about it."[6]

Stevens sent out his action request cable, stamped SENSITIVE, on July 9. It painted a dire picture of a wilderness outpost that Washington completely disregarded. From the current quota of thirty-four U.S. security personnel spread between Tripoli and Benghazi, the State Department was cutting back to twenty-seven by July 13, and down to seven on August 13, when Wood and the SST were scheduled to leave.* Rather than treat Tripoli's request with the urgency it required, the State Department filed the cable along with routine traffic coming in from the 270-odd U.S. diplomatic posts around the world.[7]

Conditions in Libya "have not met prior benchmarks" established by the embassy, the State Department, and AFRICOM for the security drawdown, Stevens wrote. Their efforts to "normalize" security operations had been "hindered by the lack of host nation security support . . . an increase in violence against foreign targets, and GOL [Government of Libya] delays in issuing firearms permits" for the locally hired guards who controlled access to the diplomatic compounds. Stevens pointedly noted that despite upgrades to physical security at

---

*Some of Wood's men flew back to Germany, the home of the C-110 counterterrorism force, and then returned to Libya under Title 10 status to train the Libyan counterterrorism proxy force. Their presence in Tripoli and their actions on the night of September 11 have been widely misunderstood.

the temporary embassy and residential compounds in Tripoli, "neither compound meets OSPOB standards." This was a shot across the secretary of state's bow, since legally she was required to personally issue a waiver to allow diplomatic facilities with substandard security to operate.

On July 13, Undersecretary Kennedy summarily turned down this latest request for additional security. "His rationale was that Libyan guards would be hired to take over this responsibility," Greg Hicks later wrote.[8] But, of course, they never had time to hire new guards, let alone train them and acquire the necessary weapons permits from the chaotic Libyan government.

Stevens sent a flurry of cables to Washington in the final weeks before the September 11 attacks, each more desperate than the last. He had Nordstrom draft a compendium of the 230 major security incidents that had occurred over the previous thirteen months, in the hope it would get someone's attention. The RSO described the climate of lawlessness, gang violence, and militia rule that had replaced any semblance of central government in Libya in his general assessment at the end. "The risk of U.S. Mission personnel, private U.S. citizens, and businesspersons encountering an isolating event as a result of militia or political violence is HIGH," he wrote. "The Government of Libya does not yet have the ability to effectively respond to and manage the rising criminal and militia related violence. . . . Neighboring countries fear extremist groups who could take advantage of the political violence and chaos should Libya become a failed state."[9]

In an email he sent to the House Oversight and Government Reform Committee staff, Nordstrom summed it up. "These incidents paint a clear picture that the environment in Libya was fragile at best and could degrade quickly. Certainly, not an environment where post should be directed to 'normalize' operations and reduce security resources in accordance with an artificial time table."[10]

Even the State Department's Accountability Review Board recognized the obvious in its after-action report.

> [T]he immediate period after the elections did not see the central government increase its capacity to consolidate control or provide security in eastern Libya, as efforts to form a government floundered and extremist militias in and outside Benghazi continued to work to strengthen their grip.
>
> At the time of the September attacks, Benghazi remained a lawless town nominally controlled by the Supreme Security Council (SSC)—a coalition of militia elements loosely cobbled into a single force to provide interim security—but in reality run by a diverse group of local Islamist militias, each of whose strength ebbed and flowed depending on the ever-shifting alliances and loyalties of various members.

If the elections were supposed to be the crucible of democracy and put an end to the violence, they failed.

Stevens got no traction with Secretary Clinton and her top deputies in his desperate quest to improve security in Tripoli and Benghazi. Instead, the managers increased the danger-pay allowance for employees in Libya from 25 percent to 30 percent on July 1.

As Representative Darrell Issa would comment during the first public hearing on the attacks, "you don't reduce security at the same time as you are increasing hazardous duty pay." To do so "sends a message that says we will pay you for the risk, we will not pay to have you made safer."

It was a cynical recognition of just how bad things had gotten.

STINGER HIT IN AFGHANISTAN

In Afghanistan, the blowback from Hillary Clinton's secret policy to arm the Libyan rebels hit home with a vengeance when Taliban fighters in Kunar province successfully targeted a U.S. Army CH-47 on July 25, 2012.

They thought they had a surefire kill. After all, their friends from Qatar had supplied them with a new-generation FIM-92 Stinger missile, and everything had worked as planned. They got word from their spotters that a U.S. assault team had boarded the large Chinook helicopter and was moving up through the Pech River Valley in the Chapa Dara district in eastern Afghanistan. When it was still a good two miles out, they could hear the engines, and the weapon commander tapped the gunner's shoulder. He depressed the aiming trigger and the infrared seeker beeped faster until it blended into a single note indicating it had locked on the target. With a cry of *Allah O-Akbar!* the gunner fired and the Stinger shot into the air, making a corkscrew path toward its target. They saw what they thought was an explosion, and cries of *Allah O-Akbar!* echoed off the mountains in the distance where their Taliban brothers had positioned themselves to fight any Americans who survived the crash.

But instead of bursting into flames, the Chinook just disappeared into the darkness as the American pilot recovered control of the aircraft and brought it to the ground in a hard landing. The assault team jumped out the open doors and ran clear in case it exploded. Less than thirty seconds later, the Taliban gunner and his comrade erupted into flames as an American gunship overhead locked onto their position and opened fire.

The next day an Explosive Ordnance Disposal (EOD) team arrived to pick through the wreckage and found unexploded pieces of a missile casing that could only belong to a Stinger. Lodged in the right engine nacelle was one fragment that contained an entire serial number. The investigation took time; arms were twisted, noses put out of joint. But when the results came in, they were stunning: The Stinger tracked back to a lot that had been signed out to the CIA *recently*, not during the anti-Soviet jihad.

Reports of the Stinger hit reached the highest echelons of the U.S. command in Afghanistan and became the source of intense speculation, but no action. Everyone knew the war was winding down. Revealing that the Taliban had MANPADS—even worse, U.S.-made Stingers—risked demoralizing coalition troops. Because there were no coalition casualties, ISAF made no public announcement of the attack.

The Taliban had been boasting since May that they had received a new shipment of surface-to-air missiles, but U.S. commanders apparently weren't taking it seriously. My sources in the U.S. Special Operations community believe that the Stinger fired unsuccessfully on July 25, 2012, against that Chinook was part of the same lot the CIA turned over to the Qataris in early 2011, initially intended for anti-Qaddafi forces in Libya. They believe the Qataris delivered between fifty and sixty of those same Stingers to the Taliban in early 2012, and an additional two hundred SA-24 "Igla-S" MANPADS.

This was the first time they had actually used them. But it would not be the last.

THE IRANIAN GAMBIT

The CIA Annex in Benghazi may have spun down its training and support operations by June 2012 and ordered the adrenaline monkeys to move on to southern Turkey to help the Syrian rebels, but they still maintained four or five operations officers, another eight to ten analysts, and seven (some sources say as many as ten) heavily armed GRS guards to protect the operators when they ventured outside the wire.

They were also supposed to defend the Annex in case it came under attack, and in an informal arrangement with the State Department, flow forces along with the mercurial 17th February Martyrs Brigade to protect the diplomatic compound in the event of trouble there. "Those guys had it all," a source who worked closely with the Annex team in Benghazi told me. "They were armed to the friggin' teeth. Here's the best stuff on the market, so buy it." [11]

Each GRS operator chose his own kit. While they all carried a NATO standard 5.56mm assault weapon, most of them preferred the Heckler & Koch 416, which the manufacturer developed in collaboration with Delta Force as the special operator's weapon of choice. In addition, some picked up Soviet-era PKMs, a lightweight machine-gun version of the venerable AK-47, from local stockpiles. Others preferred the M240, the rifleman's favorite belt-fed machine gun. For real firepower, the GRS operators relied on the M249 Minimi (French for *mini-mitrailleuse*, or light machine gun), which like the M240 was made by Belgian arms maker FN-Herstal. The Minimi used a large, boxlike magazine that held two hundred rounds. In full automatic mode, an operator could empty the entire magazine on a wave of attackers in fewer than twenty seconds, reload, and have at it again. The GRS operators packed a lot of punch.

The Annex also housed an NSA listening post that secretly monitored communications of the jihadi groups and their supporters using a dazzling array of sophisticated electronics gear, dissimulated on the roof. In some NSA facilities, these were built to look like water tanks; in others, small shacks or air-conditioning units. The outer skin of the structures was actually paper thin, so that the sensors inside could capture radio and other electronic signals. Their mission was to pick up indicators and warnings of attacks being planned against U.S. facilities in Libya or elsewhere. They also tried to keep tabs on the high-value al Qaeda targets who continued to roam the streets of Benghazi. One of these was Ezzedin Abdel Aziz Khalil, aka Yasin al-Suri, the top al Qaeda financier based in Iran.

Late in the afternoon on Monday, July 30, 2012, the Annex ears picked up chatter in Persian between a pair of Quds Force operators they were shadowing in town who were using high-end Racal VHF radios on a dedicated frequency. The NSA translator brought an English transcript of their conversation to the chief of base. He laughed when he read through it. It looks like our boys are on time, he said.

The chief of base had tasked several agents in his employ who worked for the Zintan militia that ran the Benghazi airport to track the seven Iranians scheduled to arrive that day from Tripoli. They were operating undercover as part of a Red Crescent medical team— a doctor, male nurses, medics, and administrator—and undoubtedly thought the Americans didn't have a clue as to who they really were. At least, that's what the chief of base had surmised from the intercepted coms.

But *he* knew that the Red Crescent team included operations officers the Iranians had dispatched to Benghazi to carry out an attack on the diplomatic compound. This was the big one they were all waiting for. This was the hit that Lieutenant Colonel Andy Wood had accurately warned about at the last country meeting in Tripoli with Ambassador Stevens. These officers were the "Iranian brains" who would command, fund, and perhaps actually carry out the attack.

My sources have identified the commander of the operational team as IRGC Major General Mehdi Rabbani. According to American Enterprise Institute Iran analysts Ali Alfoneh and Will Fulton, Rabbani is a member of the commanding heights of the IRGC, the inner circle of top loyalists. (My Iranian sources called this the Command Staff.) His official title at the time he went to Benghazi was deputy chief of operations for the Islamic Revolutionary Guards Corps. Alfoneh and Fulton believed he had very little, if any, combat experience from the formative Iran-Iraq War.[12] My sources tell me he was intelligence chief (equivalent to the U.S. J-2) for the southern front during the war, responsible for identifying and targeting strategic Iraqi assets. After rising to his present position in 2008, he played a major role in commanding the Bassij forces who brutally suppressed the 2009 post-election protests inside Iran.

General Rabbani was Mr. Chops, whose approval was needed to bless the operational plan to murder the ambassador and drive the Americans out of Benghazi. Among the others were Quds Force operators and civilian intelligence officers from the Ministry of Information and Security (MOIS) who specialized in terrorist operations.

The CIA chief of base knew none of this. He didn't know whether their plan was to kidnap the ambassador, kill him, hire locals to drive car bombs through the perimeter wall, or what. All he knew was, they were the bad guys and they had come to fulfill his worst nightmares.

Over the next hour or so, the chief's deputy received a steady stream of reports over the tactical radio from the Zintan militia guys on the progress of the hit team. The plane had landed at Benina International Airport. They were unloading their gear. The local Red Crescent director welcomed them. They packed into a convoy of Red Crescent vehicles (each painted white, with a large, bright red Islamic crescent painted on the doors and the roof—the Muslim version of the Red Cross). They were en route to the Tibesti Hotel. The chief had a second team waiting in separate vehicles outside to track them once they left the airport. Everything was going like clockwork.

After the Iranians freshened up at the hotel, they went out again for an Iftar dinner with the local Red Crescent guys. It was a protocol event. Because they were in the middle of Ramadan, the Muslim month of daytime fasting, they didn't actually sit down to dinner until late.

Then at one in the morning, it happened.

All of a sudden, the deputy chief jumped up from where he had been dozing off. His guys were going nuts.

The ruckus got the chief's attention. What's going on? What are they saying? he asked.

The deputy translated the excited shrieks from the trackers. It seemed the Red Crescent team had been headed back to the Tibesti Hotel when they were ambushed by a half dozen Toyota pickups with .50-caliber machine guns mounted on the beds. The militia guys forced the Iranians to get out, cuffed them, then bundled them into a pair of Jeep Cherokees and sped off.

Our guys decided it was more prudent not to follow them, he said.

So they're gone, the chief said. That's it. Kidnapped.

For the next twenty-four hours or so, the chief's network of agents in Benghazi was unable to find out where the Iranians had been taken or who was holding them. Then, according to Dylan Davies, the former British Special Operations grunt who managed the unarmed security detail at the U.S. diplomatic compound, a local fixer learned they were being held in a former Libyan army camp outside of town on the Tripoli road. "They are getting fed; they have their own beds even. They are fine," he told Davies. The fixer claimed it was a Shia-Sunni thing. "There are some who think those Iranian Shias are not welcome here," he said. "But they are perfectly okay."

Complacent (or complicit) officials in Benghazi reinforced the fixer's story. "Members of the brigade holding the Iranians are questioning them to determine whether their activities and intentions aimed to spread the doctrine of Shiite Islam," a local security official told the Benghazi stringer for Agence France-Presse. "They will be released after the investigation is concluded," he added.[13]

Davies says he learned a few days later that the CIA chief of base had tasked his former Special Forces security team to launch a hostage rescue, perhaps with the goal of interrogating the Iranians in the privacy of their armored Mercedes G wagons. At the last minute, Davies' fixer told him the Iranians had already been set free and put on a plane for Tripoli, en route to Tehran. "Mate, cancel the cavalry," he told the chief diplomatic security officer at the compound. "They left yesterday on a flight to Tehran." When the Americans expressed surprise at how he got this information, he said that his fixer's cousin worked at the airport. "He saw all seven of 'em fly out of here." [14]

But Davies—and the CIA chief of base—got played.

I learned what actually happened from two former Iranian intelligence officers, one of them a very senior operative who at the peak of his career was the chief action officer for all of Western Europe. The other was a counterintelligence officer and interrogator, who at one point worked on the protection detail of the Supreme Leader. Both of these individuals have defected to the West and now live undercover in Europe. Each has his own active network of contacts inside Iran, some of whom continue to work in senior positions in the Iranian regime. I corroborated their initial information with multiple Western intelligence sources that are not in contact with them.

The CIA chief of base was correct that the Red Crescent team included undercover Quds Force and MOIS officers who had been sent to carry out a terrorist attack against the United States. In fact, my Iranian sources said, their orders were to kidnap or kill the U.S. ambassador to Libya, to send a message to the United States that they could act against them at will anywhere and at any time in the Middle East.

But, as they were getting ready to set the plan in motion, the resident Quds Force team in Benghazi learned from its own intercepts of the Annex tactical coms that the Red Crescent cover had been blown and the CIA was onto them. So they decided to take the entire group off the streets—stage a kidnapping—in order to convince the chief of base that the danger was over.

"The team in operational command in Benghazi were Qassem Suleymani's people," the former Baghdad deputy chief of station, John Maguire, told me. "They were a mature, experienced, operational element from Iran. These guys are the first-string varsity squad." And they were playing for keeps.

Maguire had matched wits with Suleymani, the Quds Force commander, for two years in Iraq and came away with a healthy respect for his capabilities. "He is talented, charismatic. His people are competent and well trained. They have all the operational traits we used to value. And they are committed to this fight for the long haul."

Suleymani and his Quds Force operators were so successful at killing Americans in Iraq because they had penetrated U.S. operations. They didn't just randomly place an IED on a roadside, Maguire said. They placed the IED where and when they knew an American convoy was going to pass.

"They were into our coms. They were into our operational planning. That's how they were able to kill so many Americans," Maguire said. "The Iranians are a determined global service."

The faked kidnapping in Benghazi was a typical Quds Force op. They used a local militia that on the surface detested Shias, just as they used the Taliban in Afghanistan and manipulated al Qaeda. "They are very good at deception operations," Maguire told me.

And our side didn't have a clue. The CIA chief of base and his deputy fell for it hook, line, and sinker.[15]

THE CONSIGLIERE

Valerie Jarrett was President Obama's closest confidante. She was the only advisor allowed to roam the White House family quarters at night, and was so well known for her late-night sit-downs with the president that the Secret Service detail dubbed her the "night stalker."

Her official title, senior advisor and assistant to the president for intergovernmental affairs and public engagement, did not reflect her actual job, which was to conduct sensitive political negotiations on the president's behalf in Washington and overseas. Her task was to help Obama see how to maximize the damage to his political adversaries, and the gain to himself.

Journalist Ed Klein devotes two chapters to Jarrett in his 2012 critical biography of Obama, *The Amateur*. She was "ground zero in the Obama operation, the first couple's first friend and consigliere. Once asked by a reporter if he ran *every* decision by Jarrett, Obama answered without hesitation, 'Yep. Absolutely,' " Klein wrote.

Jarrett has "an all-access pass to meetings she chooses to attend: one day she'll show up at a National Security meeting; the next day, she'll sit in on a briefing on the federal budget," Klein wrote. "When Oval Office meetings break up, Jarrett is often the one who stays behind to talk privately with the president." [16]

Her father, James Bowman, was an African-American physician who moved to Shiraz, Iran, in the 1950s to help establish the Nemazee hospital. He fell in love with a childhood development specialist from Chicago and the two had their only child, Valerie, in November 1956. She was just the second child born in the new private hospital in Shiraz. Jarrett spent her first five years in Iran and claims that she still speaks some Farsi. She apparently also tells the White House staff she is Iranian-American (which she is not), since that's what I was told by an otherwise unresponsive National Security Staff spokesperson, Bernadette Meehan, when I asked her about Jarrett's involvement in the secret back-channel negotiations with Iran.

During the 2008 presidential campaign, candidate Obama pledged to conduct "negotiations without preconditions" with the Islamic Republic of Iran. It was one campaign promise he kept. The first Iranian government emissary traveled to Washington to meet with an Obama advisor in December 2008. Once he took office, the contacts multiplied.

Valerie Jarrett began meeting secretly with Iranian government representatives in Bahrain, Qatar, and Dubai in early 2009, trying to lay the groundwork for a "grand bargain" with the Islamic Republic of Iran that would resolve thirty-four years of conflict. The fact that after five years nothing came of these efforts is a testament to the deep distrust the Islamic leadership in Iran feels toward America. Many offers have been put on the table that a purportedly rational regime would have taken long ago. Jarrett was as stymied by the Iranians as Secretary of State Condoleezza Rice was in 2006, when she offered trade, aid, and even nuclear power technology to Iran, in exchange for coming clean on their nuclear research programs.[17]

A few days after the Red Crescent team was "abducted" in Benghazi, Valerie Jarrett jetted off with her Secret Service detail to Dubai, where she met with Ali Akbar Velayati, a former foreign minister and the Supreme Leader's top foreign policy advisor. For Jarrett, it was something of a homecoming, since her family had been close to the Velayatis back in Shiraz in the 1950s. Although Ali was eleven years her senior, and probably had few memories of Valerie as a child, it was a connection that Jarrett had cultivated over the years and used to insinuate herself as a player in the back-channel talks. According to Reza Kahlili, a former CIA spy inside Iran's Revolutionary Guards Corps, Jarrett met "more than ten times" with Velayati over the past twenty years.[18]

Jarrett was hoping to convince the Iranians to cooperate with Obama so they could announce a big diplomatic breakthrough—an October surprise—in time for the 2012 presidential election. Having played this game already during the 1979–81 hostage crisis, the

Iranians were attuned to the needs of U.S. politicians and our peculiar once-every-four-years timetable, which bore a faint resemblance to their own.

Several friendly foreign intelligence services reported similar accounts of the August meeting in Dubai. My sources say that Velayati told Jarrett there were elements inside Iran (such as the Quds Force) who were "out of control" and were planning to kidnap an American diplomat to show their displeasure with U.S. sanctions on Iran. Some claim that Jarrett then proposed that they transform the kidnapping into a hostage exchange, with the United States freeing the Blind Sheikh in exchange for the kidnapped U.S. ambassador. That would make Obama look like a diplomatic genius.

I have found no evidence to substantiate the allegations of a hostage exchange. Indeed, my Iranian sources believe that Velayati's goal was to reinforce the impression in the U.S. intelligence community that Iran was very concerned by the kidnapping of its Red Crescent team in Benghazi. He told Jarrett that Iran was also a victim of terrorism, and so had a lot in common with the United States. See, a team of our own doctors—doctors!—has been taken hostage in Benghazi. (Velayati, of course, had been a pediatrician himself.)

If the CIA kept hearing that the Red Crescent team had been kidnapped, from multiple sources, sooner or later they would take it to the bank and no one would be able to convince them otherwise. That was exactly what the Iranians wanted. They wanted to lull the Annex chief of base into believing that the Red Crescent/Quds Force team had been taken off the streets, and that the danger of an attack was over.

This was how the Iranians played the game of political chess. They rarely attacked directly but used indirection, feints, gambits, and complex maneuvers to distract their adversary and to hide their real strategy.

"Look at Rafsanjani and Iran-Contra," the former Iranian intelligence chief for West Europe explained. "He lied to everybody. He lied to me. He never wanted an agreement with the Americans. He wanted to negotiate a deal so he could make a big show refusing it. All the while Rafsanjani was meeting with Bud McFarlane and Ollie North in Tehran over the U.S. hostages in Lebanon, his real goal was to get U.S. HAWK missiles so he could retake the Fao Peninsula from Saddam Hussein."

National Security Staff spokesperson Bernadette Meehan insisted that Jarrett "was not involved in Iran policy, even though she is an Iranian-American." She called the story of the meetings with Velayati "unequivocally false."

"The White House people are lying on this," a senior official in a major American NGO that deals with Middle East policy on a daily basis told me. "I told them what Iranian sources told me happened at those meetings and they still denied they took place. The Iranians said, here's who sat on [Jarrett's] left, here's who sat on her right. Here's the Lufthansa flight she took out of Frankfurt. So I am absolutely certain the meetings took place. She had a relationship with Velayati. They were childhood friends. The Iranians saw these meetings as a backdoor channel to the Obama administration, so their message would reach the very top.[19]

"UNPREDICTABLE, VOLATILE, VIOLENT"

Ambassador Stevens was growing increasingly worried about his own security and the security of his employees. The more he heard from his security team about the Iranian presence in Benghazi, their support for Ansar al-Sharia, and the inability of the new government to field reliable security forces, the more worried he became. So he tasked his new regional security officer, John Martinec, and an assistant, Jairo Saravia, to draft an action request cabled on August 2.

When an ambassador sends in an action request to increase his personal security detail, you know there's trouble.

"The security condition in Libya remains unpredictable, volatile, and violent," the cable reads. "Though certain goals have been successfully met, such as the national election for a representative Parliament who will draft the new Libyan Constitution, violent security incidents continue to take place due to the lack of a coherent national Libyan security force and the strength of local militias and large numbers of armed groups." [20]

The RSO estimated that they needed eleven more locally hired bodyguards for the Ambassador's Protection Detail (APD), in addition to the twenty-four they already had. "The augmented roster will fill the vacuum of security personnel currently at post on TDY status who will be leaving within the next month and will not be replaced," Stevens wrote. The security personnel who were leaving were those he had asked the State Department previously to keep in Libya.

Just as previous cables with similar requests, this one was stamped "Routine" by the State Department and ignored.

So, Stevens took another whack at it six days later in a cable titled, "The Guns of August," on the security situation in Benghazi. It was drafted by the young political officer, Eric Gaudiosi, who was the latest of a string of diplomats to brave the Benghazi gauntlet for a short-term assignment (TDY). This time, Stevens was careful to send it with the SIPDIS caption, which meant it would be distributed through the Pentagon's Secret Internet Protocol Router Network (SIPRnet), a governmentwide clearinghouse meant to reach all U.S. security agencies. This would give it a much wider readership than otherwise available through the normal State Department channels.

Since the elections, he wrote, "Benghazi has moved from trepidation to euphoria and back as a series of violent incidents has dominated the political landscape." Kidnappings and assassinations had become rampant, as militias jockeyed for power and real estate. The Supreme Security Council (SSC), an umbrella group aimed to unite the most prominent militias to impose order on the lawless city, was an abject failure. "[E]ven in the assessment of its own commander, Fawzi Younis, SSC Benghazi has not coalesced into an effective, stable security force." The absence of any significant deterrent had "contributed to a security vacuum."

Worse still, the violence was organized and determined. "What we have seen are not random crimes of opportunity, but rather targeted and discriminate attacks" by an assortment of groups, from organized crime to former regime elements and Islamic extremists. "Attackers are unlikely to be deterred until authorities are at least as capable," Stevens concluded.[21]

Stevens knew he would be heading for Benghazi in the near future, and an inescapable dread had begun to fill him as the city where the anti-Qaddafi rebellion first took root spiraled downward into chaos. Sending a SIPDIS cable was the diplomatic equivalent of a cry of desperation. Clearly, he was hoping that *someone* would understand how dire the situation had become, perhaps the CIA or General Carter Ham at AFRICOM, and weigh in with the White House or the secretary of state personally.

This cable was also stamped "Routine" by the State Department routing officer. Help was not on the way.

INTELLIGENCE WARNINGS

On August 6, two members of the SST team were stopped at a checkpoint in Tripoli and ordered out of their armored SUV by militia members. When they refused, the situation turned confrontational, as more militiamen arrived and roughly demanded that the two Americans get out of the vehicle and leave the keys in the ignition.

The SpecOps guys were dressed in civilian clothes, and bore no militia insignia—not even boots—under the standing orders from the Department of State to downplay their presence. But they were well armed. When the militiamen refused to move away from the vehicle, the Americans fired warning shots in the air with their 9mm Glock 22s. Instead of backing off, the militiamen loosed off several bursts of Kalashnikov fire in their direction, pinging against the armor plating in the doors. This time the Americans used their weapons with purpose, and saw several militiamen go down. In the confusion, they were able to break away and speed off to the embassy compound. "We had to shut everything down after that and didn't leave the compound for over a week," said Lieutenant Colonel Andy Wood. "We felt it was probably a criminal element, but it could have been an organized al Qaeda attack."[22]

That same day, the International Committee of the Red Cross pulled out of Benghazi because of the attacks on their compound and the ongoing security threats.

Back in the United States, Ansar al-Sharia was starting to attract attention. An August 8, 2012, report from the Pentagon's Irregular Warfare Support Program of the Pentagon's Combating Terrorism Technical Support Office, published under the auspices of the Library of Congress, sounded the alarm: al Qaeda was on the move in Libya.

The report found that Ansar al-Sharia "has increasingly embodied al-Qaeda's presence in Libya, as indicated by its active social-media propaganda, extremist discourse, and hatred of the West, especially the United States."[23]

The very idea that al Qaeda was operating in Libya—far from Afghanistan—directly contradicted the Obama administration's narrative that the al Qaeda menace had been reduced to insignificance because Obama ordered SEAL Team 6 to take out bin Laden in May 2011. On the contrary, the report found that the Al-Qaeda Senior

Leadership (AQSL) in Afghanistan and Pakistan were "seeking to create an al-Qaeda clandestine network in Libya that could be activated in the future to destabilize the government and/or to offer logistical support to al Qaeda's activities in North Africa and the Sahel."

Their goals in Libya were to "gather weapons, establish training camps, build a network in secret, establish an Islamic state, and institute sharia," the report stated. "In the process, al-Qaeda will seek to undermine the current process of rebuilding Libyan state institutions as a way of preventing the establishment of strong state counterterrorism capabilities that could hinder its ability to grow in Libya." Al Qaeda's clandestine networks in Libya were "currently in an expansion phase."

The Pentagon report was widely read, according to a U.S. government intelligence analyst asked to comment on its impact and reach. "We all read the same stuff," he told the American Media Institute. "It gets circular, going through all the relevant government agencies. That report for the Irregular Warfare Office got passed around a lot." [24]

Nobody can credibly pretend that the State Department, the Pentagon, the CIA, and the White House weren't warned of the gathering threat in Benghazi. And yet, the president's reelection campaign continued to call Libya and the takedown of bin Laden a great success story.

Ansar al-Sharia gained street cred by organizing the June 7–8 gathering of pro-Sharia *katiba* in Benghazi (see chapter 11). Thirty separate brigades, commanded by fifteen different militias, participated in the show of force "and probably make up the bulk of al-Qaeda's network in Libya," the authors of the report wrote. "The al-Qaeda clandestine network has certainly stocked enough arms and ammunitions to allow it to operate independently," they added.

The authors identified Sufian Ben Qumu, one of several former Guantánamo detainees befriended by Chris Stevens during his earlier tour in Tripoli as DCM under Qaddafi, as the founder and principal leader of the group. They noted that the Ansar al-Sharia name "is also being used by al-Qaeda in the Lands of the Arabian Peninsula in so-called liberated areas of Yemen and by Salafist groups in Tunisia . . . suggesting coordination between the groups."

But the Pentagon intelligence report did not mention the Iran connection.

For a generation, the U.S. intelligence community had swallowed the fiction that Shiite Iran was locked into a deadly conflict with Sunni Islam and therefore was fundamentally incapable of cooperating with al Qaeda and other Sunni terrorist groups. That mind-set, expressed most powerfully by the CIA's top analyst in the 1990s, Paul Pillar, blinded the United States to the long-standing ties between Iran's Islamic Revolutionaries and the Palestine Liberation Organization, and later to Hamas, the Taliban, and al Qaeda. It also contributed to the CIA missing all the warning signs of Iranian involvement in the 9/11 plot. I first exposed those ties in my 2005 book, *Countdown to Crisis: The Coming Nuclear Showdown with Iran*. Based in part on my information, U.S. District Court judge George B. Daniels found that the Islamic Republic of Iran provided "essential material support, indeed, direct support [to al Qaeda], for the 9/11 attacks" on America, and awarded a $6 billion judgment to families of the victims.[25]

The Department of the Treasury has repeatedly exposed Iran for harboring top al Qaeda leaders, including Saad bin Laden, the oldest son (and titular heir) of Osama bin Laden, and members of the al Qaeda Shura (or governing) council. Treasury has also identified Iran as the center of al Qaeda's terrorist finance network. In testimony before the Senate Foreign Relations Committee in 2010, Army General David H. Petraeus, who was then in charge of the war in Afghanistan, said that al Qaeda was using Iran as a "key facilitation hub, where facilitators connect al-Qaeda's senior leadership to regional affiliates." Already in 2008, Treasury noted that Iran was providing "logistical, financial, and material support" to the Libyan Islamic Fighting Group through these al Qaeda networks.[26]

The Saudi government also understood that Shiite Iran was entirely capable of financing, training, and arming Sunni terrorist groups. They accused Iran of backing Ansar al-Sharia as a false-flag operation. "Iran supports Ansar al-Sharia financially, but secretly, because they don't want the militants of Ansar al-Sharia to break from their own leaders because of the differing views between Shiite Iran and the Sunni Ansar al-Sharia," Saudi researcher Ibraheem al-Nahas said.[27]

My Iranian sources told me the same thing. The Quds Force was using operatives recruited in Iraq, Lebanon, Yemen, Turkey, and Egypt to finance, train, and equip Ansar al-Sharia and related *katiba* in Libya, careful to show their hand only to a select few. To anyone on the ground, these Iranian operatives would look and speak just like Arabs or Turks. They helped to create a series of small, radical militias—Ansar al-Sharia was not the only one, as the June 7–8 gathering of jihadis in Benghazi showed—to block any progress toward democracy. My sources estimate that the Iranians recruited more than one thousand Libyan fighters for these militias who were "in direct contact with Quds Force officers."

What was *their* goal? "Iran wants chaos. They want to generate anti-American anger, radicalize the rebels, and maintain a climate of war," the former Iranian intelligence chief for Western Europe told me. "They are very serious about this. They want to damage the reputation of the United States as a freedom-loving country in the eyes of the Arabs. In Libya, Iran wanted to block U.S. influence, which they saw as a threat. They saw the uprising against Qaddafi—and the Arab Spring more generally—as an opportunity to accomplish this."

In other words, Ansar al-Sharia and Iran's Quds Force shared the same goals in Libya. It would soon become apparent that they shared much more as well.

THE WIZARD OF OZ OF IRANIAN TERROR

By mid-August 2012, planning for the attacks on the U.S. diplomatic compound and the CIA intelligence base in Benghazi had been in the works for two months.

My sources say the head of the Iranian team in Benghazi was a senior Quds Force officer named Ibrahim Mohammed Joudaki. Born in April 1973, he was a veteran terrorist with a long pedigree. He was the one who staged the kidnapping of his boss, General Rabbani. It was a stroke of genius.

Joudaki joined the Revolutionary Guards in 1994 at age twenty-one. He got his spurs killing Kurds as a subunit commander in northwestern Iran from 1995 to 1998. Some sources believe he was involved early in his career in liquidating Iranian dissidents in Europe.

In 2000, he was sent to Lebanon for three years to train Hezbollah fighters. He was so successful that his bosses selected him for recruitment into the Quds Force, the elite expeditionary arm of the Revolutionary Guards. They were professional terrorists in the service of the Iranian regime. At the Quds Force training academy, based in the former U.S. Embassy in Tehran, he took theoretical courses in insurgency, and political party indoctrination. He enhanced his operational skills with a course in tactical driving, where he learned to drive a chase car and to shoot from a rapidly maneuvering vehicle. At the end of the training, he was promoted to major.[28]

In 2005, Ibrahim Joudaki was sent back to Lebanon, where he helped structure the Hezbollah forces on the ground to operate against the Israeli military, and was actively involved in their successful resistance of Israel's effort to drive them out of southern Lebanon during the 2006 war. In 2007, he was rewarded for his success with a stint at the advanced officers training academy—"the equivalent of our Fort Leavenworth," a senior U.S. intelligence officer told me—and promoted to lieutenant colonel. In 2010, this Otto Skorzeny of Iran was sent to Lebanon for a third tour, this time to prepare the Iranian military mission to Syria to retrain the Syrian Army to fight a domestic insurgency. He was then promoted to deputy commander of Quds Force operations for the Levant, a theater that included Libya and Egypt.

Ibrahim Joudaki was the man in charge of Libya throughout the anti-Qaddafi uprising. He had overall control of the September 11, 2012, attacks and handled the finances and operational planning.

His deputy, Khalil Harb, was someone he had known and worked with for years in Lebanon. Harb was a top Hezbollah operative and a deputy to Quds Force chieftain Qassem Suleymani. As an Arab, it was easier for him to interface directly with the local militias and handle the logistics of the operation itself. Over those first two months, Joudaki spread money like butter on bread, while Harb and his team of hardened killers gathered intelligence, recruited militiamen, and refined the operational plan. My sources say Harb had a total of some fifty Quds Force operatives on the ground to manage the attacks. "Khalil Harb was the Quds Force chief of station for Libya," the former Western European intelligence chief said.[29]

Their initial orders were to kidnap the U.S. ambassador while he was visiting Benghazi, and to destroy the CIA Annex. They wanted to drive the United States out of Benghazi, where they believed the CIA was supervising weapons transfers to the Syrian rebels. "Iran saw the CIA presence on the ground in Benghazi as a direct threat," the former Western European intelligence chief said. The kidnapping plan was dropped after General Rabbani and the Red Crescent team arrived and were outed by the CIA. Harb and Joudaki replaced it with a straight kill order.

My sources say that a courier arrived carrying $8 million to $10 million in five-hundred-euro notes around three weeks before the attack. Joudaki distributed the money through Khalil Harb to Ansar al-Sharia leaders. The money was brought in through Tunisia, then south through the Algerian desert, and into Libya by road. It came from Quds Force accounts in Malaysia.

Andy Wood had known for months that his security team was up against trained and well-funded adversaries. But my sources say it was much worse. His real adversary was Qassem Suleymani, the Wizard of Oz of Iranian terror, the most dreaded and most effective terrorist alive.

FROM BENGHAZI WITH LOVE

Libya's jihadists saw their victory against Qaddafi as the first step in imposing Sharia law across the Muslim world and ultimately reestablishing the Islamic caliphate. Just as the Americans in the Annex were doing, they next turned their sights to Syria.

In August 2012, Abdelhakim Belhaj sent his deputy, Mahdi al-Harati, into Syria along with a contingent of Libya jihadis to join the Syrian rebels fighting Assad. A charismatic figure who had lived in Ireland for twenty years, married an Irish woman, and acquired Irish citizenship, al-Harati was famous for being wounded while fighting an Israeli boarding party on the *Mavi Marmara* in May 2010. That was the Turkish ferry, purchased by an Islamist charity with close ties to the ruling AKP party of Prime Minister Recep Tayyip Erdoğan in Turkey, which became the flagship of the Islamist flotilla trying to break Israel's naval blockade of Gaza. Erdoğan personally visited al-Harati in a Turkish hospital where he was recovering. He told the Libyan, "We are very proud of you," and kissed him on the forehead.[30]

At forty years old, al-Harati was dedicated to global jihad. And he had punched all the right tickets. After working with the Turks, who would return the favor shortly, he was spotted by the Qatari Special Forces unit that came to help the anti-Qaddafi rebels at the start of the uprising in 2011. The Qataris gave him proper military training. Some sources say he received additional tactical training from U.S. Special Forces contractors stationed at the Benghazi Annex. His connections and skills helped him to rise quickly to become the operational commander of the Tripoli Brigade of Abdelhakim Belhaj. By the end of the civil war, he led the assault on Qaddafi's compound.

This wasn't Harati's first trip into Syria. Belhaj had sent him to liaise with the Syrian Muslim Brotherhood a first time in September 2011, just after the fall of Tripoli. He returned a second time one month later along with six hundred fighters and established an international Islamist battalion in Syria called Liwaa al-Umma (Islamic People's Brigade, sometimes translated as Banner of the Islamic Nation). Hacked emails from Stratfor, the private intelligence firm, suggest that he was in contact with former CIA officer and Stratfor source, Jamie Smith, who claimed to have recruited him as a U.S. asset. If there is any truth to Smith's claim, that might explain how this die-hard jihadi acquired the assets to grow his operation in Syria so dramatically. It might also provide a clue as to how the CIA got around the legal restrictions against unreported covert actions.[31]

The media-friendly al-Harati happily met with a Reuters reporter inside Syria in August 2012 at the head of his new military unit, which he boasted had swollen to nearly six thousand fighters. The reporter helpfully told his audience that most of al-Harati's unit were Syrians, but that foreign fighters were also welcome, "including 20 senior members of his own Libyan rebel unit." Which unit that would be remained unclear. The Tripoli Military Council, which subsumed the Islamist Tripoli Brigade, had been folded into the Supreme Security Council (SSC), ostensibly the main national security force of the new Libyan government. As Reuters explained, "the Libyans aiding the Syrian rebels include specialists in communications, logistics, humanitarian issues and heavy weapons," and they "operate training bases, teaching fitness and battlefield tactics." All the skills they had learned from the Qatari and U.S. Special Forces they were now passing on to a new crop of jihadi fighters, this time in Syria.[32]

Critics of the Islamist war against Assad called al-Harati and his fighters a "foreign invasion," waging war against a sovereign state.

On August 12, 2012, just as al-Harati was taking the Reuters reporter for the rounds, the Syrian rebels shot down a Syrian Air Force MiG-23. A senior U.S. Special Operations commander told me that the shootdown occurred just weeks after major arms shipments began reaching the Syrian rebels, including from Libya.

In one YouTube video of the shootdown, labeled as a missile strike, the only apparent sounds are those of conventional anti-aircraft artillery. In a second video, however, the whooshing of a missile can be heard in the background. This video ends with footage of a Dutch military exercise using Stinger missiles purchased from Rocketstan in Turkey, which was licensed to assembly Stingers in the 1990s by Raytheon.[33]

Just days before the shootdown, a Syrian rebel posted a picture to his Facebook page of a young fighter with a SA-7 missile, claiming it was from the "first batch" of missiles to reach the rebels, with the "second batch on the way." The delivery of the missiles showed that "the international community decided to abandon the [Syrian] regime," he commented.[34] The *New York Times* reported that the photo was the first hard evidence of a complete missile system in the hands of the Syrian rebels, and gave credence to a report two weeks earlier from

NBC News reporter Richard Engel of the transfer of some two dozen shoulder-fired surface-to-air missiles through Turkey.[35]

Were the weapons now winding up in the hands of the Syrian rebels the result of a U.S. covert operation? One thing is clear: Jihadi leaders in Libya were collecting arms and supplies for their brothers in Syria. They held public events, ran ads in local newspapers, and posted billboards on the streets calling for Libyans to help the Syrian jihadis. The leaders of the weapons transfer networks were Belhaj, al-Marati, and a former 17th February Martyrs Brigade commander named Abdul Basut Haroun.

"It is just the enthusiasm of the Libyan people helping the Syrians," said Fawzi Bukatef, another former 17th February Brigade commander. "They collect the weapons, and when they have enough they send it," he said. "The Libyan government is not involved, but it does not really matter."[36]

You would have to be blind, deaf, and dumb not to see it.

What no one understood at the time was that the missile in the Facebook picture was one of the four hundred SA-7s transferred by Abdelhakim Belhaj through the smugglers' ratline to Agadez in Niger, where they were retrofitted with CIA batteries and Egyptian gripstocks. Like those missiles, this one had the dark green tube from Qaddafi's stockpile of Russian-made missiles, and telltale brown gripstock from Egypt. Yet another of these missiles turned up in the hands of an officer of the Free Syrian Army—the faction supposedly supported by the United States—three weeks after the Benghazi attacks.[37]

Taken together, I believe this is convincing evidence that the National Security Staff under the guidance of John Brennan, with possible assistance from the CIA or CIA contractors, engaged in an illegal covert action to arm the Syrian rebels through cutouts that included the transfer of surface-to-air missiles.

The game was on.

THE AUGUST 16 CABLE

Dylan Davies, the Blue Mountain employee who managed the unarmed security guards at the front gate of the Special Mission Compound in Benghazi, was not shy about giving advice. Ever since he
had taken over the contract in the spring he had been warning the
temporary State Department Regional Security Officers (TDY RSOs)
about the 17th February Martyrs Brigade. They were the local militia
that was supposed to provide the armed guards who patrolled the perimeter of the compound and provide a heavily armed quick reaction
force (QRF) in the event of serious trouble.

Early on, he met with Renee Crowningshield, the newly arrived
RSO, and gave her an earful. The 17th February Martyrs Brigade was
"putting bullets in people's heads," he told her. "The fact that the militia who formed our QRF were out there killing on the streets made
the situation at the [compound] seem all the more insane. There were
even reports that 17th February militiamen were joining the ranks of
the [Ansar al-] Shariah brigade," he said.

Crowningshield, a former cop, agreed to put the Libyans through
their paces to test their weapons skills. They failed miserably. Davies made pleas for "more boots on the ground" and a new QRF.
Crowningshield agreed. "I get what you're saying," she told him. "I'll
be writing a full security survey on the Mission, and I'll ask for the
extra funding, manpower, and equipment we need."[38]

That was in May.

Over the summer, Davies became more insistent. The 17th
February militiamen had their own bungalow, just to the right of the
main gate as you entered the compound. They could be seen lounging around in skintight yellow T-shirts, tight combat bottoms, and
flip-flops, with an AK-47 slung carelessly over their shoulders. They
couldn't even strip down their weapons. And yet, they were the only

ones besides the diplomatic security officers allowed to carry weapons inside the compound itself. The QRF has got to go, he insisted. "They need replacing with some proper, professional soldiers—like U.S. Marine Corps."

After the Brits pulled out of Benghazi, Davies says the new TDY RSO told him he had sent a cable to the State Department recommending that the QRF be replaced with a twelve-man U.S. Marine Corps detachment. If true, that cable has not been released. General Martin Dempsey later told Congress there were no Marines in Tripoli or Benghazi because the U.S. diplomatic presence in Libya was "in transition between postconflict and prewhatever it was about to become." Most U.S. diplomatic outposts operating under ostensibly normal conditions had a six-man Marine detachment whose main responsibility was to safeguard classified materials, destroying them in the event of a breach, with overall site security the responsibility of the host government, he explained.[39]

In late July, a new team of three TDY RSOs arrived in Benghazi, including David Ubben and Scott Wickland, who would still be at the compound on the night of the attack. On August 15, they sat down with Davies in the canteen of the Mission Compound to discuss the QRF at length. The lead RSO was drafting an email for Ambassador Stevens on security at the compound and asked Davies for his input. "Every RSO before you has asked for more manpower and we've been denied," Davies began. "So, presumably that's not going to change. If you can't get more men, your only alternative is more firepower. So, imagine you site a .50 caliber on top of Villa C [the VIP villa], mounted on a tripod. From there you can hit the front gate, but also swivel it around to hit the rear. A .50-cal will make even your more die-hard jihadi stop and think twice about trying to get in."

The RSO argued that Washington would never go for it. "Too aggressive," he said. But he was worried. How would they stop a full frontal assault? "Like I said, you need more men or a .50-cal. Without one or the other, you're buggered," Davies said.

The RSO said he would put that in his email to Ambassador Stevens and his bosses back at State. "I'll warn them that if the compound comes under a sustained, organized attack it will be overrun," he said.[40]

Davies' credibility was challenged after he gave an interview to CBS *60 Minutes* in October 2013, where apparently out of guilt he exaggerated his role in defending the Special Mission Compound and claimed to have personally discovered the ambassador's body in a local hospital. But his account of the August 15 security review in Benghazi and the security concerns he raised in the preceding months has been corroborated by multiple sources, including a review of the classified record by the Senate Select Committee on Intelligence. And that is why the national media jumped all over him for the fabrications in the *60 Minutes* interview: He was yet another inconvenient witness to the dereliction of duty of Hillary Clinton's State Department.

In a separate book on the Benghazi attacks, former diplomatic security agent Fred Burton recounted in nearly identical terms the emergency meeting that took place on August 15 in the Special Mission Compound cantina. "The exchanges at the meeting were direct and ominously honest," he wrote. "The TDY RSO expressed his concerns that the DS contingent would be unable to defend the post if it was subjected to a coordinated and serious terrorist attack. They cited a lack of manpower, insufficient physical security infrastructure, limited weapons systems at their disposal, and the lack of *any* reliable host-nation support," starting with the QRF. In an email back to Stevens in Tripoli, the TDY RSO "emphatically stated that he did not believe the mission could be adequately defended."[41]

With that email in hand, Ambassador Stevens convoked an emergency security meeting on August 16, 2012, at the CIA Annex in Tripoli. That meeting was "an extraordinary, not an ordinary event," commented Representative Mike McCaul, one of several House Republican committee chairs investigating the Benghazi attacks. The latest threat information provided by the CIA chief of base in Benghazi and the station chief in Tripoli convinced the ambassador that the Special Mission Compound in Benghazi was "not prepared to withstand a coordinated attack" and urgently needed additional security.

Stevens formalized these findings in a classified cable he fired off to Washington the same day. The CIA "briefed the [Emergency Action Committee] on the location of approximately ten Islamist militias and AQ training camps within Benghazi," he wrote. All present "expressed concerns with the lack of host nation security to support the U.S. Mission." A CIA officer in Benghazi "expressed concerns with the Post's relationship with [the 17th February Brigade], particularly in light of some of the actions taken by the brigade's subsidiary members," while the TDY RSO "expressed concerns with the ability to defend Post in the event of a coordinated attack due to limited manpower, security measures, weapons capabilities, host nation support, and the overall size of the compound." [42]

The key decision maker, who received all of Stevens' requests for additional security, was Undersecretary of State for Management Patrick Kennedy. When asked about the still classified August 16 cable, he made the incredible claim that he couldn't provide additional security because Stevens hadn't made clear what he needed.

Here is his exchange with Representative McCaul:

McCAUL: Did you receive that cable, the August 16th cable?
KENNEDY: Yes, sir, I did. And if I might—
McCAUL: I have limited time. Did you respond in the affirmative or did you decline that request?
KENNEDY: This cable, I did not—we did not decline the request.

McCAUL: Was additional security provided on that day, weeks before the September 11th attack?

KENNEDY: The cable, sir, and I have a copy in front of me, it closes with, "U.S. Mission Benghazi will submit request to U.S. Embassy Tripoli for additional security upgrades and staffing needs." We never received that additional request. So, there was no way I could respond to a request that had not yet been submitted.

McCAUL: Do you know if Secretary Clinton saw this cable?

KENNEDY: I do not believe so.[43]

On August 20, the embassy's weekly report on developments in Benghazi quoted a well-known women's rights activist, Wafa Bugaighis, to illustrate the worsening security situation. "For the first time since the revolution, I am scared," she told political officer Eric Gaudiosi, after being detained for attending an international conference in Benghazi. For their part, United Nations officials in Benghazi believed that the Supreme Security Council—the only official force imposing any semblance of law and order—was "fading away" in Benghazi, unwilling to take on "anyone with powerful patrons or from powerful tribes."[44]

On August 22, the TDY RSO in Benghazi sent the request to Tripoli that Kennedy referred to in the exchange quoted above. It bore the title, "Security Requests for U.S. Mission Benghazi," and included specific requirements for physical security upgrades, new equipment, and additional manpower.[45] If Undersecretary Kennedy pretended he never saw that request, he simply wasn't paying attention. Just five days later, on August 27, the State Department issued an updated travel warning on Libya, urging U.S. citizens to stay away because "inter-militia conflict can erupt at any time or any place."[46]

The flurry of reports, requests, proposals, and desperate cries for help was not enough to get Washington's attention. The message from the State Department bosses—including Secretary Hillary Clinton—to our diplomats and warriors in Libya was clear: You're on your own.

OUT OF CONTROL

On the same day that Stevens sent the secret cable begging yet again for additional security in Benghazi, a U.S. UH-60 Black Hawk helicopter was shot down in Afghanistan. My sources inside U.S. Special Forces command say it was the second Stinger hit in less than a month. Seven U.S. soldiers were killed—including two Navy SEALs and a Navy explosives expert—as well as three Afghan troops and an Afghan interpreter. The attack occurred in Shah Wali Kot, a rural area outside of Kandahar known as a Taliban stronghold.

U.S. Army Major Martyn Crighton, a spokesman for the International Security Assistance Force Joint Command in Kabul, said that the crash was under investigation but there was "no operational reporting" indicating enemy fire had brought down the aircraft.[47]

The Black Hawk went down "in the heart of Injun territory," so the search-and-rescue team hastily recovered the bodies and got out, leaving the aircraft behind, my sources report. The EOD tech that briefly inspected the charred wreckage was unable to make a definitive determination on the cause of the crash, but the flight logs showed that the aircraft had been flying well above the range of RPG fire when it went down, sources with access to the logs told me.

The administration "cannot afford to let the public or Congress know they put Stingers in the hands of the Taliban," a senior active-duty Special Operations officer told me. He said foreign intelligence sources reported that between fifty to sixty Stingers had reached the Taliban since June 2012, plus another two hundred or more SA-24s, but that U.S. forces were not actively looking for them. "We are too busy buying Mi-17s from Russia for the Afghans and leaving with our tail between our legs and claiming victory," this source said.[48]

Ten days later, Syrian rebels shot down a government military helicopter over the capital, Damascus. After an hour of firing at it futilely with AK-47s and anti-aircraft artillery, someone arrived with a more effective weapon.

Not long after this, ABC News caught up with former Navy SEAL Glen Doherty in Southern California, where he lived when not working on contract to MVM, a Beltway security firm that provided highly trained former Special Forces soldiers as security guards to the CIA and to other government agencies. According to the ABC News report, which only appeared after he was killed in Benghazi, Doherty said he was "working with the State Department on an intelligence mission to round up dangerous weapons" leftover from the Qaddafi regime.

He went on to explain that he "traveled throughout Libya chasing reports of the weapons and, once they were found, his team would destroy them on the spot by bashing them with hammers or repeatedly running them over with their vehicles." It was a desperate mission, and every day he and his colleagues at the CIA Annex in Tripoli were losing ground, as more and more missiles leaked out of the country into the hands of al Qaeda and their allies in Mali, Niger, Sudan, Gaza, and, now, Afghanistan.[49]

After inaugurating a new U.S. consular office in Tripoli on August 26, Ambassador Stevens took some time off to meet with friends in Sweden. He also traveled to Austria and Germany, where he met with European oil executives interested in investing in Libya but afraid of the security situation.

While in Germany, he stopped in Stuttgart, headquarters of

AFRICOM, to discuss with General Carter Ham the alarming security situation in Libya. General Ham had read Stevens' secret August 16 cable and had called him immediately offering to make a new SST available to him. At their sit-down in early September, he again offered to send a Special Forces team to help.

Stevens was in a bind: He needed the security, but he knew that if he fell back on the military, there would be hell to pay with his bosses at the State Department who had been trying to get rid of the SST since April. They made it clear that they didn't want a U.S. military presence in Libya, since that would detract from the narrative that al Qaeda had been defeated. Besides that, he was worried about their legal status. Now that the SST had transitioned back to Title 10 status, placing them once again under General Ham's authority— not Stevens'—they were legally subject to whatever Status of Forces Agreement was in place with the Libyan government. And there was none. If something like the August 6 attempted carjacking happened again and they were taken into custody, they would have no more legal protection than a tourist. It just didn't feel right to him to hang them out to dry.

So, he thanked the general for the offer, but declined. He had put in requests with State for them to send more DS personnel and to hire more Libyans for his personal protection detail, and was still hoping they would come up with an acceptable solution. Fingers crossed.[50]

After Stevens and his three colleagues were murdered, Admiral Michael Mullen and other administration flacks attempted to shift the blame onto Stevens for not arguing more forcefully for better security. "As the chief of mission, he certainly had a responsibility in that regard," Mullen said at a news conference when the State Department released the Accountability Review Board report he co-authored with Ambassador Thomas Pickering. "But part of his responsibility is certainly to make that case back here, and he had not gotten to that point where you would, you might get to a point where you would be considering, 'It's so dangerous, we might close the mission.' "[51]

The ARB report made the astonishing claim that, under Stevens' command, the U.S. Embassy in Tripoli "did not demonstrate strong and sustained advocacy with Washington for increased security" for the Benghazi Special Mission, a claim that flies in the face of the steady stream of reporting on the security vacuum and Stevens' multiple requests for backup, all of which were denied.

Clearly, in the eyes of official Washington, it was easier and less costly to blame the dead for not protecting themselves, than to accuse a secretary of state who wanted to become president of incompetence, callousness, and a criminal disregard for the security deficiencies of diplomatic compounds that legally fell under her personal responsibility.

13

BAITING THE TRAP

As U.S. ambassador, Chris Stevens had a full plate: He was encouraging the new government to develop democratic institutions, managing the cleanup of Qaddafi's weapons, and trying to tame the militia groups in Benghazi so Western oil companies would return. He was also working to ensure that freelancing arms dealers didn't get in the way of the Turkish, Saudi, and Qatari arms pipeline to the Syrian rebels. It was a high-wire act. And through all of it he performed without a safety net and no backup from Washington.

If that wasn't enough, things were about to get very messy.

THE FISHING BOAT

On August 25, 2012, a Libyan fishing boat, *Al Entisar* (Victory), docked in the southern Turkish port of Iskenderun, where Libyan Islamists unloaded more than 400 tons of weapons and "humanitarian supplies" that they had brought from Benghazi, ostensibly to help refugees from the Syrian conflict. They had acquired the weapons from stockpiles leftover from the fight against Qaddafi following an appeal by fellow

Islamists in Syria to help their struggle. It was the Libyan version of a charity drive.

The arrival of so many weapons and much-needed supplies created havoc between Islamist brigades close to the Muslim Brotherhood, who claimed to have paid for the goods, and representatives of the Free Syrian Army, who tried to commandeer the weapons for their own use. "Everyone wanted a piece of the ship," said Suleiman Hawari, an Australian-Syrian working with the ship's captain from Benghazi. "Certain groups wanted to get involved and claim the cargo for themselves. It took a long time to work through the logistics."

The infighting among rival rebel commanders reached a fever pitch and caught the attention of the Turkish authorities and their allies. As the rebels argued over who would get what, word leaked out and reporters began nosing around.[1]

The official recipient of the Benghazi aid package was the IHH Humanitarian Relief Foundation, the Turkish Islamist group that sponsored the infamous Gaza flotilla, where Mahdi al-Harati was wounded in 2010. The sender was an unknown organization in Benghazi, according to a UN report. On the surface, the "National Committee for the Support and Relief of Displaced People" seemed a better front than the one the Libyans used for the ill-fated shipment to the Syrian rebels in April. The same person later claimed to have organized both shipments.

"They know we are sending guns to Syria," Abdul Basit Haroun told a Reuters reporter in Benghazi. "Everyone knows." To illustrate his point, he had an associate take the reporter to a warehouse at the port where Haroun was stockpiling weapons for future shipments to the Syrian rebels.

Haroun had learned his lesson since the *Letfallah II* had been seized in April. This time he took a longer route and steered clear of Lebanon, where Iran's proxy militia, Hezbollah, wielded huge influence. Haroun was no fan of the Iranians, and he knew they were nosing around Benghazi, so this time he wasn't taking any chances.

Haroun had lived in Manchester, England, for twenty years, before returning to Libya to join the revolution. A commander with the 17th February Martyrs Brigade during the revolution, Haroun later formed his own brigade, the Abu Selim Martyrs Brigade, to commemorate the slaughter of political prisoners in Abu Selim prison in 1996. After the revolution, he became better known for his weapon-smuggling activities, and set up his charity with Saudi help. "After the end of the war of liberation, he became involved in supporting the Syrian revolution . . . sending aid and weapons to the Syrian people," said assembly member Tawfiq al-Shehabi. "He does a good job of supporting the Syrian revolution."

Haroun was also well connected to the most radical and powerful players in the new Libya, who actively supported his arms smuggling activities. Key among them was the al-Salabi brothers.

Ismail al-Salabi had been a member of the Libyan Islamic Fighting Group who cut his teeth in Abu Selim prison and became a founding member of the 17th February Martyrs Brigade, the militia that ostensibly was providing security to the U.S. Special Mission Compound in Benghazi. His brother, Sheikh Ali al-Salabi, was the spiritual guide of the Libyan branch of the Muslim Brotherhood. Both were seen as Qatari protégés. Sheikh al-Salibi joined forces with Abdelhakim Belhaj in late 2011 to form the al-Watan party in hopes of winning election to the future Libyan National Assembly. Belhaj reportedly helped organize Haroun's weapons transfers to the Syrian rebels.[2]

Haroun liked to put a positive spin on his gun-running activities. "We are doing two great things," he told Reuters. "The first is that we are taking guns off the street. The mission is so popular that we get 50 percent discounts on weapons." The second was helping the Syrian people. "Anyone who saw what I did in Syria would do the same thing," his unnamed associate added. "The water is so polluted you wouldn't even wash your hands with it. People have no clothes. I saw three births with no doctors present. People are dying without medication."

Once his weapons were unloaded in Turkey, Haroun acknowledged that he had "no control over which groups received the weapons."[3] In a separate interview with *Foreign Policy*, the forty-three-year-old unnamed rebel commander in charge of organizing the shipments (undoubtedly Haroun) lamented the lack of international support for the Syrian rebels and boasted that the *Entisar* shipment had included 120 SA-7 MANPADS. "[T]he revolution in Syria seems to have been abandoned by the world. So [in early 2012] we decided to help and send weapons."[4]

The United Nations Experts Panel on the Libyan arms embargo was quicker on the uptake than they had been when Haroun's previous shipment was seized (then mostly disappeared) in Lebanon. They called the Turkish government as soon as they learned about the arrival of the *Al Entisar* in Iskenderun, and were shown an apparently phony cargo manifest. "The Panel . . . was informed that, since the boat was carrying humanitarian cargo, in the absence of any reasonable ground to suspect otherwise, no inspection was conducted by the port authorities."[5]

So much for the vast intrusive powers of the UN. Haroun and his associate, who ran the aid organization in Benghazi, happily told Reuters that the "weapons [were] hidden among about 460 metric tons of aid destined for Syrian refugees."

THE CLEANER

On September 2, CIA Director David Petreaus made an unannounced trip to Ankara to straighten out the mess. The United States was backing the Free Syrian Army but had been stingy with direct weapons shipments, preferring to allow Saudi Arabia, Jordan, the UAE, and, especially, Qatar do the heavy lifting, with help from fellow Islamists in Turkey. These U.S. proxies had chosen to supply the hard-core Islamist brigades, including the infamous Jabhat al-Nusra, which sought to replace Assad with an Islamic state and declare Sharia law as the national constitution.

Petraeus was particularly worried by reports that the shipment on board the *Entisar* included MANPADS looted from Qaddafi's vast arsenal, which the United States was increasingly desperate to collect. Although President Obama's counterterrorism advisor John Brennan assured Petraeus they had the situation under control, the CIA director was mindful of the information provided by John Maguire, his old friend from Iraq, about the missile upgrades being done in Agadez by the brother of al Qaeda's number two. Petraeus trusted Maguire and was taken aback by the vehemence of Brennan's insistence that he steer clear of the Agadez operation.

In Iskenderun, the Libyans were boasting to journalists that they were going to be the kingmakers in Syria and would use the missiles to shoot down more Syrian Air Force jets and helicopters.

Hillary Clinton was just as worried as Petraeus, since she also had been a big supporter of the secret arms pipeline to the Libyan rebels that helped them get rid of Qaddafi. Among those weapons, as I revealed in chapter 5, were four hundred Stingers and fifty launchers, delivered by a Qatar Special Forces convoy through Chad. These were a more modern version of the deadly missiles the Reagan administration supplied to the Afghan mujahideen that helped them to defeat the Soviet army in the late 1980s.

If word got out that MANPADS were on the loose in Syria and that the United States had done nothing to stop them, it would unleash an international political firestorm. The Turks, for starters, were already hopping mad after one of their F-4 fighter-bombers had been hit by a surface-to-air missile off the Syrian coast in June. They were afraid that if the Syrian rebels got MANPADS, sooner or later the missiles would fall into the hands of the Kurds, who would give them to the PKK, the Kurdish terrorist group that had been locked in a deadly war with the Turkish government for more than twenty-five years.

The Libyans were just too chaotic, too disorganized, and too damn talkative. Someone had to get them back on the reservation. The man who knew the militias and tribal elders in Benghazi best was not a spy, but a diplomat: U.S. Ambassador to Libya J. Christopher Stevens.

Stevens knew Belhaj, the al-Salabi brothers, and most of the 17th February commanders personally. He had broken bread with them as the State Department's special envoy to Benghazi in the heady days of the anti-Qaddafi uprising in 2011. And, more recently, he had been trying to entice them to turn over their weapons, as the State Department itself acknowledged. Maguire thought of him as an ambassador of the old school. "He was like John Negroponte or Tom Pickering—the type of ambassador who understands what the Agency brings to the table and how to use it to advance the strategic goals of the United States."[6]

It is also quite likely that Stevens knew Haroun personally. At the very least, Haroun knew *him* since he was named as one of the three Libyan officials who confirmed Stevens' death in the initial Associated Press obituary, which appeared on September 12, 2012. The AP identified the arms dealer as "Benghazi security chief, Abdel Basit Haroun."[7]

Both CIA and State had a track record of calling on Stevens' services as a cleaner. In the summer of 2011, as the revolution was still facing tough resistance from Qaddafi's forces, they called on him in Benghazi to spearhead an effort to secure the MANPADS left behind by the Qaddafi regime. They sent him U.S. Army Colonel Mark Adams from the State Department's special "MANPADS Task Force" to coordinate the cleanup effort on the ground. Although the State Department touted Colonel Adams' efforts early on, and even allowed him to appear in public, they have now closeted him away from reporters. When I contacted him in the course of writing this book, he insisted that his remarks be cleared by a State Department flack, who, after multiple emails and phone calls, declined to make Adams available.[8]

Petraeus and Clinton wanted Stevens to leverage his close relationship with the former rebel military leaders to get the MANPADS back or, at the very least, to prevent any further shipments. They wanted him in Benghazi. As of yesterday.

In his personal diary for September 9, 2012, just before departing for Benghazi, Stevens noted: "Stressful day. Too many things going on, everyone wants to bend my ear. Need to pull above the fray."[9]

HILLARY'S PHOTO OP

The official reason Stevens went to Benghazi was to certify that the Special Mission Compound could be opened as an official U.S. Consulate before September 30, the end of the fiscal year.

According to his deputy, Greg Hicks, he went there on direct orders from Secretary of State Clinton. "Chris told me that in his exit interview with the secretary after he was sworn in, the secretary said we need to make Benghazi a permanent post. And Chris said I will make it happen," Hicks told the House Oversight and Government Reform Committee.[10]

Clinton planned to travel to Benghazi that autumn, just before the U.S. presidential elections, on a victory tour to tout the achievements of her four years as secretary of state. She wanted to announce the opening of the new consulate in the birthplace of the Libyan revolution during that trip.

However, as the security situation deteriorated, with no reinforcements from State, Stevens put Benghazi on the back burner. "[T]he Ambassador had not indicated any sort of desire to travel to Benghazi," said Eric Nordstrom, his chief security officer until late July. He vaguely mentioned that he might get out there sometime in October, once the security situation had been resolved.

The push to move up the visit to Benghazi was not the ambassador's decision, as the ARB report tried to suggest. It came from assistant for Near Eastern affairs Beth Jones, who was acting on Clinton's orders. She sent instructions for Stevens to certify that Benghazi could be converted into a "permanent constituent post" by September 30, because her bureau had funds leftover from Iraq that could be used for Benghazi. The funds "had to be obligated by September thirtieth," Hicks testified. Jones insisted that Stevens get to Benghazi as soon as possible so that they could fund the new consulate without having to go through the more complicated annual budgetary process, which involved congressional review.[11] Given all the "negative" reporting about security in Benghazi, Congress might insist that they build a new facility to the Inman standards, as required by law, rather than just upgrade the existing villa compound. With money scarce, Secretary Clinton preferred spending new funds on other pet projects, such as funding abortion clinics in the Third World. Congressional attention to Benghazi might also reveal the ongoing screw-up of the missing MANPADS, another "success story" touted by the administration that had gone badly wrong.

As they started to start planning the trip in late August, just before Stevens left for Europe, Nordstrom's replacement tried to slam on the brakes. "Regional Security Officer John Martinec raised serious concerns about his travel. Because of those concerns, the ambassador adjusted his plans for that trip," Hicks said. "First, he agreed that he would go in a low-profile way, that his trip would not be announced in advance. . . . And second, he eventually decided also to shorten his trip."[12]

While my sources believe Stevens was sent to Benghazi primarily to "clean up the mess" created by the publicity over the unauthorized arms shipments to the Syrian rebels, the revelation that Hillary Clinton also wanted him there so that she could stage a photo op caused a firestorm when whistle-blower Greg Hicks first revealed it at the May 8, 2013, HOGR hearing. Why? Because it contradicted the State Department's official version in the ARB report that claimed Stevens made the decision on his own, with no prompting from Washington.

Yet another inconvenient truth.

"ME TARGETED"

"Islamist hit list in Benghazi. Me targeted," Chris Stevens wrote in his diary before departing for his last mission.

The seven-page diary was discovered in the charred ruins of the Special Mission Compound by a CNN reporter one week after the attack. When CNN published portions of it, including Stevens' comment, the State Department went ballistic. Philippe Reines, identified as a "senior advisor to Secretary Clinton," called CNN's actions "indefensible." CNN countered that Stevens' fears of a terror attack were newsworthy and "are now raising questions about why the State Department didn't do more to protect Ambassador Stevens and other U.S. personnel."[13]

As the embassy team was nervously preparing his trip, Libyan Interior Ministry officials declared a state of maximum alert following gun battles between local police and militia forces in Benghazi. The only good news: that meant the police had "established a 24/7 police presence at the Mission in response to our long-standing request," political officer Eric Gaudiosi wrote in the weekly Benghazi update for Washington.[14]

But that protection never materialized. That was one of the first things Alec Henderson, the regional security officer who had arrived in Benghazi just one week earlier, told Stevens after his flight from Tripoli touched down at Benghazi's Benina Airport at around noon on Monday, September 10, 2012. Henderson whisked Stevens and the two African-American DS officers who had accompanied him from Tripoli into armor-plated Toyota Land Cruisers for the twenty-minute drive to the compound. Also greeting him was David McFarland, the TDY political officer who was heading back to Tripoli the next morning, who gave him a rapid-fire update in the car. A Toyota Hilux pickup with a .50-caliber machine gun from the 17th February Martyrs Brigade led the small convoy as it rapidly weaved in and out of the mid-morning traffic and around the many traffic circles to the Fourth Ring Road in the upscale Fwayhat district.[15]

For Chris Stevens, returning to the Benghazi compound, where he had lived for most of the heady days of the anti-Qaddafi uprising, was like coming home. But this time, the atmosphere felt different. What was the deal with the 17th February Martyrs Brigade, he asked? Why had they provided just one gun truck as an escort, and why were they driving like they were afraid of being pursued? During the revolution, they had chaperoned his every movement proudly and openly with several vehicles, honking through the checkpoints, sirens blaring. Now it was as though they were afraid to be caught dead with an American.

They were lucky to have even one gun truck, McFarland told him. The 17th February guys essentially had gone on strike last week to protest salary and working hours. Thanks to Alec, things were getting back on track.[16]

For an ambassador, visiting a distant outpost in a country as secretive as Libya was the only way of getting a feel for the situation on the ground, unfiltered by politics and make-nice reporting. Chris Stevens was about to get lots of it.

Once Stevens had stowed his gear in the master bedroom of the sumptuous VIP villa, Alec Henderson briefed him and the two DS agents he had brought with him from Tripoli on the emergency REACT plan, in case of a terrorist attack. They'd made a number of security improvements since Stevens had last lived here. They'd raised the perimeter wall, installed Jersey barriers to prevent car bombs, and put up guard booths outside protected by sandbags. They now had security cameras covering the inside perimeter wall, which spanned more than three hundred yards along the front and back and nearly one hundred yards on the sides. While some of them were still in boxes, waiting to be installed, they had all three access gates covered, two in the front and one at the rear of the compound. They also had an operational camera showing the outside of the main gate.

The unarmed Blue Mountain guards at the main gate were there to verify the identity of visitors only, not to provide actual defense in the event of an attack. They controlled the drop-down steel pole and the double-steel doors to allow authorized vehicles to enter the compound. In the event of an emergency, they were supposed to activate the Imminent Danger Notification System and get the hell out of the way. If that ever happened, they would know it, Henderson said. Wailing sirens would go off all over the compound. If he was in the Tactical Operations Center (TOC), which was housed in a separate villa across a narrow interior driveway from the ambassador's residence, they'd probably hear his voice as well over the loudspeakers.

They called it the duck-and-cover alarm. If you hear that, he told the DS agents from Tripoli, it is not a drill.

The most important thing was to get the VIP package (in this case, the ambassador and facility manager Sean Smith) into the safe room, where they were supposed to shelter in place until reinforcements arrived. Scott Wickland, one of the DS agents who had arrived in Benghazi a few weeks earlier, had the keys and would relieve whoever was with the ambassador to make sure he was secure. Then they would evacuate to the more heavily fortified CIA Annex. Another DS agent, David Ubben, gave them the tour and showed them how the security grille with its heavy steel bars was designed to prevent entry into the living quarters, where the safe room was located, then he demonstrated how to operate the special emergency release windows so that they could escape to the roof if need be. If things got really bad, and they had to abandon Benghazi altogether, or if they were attacked on the road, their standing instructions were to head for the Egyptian border by road—eight hours by the coastal highway.

It was a sobering briefing. Everyone who spent the night at the Special Mission Compound got it. If they were ever attacked, God help them.

Stevens, ever the diplomat, wanted to meet the whole team at the Mission Compound, so after a brief lunch in the sumptuous living-dining area of the VIP residence with Smith and McFarland, he walked over to the canteen in the commons villa. There he briefly met Dylan Davies, the British contractor who managed the unarmed security guards at the gate.

In his somewhat confused account of these events, Davies says that Stevens asked him what he thought of the security arrangements at the compound. "Put it this way, sir—more could be done," Davies said. Then he repeated what he had been telling the various RSOs for several months: They needed more physical barriers, more DS agents, and especially, they needed to get rid of the 17th February Martyrs Brigade as the QRF, because they were unreliable. They simply wouldn't come if needed.

Stevens went outside and chatted in fluent Arabic with the Blue

Mountain guards, asking them about their families and making small talk about Benghazi. They were impressed, Davies recalled.[17]

What Davies didn't know was that, just the night before, Henderson had emailed John Martinec, his boss in Tripoli, warning that the 17th February Martyrs Brigade had told him they "would no longer support U.S. movements in the city, including the Ambassador's visit." They were flying by the seat of their pants.[18]

At two in the afternoon, Henderson and the full complement of DS officers saddled up and drove the ambassador a kilometer down the road to the CIA Annex for a two-hour briefing. Stevens was met by the CIA chief of base, who picked up where Henderson and Ubben had left off.

The Annex was their Fort Apache. Although it was just over half the size of the diplomatic compound, it contained four large villas set at diagonals around an inner courtyard and parking area. It had much higher walls and much, much better security. For starters, they had fewer neighbors. With a warehouse complex and empty lots on one side, walled villas on two others, their main vulnerability was the street side, and that was protected by impenetrable blank walls with no telltale sign of what lay behind them. Plus, they had a height advantage. From the rooftops, they dominated the entire neighborhood and no one could look down on them. Anyone trying to approach from the street was dead meat.

The chief of base introduced some of his GRS operators, including former Spec Ops warriors Tyrone Woods and Mark G——, and had them give a brief display of the firepower they had at their disposal in case something went down. (Mark G——'s full name, which is known to the author, had not been released at the time this book went to press.) He repeated the evacuation plan Henderson had sketched out back at the compound. If anything happened, they were supposed to hightail it to the Annex. Once they got there, the GRS guys could hold off the bad guys until more serious help arrived. Or they evacuated by road to Egypt.

It wasn't a bad plan, all things considered. But Ibrahim Joudaki and Khalil Harb, the Quds Force operatives in charge of the attack, had already analyzed it inside and out and identified its inherent weaknesses. And they intended to exploit those weaknesses to the fullest. Their men had been conducting surveillance on both American compounds for nearly three months by this point. As for the REACT plan itself, security at the compound was so lax that copies of it had been left lying around, so the Quds Force men had had no problem acquiring it for a fee from a disgruntled local security guard.*

WAS STEVENS INVOLVED IN GUN-RUNNING?

The CIA chief of base next got down to the real business that had brought Ambassador Stevens to Benghazi: their ongoing effort to track down chemical weapons, SCUD missiles, and other WMD; and the hunt for the missing MANPADS from the Qaddafi regime.

The contractors working on MANPADS collection had left Benghazi in June, so now the effort was focused on keeping tabs on the Libyans who were buying weapons and sending them to Syria. The *Al Entisar* deal was an absolute mess. As more news on that shipment came out, it was bound to attract wannabe arms dealers who would make Abdul Basit Haroun and his guys look like pros. They had to shut this thing down before it went completely out of control.

Stevens had two meetings the following day, September 11, that were highly unusual for an ambassador, and that contributed to rumors that the main purpose of his trip to Benghazi was to facilitate arms shipments to the Syrian rebels.

---

*Even foreign journalists found a copy of the REACT plan, along with Ambassador Stevens' schedule, in the charred ruins of the VIP villa weeks after the attack.

The first was with Mahmoud El-Mufti, the marketing director of the Al Marfa Shipping and Maritime Services Company, a private Libyan shipping company. El-Mufti was coming to meet him at the compound in the afternoon, according to Stevens' official schedule. The subject was not mentioned.

Why would an ambassador meet with an official from a local shipping company? Was he seeking a better, more discreet shipper for the weapons the White House had publicly stated it wanted to see reach the Syrian rebels? Someone who could maintain control over the goods, and make sure they reached their authorized destination?

El-Mufti appeared to have the credentials. Already in March 2011, at the start of the rebellion, he attracted the attention of a *USA Today* correspondent for his political enthusiasm. "We are Arabs. We are Muslim," he told the reporter as they drove through the main gate of the Benghazi port to greet a Turkish and Emirati aid ship as it docked. "We continuously support the Palestinian cause." Then he caught himself and added, "But Israel is not our priority. We're not interested in war. We want to see prosperity first. We can be the Dubai number two."

Those aid ships arriving in Benghazi in early 2011 were undoubtedly bringing weapons to the Libyan rebels. Al-Marfa Shipping advertises its services as a "port agent" engaged in "drawing up, initiating and delivery the required documents (booking lists, shipping permits, delivery orders) related to the cargo." They would be qualified to organize arms shipments to the Syrian rebels. But why would an ambassador engage in such discussions? If the president had authorized covert arms shipments to the Syrian rebels from Libya, wouldn't that be a task for the CIA chief of base?

Lieutenant Colonel Andy Wood mentioned in our conversations that the embassy in Tripoli had encountered difficulties in getting military supplies and even containers with personal belongings cleared through customs in Tripoli and Benghazi. But again, would the U.S. ambassador travel all the way from Tripoli to discuss such a mundane subject with a midlevel shipping company official? Unlikely.

Ambassador Stevens' diary includes a brief note about this meeting. In it, Stevens identified al-Mufti as "shipping Co. owner & broker of PM contender Dr. Moh Mufti." The State Department has never concealed Stevens' close ties to political power brokers in Libya, or that he intended the Benghazi trip as an opportunity to reconnect with some of them. Libya was, in fact, in the throes of selecting a new prime minister at precisely this time, and Dr. Mahmoud El-Mufti—a relative of the Benghazi shipping company official—was considered a "minor contender."

Mahmoud El-Mufti still works at the shipping company and speaks good English. When I reached him on his cell phone and told him I was writing a book on Benghazi, and understood he met with Ambassador Stevens on the day he died, he cut me off. "No comment. No comment, please," he said and hung up. Despite his nervousness, I believe that this meeting was one of Stevens' planned political consultations. It's inconceivable that the ambassador didn't take the opportunity to pick El-Mufti's brains on the *Al Entisar*, and what he would have done to keep the whole thing from coming unglued. That was simply part of his job, and part of his character. But it wasn't his primary mission in meeting El-Mufti.

The second unexplained meeting that has drawn significant attention was with Turkish Consul General Ali Seit Akin. His role remains more murky. In a message of condolence he posted to the U.S. Embassy in Tripoli's Facebook page, he said that he and Stevens "knew each other well and were good friends already" because they had both been posted to Benghazi during the uprising.[19] In his diary entry previewing the day, Stevens mentioned the upcoming meeting with Akin, "who had helped me land in Benghazi last year."

Turkey was a major player in Libya. Initially, the Turkish government of Islamist premier Tayyip Recep Erdoğan opposed the NATO bombing attacks and, despite prodding by Secretary of Defense Robert Gates and other top NATO officials, refused to take part in it. Turkish businesses had invested heavily in Qaddafi's Libya and opposed the bombing, knowing that their assets would be among the targets. Tens of thousands of Turkish construction workers lived in Benghazi and in other major cities, and had to be evacuated at the beginning of the conflict. Later, Erdoğan's government sent humanitarian aid and assistance to the rebels in Benghazi, and held extensive political consultations with them in Ankara, Doha, and elsewhere. Ultimately, Turkey supported the new government.

Although Akin posted a CV to the consul general's website, it suggested that he was an intelligence officer under diplomatic cover, or just a very mediocre diplomat. After a twenty-year career that began at the Department of Defense (where many officers in Turkey's National Intelligence Organization, MIT, start their careers) and took him to Tehran, Frankfurt, New York, and Riyadh with the Foreign Ministry, he ended up as the deputy Turkish representative to the World Trade Organization, and then as a deputy consul in Benghazi.* Really? Although I contacted Akin when writing this book, he refused to answer questions by email or telephone.

Given the mess created by the *Al Entisar* shipment, it's hard to believe the two diplomats didn't discuss it. Erdoğan's government made clear their displeasure with the publicity the shipment had received and the lack of discretion of the Syrian rebels and their Libyan supporters. They were also unhappy with the way in which the weapons were distributed and felt they should have been in control, given the involvement of Erdoğan's favorite "NGO," the IHH Humanitarian

---

* The Turkish consul general in Benghazi at the peak of the civil war in the spring of 2011 was career diplomat Ali Davudoglu.

Relief Foundation, a group brazenly used as a cutout by the Turkish government.*

Many have speculated that Chris Stevens went to Benghazi to oversee a secret weapons pipeline to the Syrian rebels. That speculation was fueled by Secretary of State Hillary Clinton, in her only public testimony before Congress on the Benghazi attacks. Her exchange with Senator Rand Paul is worth quoting in its entirety:

> **SEN. PAUL:** Now, my question is, is the U.S. involved with any procuring of weapons, transfer of weapons, buying, selling, anyhow transferring weapons to Turkey out of Libya?
>
> **SEC. CLINTON:** To Turkey? I will have to take that question for the record. That's—I—nobody's ever raised that with me. I—
>
> **SEN. PAUL:** It's been in—it's been in news reports that ships have been leaving from Libya and that they may have weapons. And what I'd like to know is, the annex that was close by, were they involved with procuring, buying, selling, obtaining weapons, and were any of these weapons being transferred to other countries, any countries, Turkey included?
>
> **SEC. CLINTON:** Well, Senator, you'll have to direct that question to the agency that ran the annex. And I will—I will see what information is available and—
>
> **SEN. PAUL:** You're saying you don't know.
>
> **SEC. CLINTON:** I do not know. I don't have any information on that.[20]

* For example, after their deadly encounter with militants on board the *Marvi Marmara*, the Gaza flotilla ship, the Israelis learned that the Turkish government had trained IHH volunteers in military assault tactics so they could inflict maximum casualties on Israeli commandos.

Secretary Clinton of course knew exactly why she had sent Chris Stevens to Benghazi, and it was not to facilitate freelance gunrunning to the Syrian rebels, but just the opposite. My investigation has found that Stevens was sent to ferret out unsolicited weapons transfers by overzealous jihadis and do what he could to shut those down, so that professionals—most likely under John Brennan's control—could do the job. "The Turks were squawking about what a shitty job we did at controlling weapons distribution," a senior U.S. Special Operations officer told me. "Stevens' job was to get the MANPADS back, or at least out of the hands of al Qaeda–affiliated groups."

Larry Johnson, a former CIA officer who supported Hillary Clinton in her 2008 presidential bid, told me his sources believed Stevens was sent to Benghazi "to support the White House arms smuggling operation, run by John Brennan. Brennan was responding to Saudi and Turkish pressure to do something to halt the spread of Iranian influence, and was assisting them in acquiring weapons that could be used by the rebels."[21] Both of these sources agreed that the CIA was not running guns out of Benghazi, but was working with others to do so. "That way, they didn't have to report the activity to Congress," Johnson said.

But they were also trying to buy back the four hundred or so Stingers that had gone loose. "The MANPADS buyback was key to why Stevens was in Benghazi," a third intelligence source who was directly engaged in the MANPADS buyback operations told me. "There was a meeting planned that night on the buyback."

If Stevens didn't raise the subject of the *Al Entisar* shipment with Akin, the Turkish official probably did. Akin's role in the Benghazi attacks is considered so sensitive that, until now, the State Department and the CIA have denied Freedom of Information Act requests relating to his meeting with Stevens.[22]

Their meeting was critical to this story for another reason as well, which I will reveal in the next chapter.

CAT AND MOUSE

After his briefing at the CIA Annex, Stevens went for a scheduled meeting at the Fadeel Hotel downtown to confer with members of the Benghazi Local Council. Although touted as a four-star hotel, the Fadeel was long in the tooth, left dusty and a bit beaten up by the revolution. Twenty of the council's forty-one members showed up for the closed-door meeting, including President Jumaa al-Sahli. This was Stevens' opportunity for more ground truth.

The council members were "frustrated at the slow pace of reforms instituted by the Transitional National Council and its successor, the General National Congress," Stevens reported. They feared that the U.S. candidate for prime minister, Mahmoud Jibril, would focus national resources to develop Tripoli and the western half of the country, leaving Benghazi and the east to its own devices. They urged Stevens to get the U.S. government "to 'pressure' American companies to invest in Benghazi," not a good sign. It is quite likely that Stevens laid out his concerns about loose weapons and the need to collect them in a responsible manner, as they reorganized and integrated the many militias vying for control of the city and the surrounding region.[23]

Stevens had real anxiety about the security situation in Benghazi, especially as multiple militias, some of them commanded by known al Qaeda operatives, were claiming to be in charge of the city's security—and even his own. David McFarland had briefed him on a disturbing meeting just the day before with Wisam bin Hamed and Muhammad al-Gharabi, commanders of rival Libya Shield brigades. Just as the council members would the next evening, they pledged "they would not continue to guarantee security in Benghazi" if the United States continued to back Jibril for prime minister over the Muslim Brotherhood candidate, Awad al-Barasi.

Making the threat even more alarming was the fact that Wisam bin Hamed reportedly fought against U.S. troops in Iraq and Afghanistan before returning to Libya in 2011 to fight Qaddafi. His Libya Shield brigade regularly flew the black flag of al Qaeda at rallies and in battle. According to the Library of Congress report, he worked hand-in-glove with Ansar al-Sharia. As McFarland had reported, "blurry lines defined membership in Benghazi-based brigades," so it was often difficult to know who was who.[24]

One of Stevens' jobs during his stay in Benghazi was to smooth over relations with the 17th February Martyrs Brigade, which had supposedly quit as the mission's quick reaction force a few days earlier. Alec Henderson had been working day and night to hammer out a new agreement with them. But who were they really? Were they making a pact with jihadis? Or even al Qaeda?

The meeting at the Fadeel Hotel was the first avowed security breach of Stevens' trip. Someone on the local council had invited the press, who "showed up unexpectedly, despite U.S. efforts to keep the Ambassador's program and movements from being publicized," the State Department review board found.

It worried Alec Henderson, the RSO. And it worried Stevens.

The cat was out of the bag.

But Khalil Harb and his Quds Force team didn't learn that Stevens was in town from journalists who attended that dinner. Rather, they sent journalists to the dinner as a subtle warning to the Americans: We're watching you. We know where you are.

That was how the Iranians worked, like a cat playing with a mouse. They were toying with the American ambassador before they killed him.

●　　●　　●

The ARB report, which acknowledged the security breach, also propagated the lie that Stevens "met with the City Council at a local hotel *for dinner*" (italics added). According to Stevens' schedule and my sources, the City Council meeting at the Fadeel hotel took place in late afternoon. Stevens had other plans for dinner that the State Department was intent on keeping from the public eye.

In his diary, Stevens noted that he dined that evening after the Local Council meeting with a Libyan friend and the friend's partner. He described his friend, whose name I have voluntarily withheld, as a "hotelier and caterer extraordinaire" who came from Jalu, an oasis town in the desert south of Benghazi. Although this was clearly a private dinner, Stevens noted that the couple gave him "intelligent comments on the history of the revolution and the [Muslim Brotherhood's] influential role in supporting it from the beginning." Nevertheless, when Stevens showed his sympathy for the Muslim Brotherhood, they offered "heated words" that were "very anti-MB."

In a comment to himself, Stevens noted that Habib Bubaker, a local translator he had used for official meetings since first arriving in Benghazi during the revolution, also blamed TNC chairman Abdul Jalil "for handing the revolution over to the MB."

Clearly, Chris Stevens was struggling with his own pro–Muslim Brotherhood leanings, which he shared with the State Department hierarchy. However, unlike his bosses, he was inquisitive and challenged his prejudices.

14

THE ATTACKS

The next day was the eleventh anniversary of the September 11, 2001, terror attacks on America, and the Obama White House wanted everyone to know they were on top of it. They announced that White House counterterrorism advisor John Brennan was the man in charge. For more than a month, Brennan had been convening meetings and reviewing security measures to make sure this most auspicious of anniversaries came and went peacefully.

> During the briefing today, the President and the Principals discussed specific measures we are taking in the Homeland to prevent 9/11 related attacks as well as the steps taken to protect U.S. persons and facilities abroad, as well as force protection. The President reiterated that Departments and agencies must do everything possible to protect the American people, both at home and abroad.[1]

Chris Stevens, Greg Hicks, and Lieutenant Colonel George Bristol— Andy Wood's successor in command of the scaled-down Special

Forces team that remained in Tripoli—must have been pleased to know that John Brennan was taking charge of their security, since no one else was.

Apparently, Brennan's idea of security preparations was to lie low, hunker down, and hope for the best. Astonishingly, given the number of vulnerable U.S. diplomatic and military facilities in hot areas of the Middle East, not a single military asset was put on readiness alert for that day anywhere within an eight-hour flight of Libya. No one thought to contact AFRICOM commander General Carter Ham with orders to beef up security for U.S. facilities in the fifty-four countries in his area of responsibility; no one called Admiral James G. Stavrides, the Supreme Allied Commander–Europe (SACEUR), to put his Commander's In-Extremis Force (CIF)—then in Croatia—on standby, or to make sure that U.S. fighter jets and refueling aircraft were placed on strip alert. They didn't even give a heads-up to the interagency Foreign Emergency Support Team (FEST), a special unit created to respond to terrorist incidents at U.S. embassies worldwide, to be prepared in case they were needed.

In fact, as General Martin Dempsey, chairman of the Joint Chiefs of Staff, subsequently told the Senate Armed Services Committee, the September 10, 2012, meeting referred to in the White House press release was a "routine monthly review of counterterrorism operations, worldwide," not specifically devoted to the September 11 anniversary.[2]

It was another seemingly petty lie from the Obama White House, but one that carried deadly consequences.

THE CIA WARNING

And yet, the CIA had actionable intelligence identifying credible threats to U.S. diplomatic facilities in the region.

In Egypt, prominent jihadis had issued a call for supporters to burn down the U.S. Embassy in Cairo and kill everyone inside, to pressure the U.S. government to release the Blind Sheikh and Muslim detainees in Guantánamo Bay.[3] Supporters of the Blind Sheikh had been camped out in front of the embassy in Cairo for days before the 9/11 anniversary, claiming to be conducting a peaceful protest. They were led by Mohammad al-Zawahri, the brother of al Qaeda leader Ayman al-Zawahri. Mohammad had been in an Egyptian jail on terrorism-related charges until just a few weeks earlier, when he was pardoned by President Mohamed Morsi.[4]

You would think that the presence of the brother of the Supreme Leader of al Qaeda outside the U.S. Embassy would set off alarm bells back in Washington and perhaps prompt someone to ask CENTCOM commander General James Matthis, who had direct responsibility for Egypt, to put his CIF on alert. Didn't happen.

CNN reporter Nic Robertson took notice. He interviewed Zawahri in front of the U.S. Embassy on September 10 about his demand. "And when you call for the release of prisoners, you're talking about Sheikh Abdul-Rahman," he said.

"Of course" al-Zawahri replied, "he is the first one we ask about." The Blind Sheikh's son was standing next to him and nodded his head. CNN reportedly caved in to Obama administration pressure and stopped referring to Robertson's astonishing report later on, since it contradicted the official line that the attacks came in response to an Internet video.[5]

Egypt specialist Raymond Ibrahim, who scours the Arabic-language press and social media to reveal information that often gets shunted aside by the national media as politically incorrect, reported on September 10 that two prominent jihadi groups tied to al Qaeda were also calling on their supporters to storm the U.S. Embassy. They went a step further, threatening not just to burn down the embassy but to take hostage anyone left alive "unless the Blind Sheikh is immediately released."[6]

The CIA took these threats seriously and, on September 10, "notified Embassy Cairo of social media reports calling for a demonstration and encouraging jihadists to break into the Embassy," a draft of the post-Benghazi administration talking points reveals. That sentence was subsequently deleted, apparently by Hillary Clinton's deputy chief of staff, Jake Sullivan.[7]

In response to these direct threats and to the explicit CIA warning, U.S. Ambassador Anne Patterson, who boasted of her close ties to Morsi and her sympathy for the Muslim Brotherhood, ordered the evacuation of all U.S. Embassy personnel by noon on September 11 and issued a warning to Americans living in Egypt not to visit the embassy that day. She also ordered the U.S. Marine guards to remove the magazines from their duty rifles, so as not to appear too aggressive. That order would have near-fatal results.[8]

But even those dramatic actions did not prompt U.S. military leaders or their civilian bosses to put U.S. military forces in the region on alert, as Admiral Mullen, General Dempsey, and AFRICOM commander General Carter Ham have all testified.

According to the official version, there were no "indicators and warnings" of a specific attack. "We were blindsided," the former director of operations (J3) for AFRICOM, Rear Admiral Richard Landolt, told me.[9]

RECON

In Benghazi, the day began with a bad omen.

At 6:43 AM, not long after sunup, one of the unarmed Blue Mountain security guards at the Special Mission Compound noticed that a Libyan policeman had parked his car by the front gate and climbed to the roof of a building under construction just across the street, where he was taking photographs inside the mission's thirteen-acre walled compound. At the same time, he was talking to someone on a cell phone, apparently getting instructions. As he was getting back into his car, where two other individuals were waiting for him, the guard went over and asked what he had been doing. The policeman angrily told the guard to get lost and sped off.

When the guard told David Ubben and Scott Wickland, the two assistant regional security officers (ARSOs), they went nuts. According to the State Department review board, they asked the guard to take them across the street and show them exactly where the man had been standing. To both of them, it smacked of a recon mission. The policeman clearly had been in a position to take pictures of all four of the villas inside the compound, which otherwise were hidden from view by the perimeter wall.

That was the second warning from the Quds Force team. They already had all the pictures they needed of the Special Mission Compound and of the Annex. Indeed, they had even taken measurements with a laser rangefinder, as became apparent later.

The incident so disturbed Ubben and Wickland that the mission drafted a complaint to Mohammad Obeidi, the head of the Ministry of Foreign Affairs office in Benghazi, and to the local police chief. The complaint noted that the person photographing the compound "was part of the police unit sent to protect the mission," and had arrived in official police vehicle number 322. It also reminded Obeidi of the ongoing security concerns at the mission, and the failure of the Libyan authorities to provide the normal assistance accorded by host countries to accredited diplomats:

> On Sunday, September 9, the U.S. mission requested additional police support at our compound for the

duration of U.S. ambassador Chris Stevens' visit. We
requested daily, twenty-four hour police protection at
the front and rear of the U.S. mission as well as a roving
patrol. In addition, we requested the services of a police
explosives detection dog.

We were given assurances from the highest authorities in the Ministry of Foreign Affairs that all due support would be provided for Ambassador Stevens' visit to
Benghazi. However, we are saddened to report that we
have only received an occasional police presence at our
main gate. Many hours pass when we have no police
support at all.[10]

The State Department review board commented laconically that the
complaint was never actually submitted to the Libyan authorities "due
to the typically early closure of Libyan government offices." Nor was it
provided to Congress despite repeated requests that the State Department provide all relevant documents relating to the Benghazi attacks.
Congress only saw it after a reporter stumbled upon copies of both
letters in the ruined compound six weeks after the attacks. "Given
the location where they were found, these documents appear to be
genuine and support a growing body of evidence indicating that the
Obama administration has tried to withhold pertinent facts about the
9/11 anniversary attack from Congress and the American people,"
Representative Issa wrote Secretary Clinton.[11]

EYES ON

With the continued uncertainty about their agreement with the 17th
February Martyrs Brigade, RSO Alec Henderson asked Stevens to re-
strict all of his meetings to the compound that day. He didn't want to

risk moving around town, especially without a gun truck escort from the 17th February Martyrs Brigade.

Stevens' draft schedule, found by reporters in the wreckage of the compound, had him visiting the offices of the Arabian Gulf Oil Company (AGOCO) that morning. AGOCO was a Benghazi-based oil producer and exporter that controlled roughly one-quarter of Libya's oil. They had played a critical role in the rebellion by pledging to repay the government of Qatar and the private oil trading giant, Vitol, for fuel they provided to the rebels. AGOCO operated the largest oil export terminal in the country, in Tobruk, which could easily be shut down in the event of political protests. AGOCO undoubtedly wanted Stevens' assistance in taming the rivalries among warring militias, so that they could continue to export oil. That meeting got shifted to the following day.

David McFarland, the TDY political officer, returned to Tripoli on the early morning commercial flight. That left Stevens, Sean Smith, and the five DS agents as the only Americans at the compound. Because it was the September 11 anniversary, flags at the compound were flown at half-mast.

From Stevens' diary, we know that he was happy to be back in Benghazi, but also apprehensive. "It's so nice to be back in Benghazi. Much stronger emotional connection to this place—the people but also the smaller-town feel and the moist air & green & spacious compound," he wrote that morning.

He had breakfast with Habib Bubaker, his fixer, translator, and friend. Bubaker ran an English school in town, and first met Stevens during the heady days of the insurrection the year before. While Stevens spoke fluent Arabic, "he preferred English in his official meetings." Bubaker made introductions and became a confidant. Bubaker accompanied Stevens throughout most of that fateful day.[12]

He had set up meetings with appellate court judge Naeem Jibril, and with the TNC's local representative, Dr. Fatih Baja. Both were trusted personal sources he had cultivated during his earlier stay in Benghazi. He felt he could rely on them to give him ground truth on the political lay of the land, without the posturing and breast-beating he'd endured with the local council the night before. He also met as scheduled with Mahmoud El-Mufti of the Al Marfa Shipping and Maritime Services Company. These meetings were devoted to the intense jockeying for the prime minister's position, where different militias, clans, and corporate interests were all vying for power, influence, protection, and jobs. "[Mahmoud] Jibril reportedly fared well. Possibly also Mufti," Stevens wrote in his diary that morning. "Baja says Jibril should win. But they're worried about the MB [Muslim Brotherhood] and extremists denying him his rightful place or making life difficult if he wins."

Stevens also met at least briefly with some of the local security guards, including those from the 17th February Martyrs Brigade he had known from the old days, but didn't yet have the down-and-dirty negotiating session with brigade commanders Fawzi Bukatif and Fathi Obeidi. He was also planning to meet with the Egyptian, Swedish, and Italian consuls, whose contact information was conveniently included at the end of his five-page schedule.

His last meeting of the day was with the Turkish consul general.

He had every reason to be meeting with the Akin. Turkey played a major role in Libya both politically and economically. Plus, there was Turkey's direct involvement in the Syrian civil war. Turkey was both the staging area for the Syrian rebels, providing them safe haven and protection, and an active participant in the weapons pipeline. Akin would be a tremendous source of information for who was doing what for the Syrian rebels in Benghazi. If he didn't know the players, nobody did.

It should come as no surprise that the ambassador was collecting intelligence. After all, it was his job. But, unlike the CIA, he gathered information openly, like a reporter, then used it to form the political judgments that he sent back in official cables to his superiors in Washington.

An initial State Department "tick-tock" of events that night wasn't released until October 9. The briefer warned reporters he was giving them "as much granularity as I possibly can. This is still, however, under investigation. There are other facts to be known, but I think I'm going to be able to give you quite a lot, as far as I know it. I have talked to the—to almost all the agents that were involved, as well as other people." [13]

And yet, for some reason the briefer apparently got the hour of Akin's visit wrong. He had him arriving at the compound at seven-thirty, and departing at around eight-thirty with Stevens and a DS agent accompanying him to his car, which was parked in the street just beyond the front gate. The State Department Accountability Review Board corrected this, and had Akin arriving at six-thirty and leaving at around seven-thirty. Akin repeated the earlier times in an email exchange with columnist Diana West.

The State Department briefing gave rise to speculation that Akin left *after* Ansar al-Sharia had set up the first roadblocks, and accusations that he had failed to warn Stevens. I have found no credible evidence to support those claims. I believe the State Department initially was hoping to cover up the arrival of a British security team, who dropped off an armored vehicle and weaponry at around eight-thirty after supporting a VIP visit for the day. Ever since the assassination attempt on the ambassador in June when they pulled out of Benghazi, they had arranged with the RSOs to stow their gear at the American compound, because Benghazi was too dangerous for them to stay.

Changing the hour of Akin's departure to mask the arrival of the Brits may sound petty, even stupid, but it fits with the overall strategy of Hillary Clinton and the Obama administration in the run-up to the U.S. elections to pretend that the United States faced no security threat from al Qaeda because they had killed bin Laden. There were several other details they changed in the chronology.

Akin's visit was critical for a very different reason that I can reveal here for the first time: It gave Khalil Harb and his Quds Force team the "eyes on" confirmation they needed in order to pull the trigger on the attack. As is standard in any high-risk intelligence operation, they wanted direct, physical proof that their target was indeed still in place. My sources say that someone on Akin's staff—a driver, a bodyguard, or a personal assistant—had been co-opted by a Quds Force operative and made a cell phone call to tell them Akin had indeed met with Stevens.

"The Iranians used the Turkish consul for their own hidden purpose," the former Iranian intelligence chief told me. "They wouldn't have attacked without confirmation that the ambassador was inside."

In addition, of course, Harb had watchers surrounding the compound. But he hadn't expected Stevens' willingness to expose himself by escorting the Turkish diplomat all the way to his car, where they could see him with their own eyes.

It was a gift.

THE CAIRO DEMONSTRATION

Stevens used the extra time between Akin's departure and the arrival of a friend he was expecting later that evening to catch up on paperwork and the day's events. He retired to his work area inside the residence, which the State Department refers to as Villa C, and began emailing back and forth with embassy staff in Tripoli. He still had the matter of the new contract with the 17th February Martyrs Brigade to resolve, which was especially troubling, given the latest reports from David McFarland and Eric Henderson. That's one reason he had accompanied Akin out onto the street: He wanted to take the opportunity to say a few words to the guards outside, let them feel a little love.[14]

Greg Hicks had been sending him text messages during the day about the alarming turn of events in Cairo. Ambassador Anne Paterson had pulled the trigger on the REACT plan by midday and evacuated all embassy personnel. The embassy also put out a warning to Americans in Egypt not to visit the embassy that day.

By midafternoon, the Egyptian police had abandoned their posts. Out in the street, protesters were shredding and stomping on the American flag. Then a bit after five o'clock, the organized jihadis came onto the streets. They brought up a ladder and started to scale the perimeter walls around the embassy grounds. Soon, they were tearing down the American flags flying at half-mast inside the compound itself and replacing them with the black flag of al Qaeda. Video footage showed them chanting, "Obama, Obama! We are all Osama!"[15]

Because Cairo was the largest U.S. embassy in the Middle East after Baghdad, with hundreds of employees, it had both a Marines Corps detachment and a State Department Mobile Security Guard contingent. They responded as they had been trained, donning helmets and body armor and withdrawing to defensive positions to protect the sensitive areas of the embassy containing classified files and communications gear, protectees behind them, ready to shoot the next person who came through the door. It was a full-blown crisis, and as Stevens skimmed the cable and email traffic of the afternoon, it just kept on getting worse and worse. They would be having a sleepless night in the State Department Operations Center.

Despite the chaos in Cairo, which at one point seemed "just minutes away from a hostage situation," no senior official in the Obama administration thought to put military forces in the region on alert.[16]

There was no call from Secretary of State Hillary Clinton asking for help, no secure video-teleconference with regional commanders called by the chairman of the Joint Chiefs of Staff, General Martin Dempsey, no panicked phone call from the president's counterterrorism advisor, John Brennan, that their worst 9/11 nightmares appeared to be coming true.

Later, varying accounts of what happened in Cairo began to emerge and it became evident that even *that* attack was a planned jihadi operation, not a spontaneous protest over an Internet movie.

However, the important thing for Chris Stevens that evening was the knowledge that both Tripoli and Benghazi remained calm. As Greg Hicks told Congress, there was not a hint of a protest in either Libyan city, just the constant grinding knowledge that they were being watched and that the jihadis were on the rise.

The last words in Stevens' diary were ominous. "Never ending security threats . . ."

THE GAMER

Sean Smith was looking forward to returning to his wife, Heather, and their two children, in The Hague. After the bad days in Baghdad, where he had been posted in 2007–2008, he was sent to Montreal and was now on a three-year posting to boring Holland. And it was great. He had come to Benghazi on temporary duty (TDY) on September 3 for a four-week assignment as the communications specialist and building manager, but vowed it was his last hazardous duty assignment. It just wasn't fair to his family, he said.[17]

The thirty-four-year-old former Air Force sergeant was gregarious and easygoing. He was also a computer freak. His mother, Patricia Smith, later told their hometown paper, the *San Diego Union Tribune,* that it was hard to imagine him away from a keyboard. "Computers were a part of him. You couldn't have one without the other," she said.[18]

Sean Smith led a double life: diplomat by day, computer gamer by night. In the EVE Online community, he was known as Vile Rat, or VR for short. EVE was a gigantic, ever-evolving space warfare game world that Vile Rat had helped to shape. "Vile Rat was a spy for the Goonfleet Intelligence Agency," said his online "boss," Alex "The Mittani" Gianturco. "If you were an alliance leader of any consequence, you spoke to Vile Rat."[19]

The gamers knew that Vile Rat was in Benghazi. "He commented on how they use guns to celebrate weddings and how there was a constant susurrus of weaponry in the background," the Mittani wrote. Because Sean Smith's job was to provide IT services for the Special Mission Compound he was online all the time, "hanging out with us on Jabber as usual and talking about Internet spaceship games." He also moderated a blog for online gamers and computer nerds called Something Awful, to which many fellow State Department employees belonged.

After Stevens had retired to his room, Smith remained at his workstation out in the commons area of Villa C, where he had rigged up a CCTV video feed of the security cameras, in addition to his personal laptop, where he could play EVE.

Earlier that evening, he had sent out a message to Gianturco with typical gallows humor. "Assuming we don't die tonight. We saw one of our 'police' that guard the compound taking pictures."

He was online with the Mittani at 9:42 PM local time when sudden movement on the CCTV cameras caught his attention. "He was on Jabber when it happened, that's the most f—ked up thing," Gianturco wrote. "In Baghdad the same kind of thing happened—incoming sirens, he'd vanish, we'd freak out and he'd come back ok after a bit.

This time he said 'F—K' and 'GUNFIRE' and then disconnected and never returned."[20]

A State Department gaming colleague speculated on the blog site Smith moderated that given his dedication and cool-headedness, "it is likely that he spent his last moments completing the destruction of cryptographic technology and classified materials that he was responsible for."[21]

The assault happened so quickly and with such extreme violence there was no time to go through the complex REACT drill. Whatever classified equipment was there fell into the hands of the enemy.

But there was still time to die.

INSIDE THE TOC

Alec Henderson was monitoring the CCTV cameras inside the Tactical Operations Center, the TOC. This was a heavily fortified one-story cement building in a separate part of the compound, about seventy-five feet to the east of Villa C, where the ambassador and Sean Smith were headquartered. The DS half of the compound had its own entrance, known as Bravo-1, a bit farther down from the main gate. A driveway led directly to the DS barracks and the TOC behind it.

The DS half of the compound was actually a separate property, set off from the residence by a north-south alleyway enclosed in nine-foot masonry walls that spanned the entire 300-yard length of the property.* When Stevens first rented the property the year before, they had opened a lateral drive through the wall that gave access to the

---

* The owners of the compound had a separate gate at the far end giving direct access to the Fourth Ring Road. This gate (referred to as the "C3 gate" in the ARB report), just wide enough for a single vehicle, was secured by a full-height metal gate. Just inside the gate, two of the SMC's armored Toyota Land Cruisers were parked as obstacles.

pool at the back of Villa C. This is where Scott Wickland, David Ubben, and one of the ambassador's personal security detail from Tripoli now were lounging and talking, enjoying the Mediterranean summer evening.[22]

At approximately 9:02 PM, a tan Toyota Hilux gun truck with markings from the local SSC drove up and parked on the gravel road in front of the main gate and turned off its lights. This was a welcome addition, given Henderson's repeated demands to the SSC for round-the-clock police protection. All that could be seen of the two men inside the darkened vehicle was the glowing embers of their cigarettes.

None of the working security cameras showed any other activity out on the street or along the perimeter wall. The Blue Mountain security guards out front were smoking and chatting. All was calm—or so Henderson thought.

If the security cameras at the rear of the compound had been working, Henderson would have noticed gun trucks pulling up along the Fourth Ring Road, also known as Venice Road, at around 9:30 PM. About a dozen armed men jumped out and began milling around outside the rear gate of the compound, according to diners at the nearby Venezia Café, who were interviewed by a reporter from the *Guardian*.

"One of the militia jeeps bore the black banner of a local Islamist militia, Ansar al-Sharia. The militiamen made no attempt to hide. . . . Neighbors saw militia 4X4s blocking streets leading to the compound. All were surprised there was no reaction from the compound," the *Guardian* reported.[23]

At 9:40 PM, Henderson noticed the lights on the police car go back on, then it drove off. Then all was dark once again in the CCTV monitors.

Two minutes later, all havoc broke loose.

A swarm of heavily armed assailants stormed the guard booth. Two of the Blue Mountain guards ran off. The other two were trapped by the lead assailant, who demanded that they open the main gate and the steel drop bar. When they hesitated, he forced the first one to his knees and shot him in the leg. The other one opened the gates and the attackers let the two men go. "We're here to kill Americans, not Libyans," they reportedly said. The two men, both wounded, were later taken to the hospital.

As soon as the gates opened, armed men emerged from the darkness and rushed into the compound in a rapid, disciplined, military assault. Some of them were wore long beards and Afghan-style *shalwar kameez*, the long baggy tunic with loose-fitting pants beneath.

Henderson heard the shots and turned back to the monitors, where he now saw upwards of fifty armed men starting to rush through the main gate into the compound. He hit the duck-and-cover alarm and, for good measure, activated the public address system. "Attack! Attack!" he shouted. Then he picked up the separate VHF radio he and the other DS agents used to communicate directly with the Annex and told them they were under attack.

The main gates were on the north side of the compound. One group of attackers jumped the low hedge and ran off to the left through the olive orchard to the Bravo-1 gate along the front wall. After a bit, they were able to open it and four vehicles screeched in, two Nissan Pajeros without license plates and two gun trucks. Another dozen or so fighters jumped out of the vehicles and headed directly to the Villa C to find the ambassador. The gun trucks "took up strategic firing positions on the east and west portions of the road to fend off any unwelcome interference. Each vehicle flew the black flag of jihad."[24] Pajeros were the favorite vehicle of Hezbollah, the main proxy of the Iranian Revolutionary Guards Quds Force. Hezbollah military commander Imad Mugniyeh was assassinated in February 2008 when the headrest of his Pajero blew up just after he got into the vehicle following a meeting at the Syrian intelligence headquarters in downtown Damascus.

Another group moved to the bungalow on the other side of the main gate that served as a dormitory and rec room for the 17th February Martyrs Brigade guards. They found the diesel jerry cans exactly where they expected them, stockpiled by the new generator that was due to be installed in the near future. They poured diesel fuel over the cars parked nearby and set them alight. Two of the three militiamen on post barricaded themselves in a closet and later told the FBI they fired their weapons at the intruders. The third ran for his life toward the DS barracks, where he knew more weapons were stockpiled.[25]

The two Blue Mountain guards who had run away earlier managed to sound the alarm right at the start using their VHF radios. That gave the DS agents a few precious moments of advance warning—just enough to get out of the way. In accordance with the REACT plan, Scott Wickland, who was at the pool with the others, rushed into the main villa to take charge of the ambassador and Sean Smith, and relieved the agent who had accompanied Stevens from Tripoli, who was lounging about barefoot in his underwear and a white T-shirt. Without his quick thinking, Ambassador Stevens and Sean Smith most likely would have been slaughtered immediately.

Wickland had his protectees don flak jackets, stored in their bedrooms, and rushed them into the safe haven, leaving all other belongings behind them. As soon as he locked the security grille behind them, he radioed the TOC to tell Henderson that he had the ambassador. The security grille blocked access to roughly half of the ground floor of the large, luxurious residential villa, including multiple bedrooms, a bathroom, and a large walk-in closet with no windows, which was the designated safe haven. Wickland found his own kit, which consisted of body armor, an M-4 carbine, spare magazines, a Remington 870 12-gauge shotgun, a ballistic helmet, smoke grenades, a compass, whistle, and emergency medical gear. He would need it all.

The DS agent gave his cell phone to the ambassador, who began making calls to local consulates and to the embassy in Tripoli, as well as to the 17th February Martyrs Brigade. As soon as the ambassador and Sean Smith were secured inside the closet, Wickland "took up a defensive position inside the Villa C safe area, with line of sight to the safe area gate and out of view of potential intruders," the ARB report states.

Outside was pandemonium, with gunshots and loud explosions from RPGs as the assailants opened a breach in the boundary wall between the two compounds just inside the Bravo-1 gate. David Ubben and the second DS agent who came with Stevens from Tripoli pulled out their sidearms and sprinted down the short driveway connecting Villa C to the DS side of the compound. Initially they went to the Tactical Operations Center (TOC), where Ubben's battle kit was stored and where the other agent thought he had last seen the ambassador.

The sheer physical challenge of moving from place to place within the large compound, with gunfire rapidly getting closer, is forgotten sometimes in the telling of the events of this night. Former DS agent Fred Burton described the movements of the four ARSOs, relaying one another and sprinting to the other side of the compound to retrieve their kit, as "an elaborate spider web." I think of it more as a desperate race for home plate. Everyone was running for their lives.

ARSO 3 and ARSO 4—the two men from Tripoli—moved to their own barracks in front of the TOC, which the State Department has called Villa B. Since they were visiting, that's where their own weapons and battle kit were stowed.

In the meantime, the attackers had driven up from Bravo-1 gate in the gun trucks and taken up blocking positions, while foot soldiers were now moving toward the residential villa where the ambassador, Sean Smith, and Scott Wickland were holed up. They moved silently, carefully, using military hand signals as they checked each corner to make sure it was clear. While surveillance footage viewed by members of Congress showed that many of the attackers were "just milling about, trying to figure out why nobody was shooting at them," this group knew exactly what they were doing.[26]

The ARB claims the two visiting ARSOs tried to reach the ambassador's villa via the lateral driveway but saw they were outnumbered and outgunned so they turned back and barricaded themselves into the DS villa. Later, the five GRS contractors who survived the rescue attempt said they had found the DS agents from Tripoli hunkered down in a closet, still barefoot and unarmed.

Asked if he felt the DS agents had done their jobs, Representative Lynn Westmoreland said, "Absolutely not. I think this will come back to the State Department," he said after a closed-door Intelligence Committee hearing with the CIA contractors. "They were not armed. One of them was barefooted. I believe they were totally unprepared for any kind of attack."[27]

In defense of those two agents, who essentially took themselves out of the fight that night, they would have had to cross the north-south alleyway that cut the large compound into two, with the DS villa on one side and the ambassador's residence on the other. With nine-foot walls on both sides, that alleyway was a free-fire zone that now belonged to Khalil Harb and his attackers. And they were firing at will.

So, they radioed Henderson in the TOC to tell him they were safe, but immobilized. That left Ubben and Henderson in the TOC, and Wickland with the ambassador and Sean Smith, versus somewhere between sixty to a hundred heavily armed attackers. As the ARB noted, there was "no line of sight" between the two halves of the compound, so all Henderson and Ubben knew was what they could see on the security cameras.

The first sign that things were getting really bad were the flames.

THE INITIAL RESPONSE

Greg Hicks was relaxing in his villa in Tripoli after what he described as a "routine day" when John Martinec ran in at around 9:45 PM. "Greg, Greg, the consulate's under attack!" he yelled.

Hicks picked up his cell phone, which was on mute, and saw that he had two missed calls: one from Ambassador Stevens and the second from an unknown number. When the ambassador's number didn't answer he punched the unknown number and Stevens picked up. "Greg, we're under attack," he blurted out. Then the line went dead.*

The Embassy Compound in Tripoli also had residential villas, barracks, and a Tactical Operations Center within its walls. As Hicks went over to the TOC, he kept trying to call Stevens back on both numbers, without success. Martinec was on the phone with Alec Henderson in Benghazi, who told him that the Special Mission Compound had been breached by "at least 20 hostile individuals" with weapons. Hicks next called the CIA chief of base at the Benghazi Annex, to see if he had already dispatched reinforcements to evacuate the mission, as required in the REACT plan. The CIA chief said he was "mobilizing" reinforcements, but that they hadn't yet moved out.

It was now about 10 PM in Benghazi, and 4 PM in Washington, D.C. Hicks then phoned the State Department Operations Center with the information. Five minutes later, the Ops Center sent out a brief notice to the secretary of state, the White House, the secretary of defense, the national security advisor, the director of national intelligence, the FBI director, and their operations staff,

---

*As Hicks testified on May 8, 2013, he hadn't recognized the number because Stevens was using DS agent Scott Wickland's cell phone. Hicks' statement in the following account is taken from his congressional testimony.

with the subject line: "U.S. Diplomatic Mission in Benghazi Under Attack."

> (SBU) The Regional Security Officer reports the diplomatic mission is under attack. Embassy Tripoli reports approximately 20 armed people fired shots; explosions have been heard as well. Ambassador Stevens, who is currently in Benghazi, and four COM personnel are in the compound safe haven. The 17th of February militia is providing security support.
>
> The Operations Center will provide updates as available.

Hicks was moving quickly, and hadn't had time yet to discuss all the details with Henderson in Benghazi. The important thing was, Washington was now in the loop, and Henderson was patched into the State Department Operations Center, where Deputy Assistant Secretary Charlene Lamb was able to follow events in Benghazi through a live audio feed.

Meanwhile, Hicks started to work the phones, just as Stevens was trying to do from the closet in the safe haven. He began by calling President Magariaf's chief of staff to report the attack and request immediate assistance. Then he called the prime minister's chief of staff and the Ministry of Foreign Affairs to make the same request. Hicks also told the CIA chief of base in Benghazi that he was organizing a response team from Tripoli via a chartered plane for additional backup, while they kept pressing the Libyan government to get the Benghazi SSC or the militias to respond.

Glen Doherty, the former Navy SEAL who had spoken to ABC News in Southern California just two weeks earlier, was the first to volunteer.

THE FATWA

As the cars in front of the 17th February bungalow were burning, a crowd of attackers was attempting to storm Villa C, pounding on the massive, reinforced front door, trying to break in. Some ran around to the side of the villa, smashing windows with their rifle butts, but were unable to get through the security grilles. Finally, at around 10 PM local time, they backed off, and someone blew the front door open using an RPG. The explosion rocked the walls where the three Americans were huddled. They could hear bits of wood and metal slam into the walls, and then in the brief silence, a pattering like rain as small chunks of masonry fell back onto the marble floor.

And then, utter chaos erupted. The intruders swarmed inside, shouting and hooting *Allah-o Akbar!* Some of them began ransacking the elegant reception and dining area, slicing the overstuffed furniture, smashing vases, mirrors, picture frames, and everything else in sight. In their fury, they even ripped the kitchen cabinets from the walls. Former DS agent Fred Burton said their rampage had "all the charm of an urban America blackout and the bloodlust of tribal genocide."

But Khalil Harb was not motivated by anger or revenge. He had a plan, and that was to get the ambassador and kill him.

My Iranian sources revealed that Ayatollah Mesbah Yazdeh, a senior cleric in Tehran who had been the mentor of President Mahmoud Ahmadinejad, issued a written Sharia law decree, or fatwa, authorizing the murder of Ambassador Stevens a few days before the attack.

Some jihadis believed they didn't need a religious edict in order to kill American civilians. American-born imam Anwar al-Awlaqi famously told his al Qaeda followers in 2010, shortly before he himself was killed by a U.S. drone strike, that they didn't need to consult with anyone to "fight the devil." For Muslims, he said, "it is a question of 'us or them.' "[28]

But the Iranians still did. The fatwa gave Ibrahim Joudaki, Khalil Harb, and the Quds Force team the certainty that if they died in this battle, they would soon be transported to paradise where they could drink from rivers of wine and enjoy the ministrations of seventy-two virgin girls.

Khalil Harb's intelligence on the compound was detailed and accurate. While the Ansar al-Sharia pickup boys ransacked the main room of the residence, his team went to verify the location of the ambassador. From the REACT plan, they knew exactly where he was supposed to be.

At a few minutes after 10 PM local time, his men reached the steel grille blocking off the residential quarters and the safe haven from the reception areas of the ground floor. Scott Wickland saw them coming and moved back around a corner, so that they couldn't see him. They banged their rifle butts against the steel bars. They grabbed furniture and anything they could find in a futile effort to break the locks. Wickland slipped into the closet and told Ambassador Stevens and Sean Smith to "prepare for the intruders to try to blast the safe area gate locks open." He most likely gave his shotgun to Sean Smith, who had been trained to shoot in the military. Perhaps he offered his SIG Sauer P229 semiautomatic pistol to the ambassador. If they were going down, they might as well take some of the bad guys with them.

And, then, a miracle happened. Someone gave an order, and the attackers went away.

TY WOODS SADDLES UP

Ty Woods, a twenty-year veteran of the Navy SEALs, was chomping at the bit. As soon as the alert that the diplomatic compound was under attack came over the dedicated VHF frequency the men at the Annex used to communicate directly with the TOC at the compound, he began throwing his kit—always at the ready—into one of their armored SUVs.

The Annex had an impressive armory, and Woods, Mark G—, and the five other former SpecOps guys who were working as babysitters for the CIA in Benghazi plunged right in, pulling out their HK 416 assault rifles, a M249 Minimi light machine gun, sniper rifles, web kits, body armor, and more.

Within minutes, the seven Global Response Staff had suited up and began running back and forth to stow their gear in two of the armored Mercedes G-wagons in the courtyard. They commandeered a translator, just in case they encountered trouble along the way.

Ty's former wife, Patricia Ann So, was not surprised that he was first to volunteer. "He was balls to the wall," she said. "He loved life, loved adrenaline." Rescuing hostages and responding to emergencies with highly focused violence what he had done for most of his adult life.[29]

As they were loading up, the CIA chief of base intercepted him, cell phone glued to his ear, and told him to hold off. Woods couldn't believe it. They could hear the explosions at the compound less than a mile away.

The chief was a former U.S. Army officer who had been recruited into the agency after 9/11 as the CIA became increasingly focused on paramilitary operations, so he wasn't some Ivy League desk jockey who had never fired a weapon in anger.[30] He said he wanted backup. He wanted the 17th February boys to rally the troops. The situation at the compound was way out of control. He wanted gun trucks with .50 calibers. And Libyans out in front. And he wanted one of the GRS shooters to stay behind just in case. Mark G— got the short straw and remained at the Annex, where he was critically wounded later.

Woods and his fellow operators continued loading up. When they were finished, he went back to the chief to get the okay to depart and was told a second time to wait.

Ty's father, retired attorney Charles Woods, was instrumental in setting up a Citizens Commission to investigate the Benghazi attacks and the U.S. government's failure to protect or rescue the Americans who were posted there. "Three times Ty and several of his fellow defenders asked for permission to run to the gun," he told me at the first public session of the commission, "but they were told to stand down."

Ty Woods didn't ask a fourth time. He just said, screw it, and took off. By the time they rolled out of the Annex gate, it was 10:07 PM.

The CIA later issued a formal statement denying that anyone at the agency had ever given a "stand down" order. The chief of base was merely trying to assemble a more potent response team, especially the gun trucks. He was also worried that once Ty and the other GRS operators departed, there would be no one left to defend the Annex except the four case officers and ten analysts left behind.[31]

But if Ty Woods and his teammates had been allowed to depart when they were initially ready to roll, the odds are good that they could have saved the ambassador and Sean Smith. He knew it then. We know it now.

FIRE!

The brief calm that ensued once the intruders gave up their attempt to break through the steel grille protecting the safe haven soon gave way to a sense of dread and foreboding. The lights went off, then Scott Wickland, Ambassador Stevens, and Sean Smith heard the ominous *whoosh!* as someone set a match or a lighter to the diesel-fuel-soaked furniture in the residence.

It was around 10:05 PM.

Within minutes, thick noxious fumes began pouring through the bars of the security grille, and Wickland said they had to move. He led the ambassador and Sean Smith on all fours in the darkness into the bathroom, since it had an exterior window. Once they got inside, they tried to seal off the door with wet towels to keep out the thick smoke. As it became more difficult to breathe, Wickland mistakenly decided to open the window, thinking he could ventilate the noxious fumes. Instead, the open window created a vacuum effect that drew more smoke inside.

Wickland "determined they could no longer stay in the safe area and yelled to the others, whom he could no longer see, to follow him to an adjacent bedroom, where there was an egress window," the ARB report states.

The smoke by this point was so thick that Wickland resorted to "banging on the floor" to guide the ambassador and Sean Smith to his position. He eventually reached the specially designed emergency exit window, which was protected by an exterior grille to forestall intruders but had a latch on the inside that released both the window glass and the steel grille. When he forced himself through the open window, he collapsed on a small patio that was shielded from the attackers by a wall of sandbags, thankfully erected during an earlier visit by one of the MSD teams. The intruders were firing their AK-47s in wild bursts in all directions. Bullets were impacting all around him, and cement chips were flying. But he couldn't tell for sure if they were actually aiming at him.

Then he realized that neither the ambassador nor Sean Smith had followed him out the window.

Wickland could hardly breathe, let alone talk. His face was blackened, and he was gagging from the smoke. As Greg Hicks later testified, petroleum-based fires give off large amounts of hydrogen cyanide gas, especially in the initial stages, so he was lucky to be alive. But he went back inside several times trying to find his protectees, until the heat and the smoke forced him to exit again.[32]

Finally, after several attempts, he realized he was too weak to go back in. The villa had become a death trap, so he climbed a ladder positioned near the emergency exit window and climbed to the roof, where he radioed to Henderson in the TOC for assistance. With all the smoke and noxious fumes he had inhaled, his voice was unrecognizable. It took several attempts to croak out his status before Henderson and Ubben understood what he was saying.

The ambassador and Sean Smith were missing.

THE POSSE ARRIVES

David Ubben decided to make a move. Grabbing smoke grenades from the armory in the TOC, he cracked open the metal reinforced door, tossed a grenade outside to obscure his position, and then made a run for the DS barracks just across the way. He had radioed ahead of time, so the two agents who had accompanied Stevens from Tripoli were waiting to let him in.

The ARB report claims that the three of them jumped into one of the armored Toyota Land Cruisers parked outside, crossed the north-south alleyway, and drove the hundred feet or so of the lateral driveway to the residential villa, where they joined Wickland. But when the surviving GRS agents were finally allowed to talk to members of the House Intelligence Committee in November 2013—more than a year after the attacks—they insisted they had found the two agents from Tripoli barricaded in their barracks, barefoot and unarmed.

Ubben was shocked to find Scott Wickland in such a state on the roof. He was vomiting and gasping for breath. Nevertheless, he pointed to the window down below that led to the safe haven, so Ubben climbed down the ladder and himself entered the burning villa five, six, seven times, or more in search of the ambassador and Sean Smith.

By this point, the first wave of attackers began to leave the compound, having accomplished their goal. There was no way that the ambassador could have survived the fire.

Ty Woods and his men were driving lights out, at breakneck speed along the Fourth Ring Road along the south side of the compound, and turned right onto the main road heading north. As they tried to turn right again, hoping to reach the front gates, they were stopped by militiamen from the 17th February Martyrs Brigade—the same unit that was supposed to be protecting the compound and providing a quick reaction force in case of attack.

Instead of protecting the Americans inside the compound, they blocked the Americans who had come to rescue them. The ARB report tried to paper over the interference, saying that Woods and the GRS team had voluntarily stopped in an attempt "to convince militia members there to assist."

Deputy Assistant Secretary of State Charlene Lamb introduced the fiction in her sworn testimony before the House Oversight and Governmental Affairs Committee on October 10, 2012, just one day after her colleagues had briefed the press (but not Congress) on background. She claimed the Annex fighters "arrived [at the diplomatic compound] with approximately 40 members of the Libyan 17th February Brigade." In other words, the State Department security procedures that relied on the local militia were working.

That simply was not true. After extended palavers, reportedly including glasses of tea, all Ty Woods and his men could scare up were three Libyan militiamen who were willing to join them—as volunteers. On the excuse that Ambassador Stevens had not yet finalized their new contract, the militiamen weren't budging.[33]

The delay had stretched out the five-minute drive from the Annex to the diplomatic compound to nearly twenty minutes. When they finally roared onto the gravel road along the front side of the compound at 10:25 PM, they nearly drove into a crowd of men milling around outside. As soon as the militiamen realized who they were, they opened fire. According to the timeline provided by the CIA, it took nearly fifteen minutes for three of the GRS men to fight their way into the compound from the front. At the same time, three others entered on foot from the gates at the rear of the compound to reconnoiter the situation.

By the time Ty Woods and his team reached Villa C and found Wickland and Ubben on the roof, it was 10:40 PM. The fire had died down, but smoke was continuing to pour out of the smoldering building. In addition to his twenty years as a sniper with the Navy SEALs, Woods was trained as an 18-Delta combat medic, and began immediate emergency treatment on the two DS agents for smoke inhalation.

The GRS operators decided to split up into three teams, two men each. Ty's teammate made repeated trips back into the safe haven, and eventually found Sean Smith and brought him out. Most accounts, including the ARB report, say he was already dead. However, an early rendering by former Special Forces operators Jack Murphy and Brandon Webb says that Smith "was unconscious and would later be declared dead."

A second two-man team set up a sniper's position on the roof, protected by sandbags; one shooter, one spotter. Two others went to evacuate Alec Henderson from the TOC and the other two DS agents from the barracks. According to former DS agent Fred Burton, this gave Henderson enough time "to secure weapons and equipment, including the firearms of the British specialists, and make sure that the laptops were destroyed and any sensitive material on them would be impossible to retrieve." Much of the destruction was done "old-school," using a hammer to smash the hard drives.

There was no question anymore but that they were evacuating Benghazi. Leaving for good, shutting down.

At around 10:45 PM, fresh crowds of jihadis began gathering at the rear gate, Charlie-3, along the Fourth Ring Road. They now started to advance on Villa C. From the roof, the spotter pointed them out and the shooter picked them off with two-tap shots to the head. After a number of the attackers went down, they fell back to regroup. The brief respite was enough for the two men who had gone to retrieve the DS agents to rejoin the GRS team and the three 17th February militiamen at the main villa. They all took turns crawling back into the smoldering villa to find the ambassador.

Henderson tried to go in through the front door of the villa, but no sooner had he reached the main room than the ceiling collapsed. Above, the roof was getting hotter beneath the GRS snipers. It was only a matter of time before it burned through or melted. They could see crowds of jihadis moving through the vineyard and the guava trees near the rear of the compound, and more of them milling near the front gates. They knew they had to evacuate. But they still hadn't found the ambassador.

By around 11 PM, the jihadis had regrouped and launched a second wave of attacks. This time, all the Americans were in one location—Villa C—and made for a formidable force: six GRS shooters, five DS agents, plus the three Libyan militiamen. They had better training and more discipline than the attackers, who had been joined by wannabes and adventure seekers from around town. The attackers approached through the orchard to the south, where the GRS snipers picked them off one by one at a distance of around four hundred feet. Others tried to reach the villa from the main gate to the north, fewer than two hundred feet away. They were well within range of the DS agents and their M-4 carbines and the GRS shooters with their HK 416s and the

M-249 Minimi light machine gun. The Americans had the advantage of shooting from behind sandbags. Those sandbags saved their lives.

THE WHITE HOUSE MEETING

Rear Admiral Richard Landolt, the AFRICOM J-3 (director of operations), was the man on duty station that night at Kelley Barracks in the Stuttgart, Germany, headquarters of U.S. Africa Command. "I took the call at 9:42 PM and ran down to my man cave with all the classified equipment," he told me. "After I spoke to our Ops Center, I immediately Tandberged with General Ham in the Pentagon." The Tandberg was a secure video-teleconference (VTC) system widely used by the military and by the national command structure. "I spoke to him again about three times over the next thirty minutes."

Stuttgart was on the same time as Benghazi in mid-September, and both were six hours ahead of Washington, D.C.

RADM Landolt's next call was to his immediate boss, AFRICOM deputy commander for military operations Vice Admiral Charles J. "Joe" Leidig. "He decided we needed to set up a command center right now, so we did."

The problem was that AFRICOM had no assigned forces. "We were due to get our own [Commander's In-Extremis Force] on October 1, but without any enablers," Landolt said. That was the same CIF that was in Croatia on September 11, 2012, on an exercise. Before requesting that the CIF deploy, they needed more information. So, at 9:59 PM, as soon as he got off the Tandberg with General Ham, Landolt gave orders to the Predator drone then loitering over suspected terrorist camps in Derna, redirecting it to Benghazi.[34]

At the Pentagon, visiting AFRICOM commander General Carter Ham sprang into action. "My first call was to General Dempsey—

General Dempsey's office to say, hey, I am headed down the hall. I need to see him right away."

General Ham briefed the chairman of the Joint Chiefs of Staff—the highest military leader in the nation, and the president's top military advisor—on what he knew. "We immediately walked upstairs to meet with Secretary Panetta," Ham recalled.[35]

In that initial briefing with Panetta, Ham said, there was no discussion of a demonstration. "It became pretty apparent to me, and I think to most at Africa Command pretty shortly after this attack began, that this was an attack," Ham recalled. As reports began to stream in through intelligence channels and the National Military Command Center, "we started very quickly to think about, you know, the possibility of a U.S. Ambassador being held hostage in a foreign land and what does that mean," Ham said.[36]

It was now approximately 4:40 PM in Washington, D.C. Equipped with that knowledge, Secretary of Defense Leon Panetta and General Dempsey rushed downstairs to a waiting convoy of armored SUVs, waiting to take them to a previously scheduled five o'clock meeting with President Obama.

While they were briefing the president in the Oval Office on what they knew about the attacks, they received word that the unarmed Predator drone that RADM Landolt had diverted from its original mission over Derna had arrived on station over the Special Mission Compound in Benghazi. It was equipped with a complete surveillance package, including a night vision optical suite.

It was now 5:10 PM in Washington, 11:10 PM in Benghazi. From that moment on, Washington had eyes on the compound. The video uplink went to the Pentagon and to numerous operational headquarters in the United States and around the globe, and was immediately accessible to the White House Situation Room.

Nobody at senior echelons of the U.S. government could pretend from that moment on that they were unaware of the extreme violence of the events taking place on the ground.

Although the drone launched from Sigonella Naval Air Station in Sicily, fewer than five hundred miles (and a one-hour flight) from Benghazi, it was being piloted by remote operators at a base in the United States. A man claiming to be the camera operator on the drone called into the *Sean Hannity Show* in May 2013 and gave listeners a glimpse of what the decision makers were seeing. "There were dozens if not hundreds of people surrounding the [compound]," he said. He said the buildings were "already on fire" when the drone came on line. Although the fire partially blinded the cameras, which were in infrared mode, he could see people and cars constantly stopping, apparently bringing up reinforcements. And he could see muzzle flashes just about everywhere. "It was already under attack," he said.[37]

Did Panetta suggest to the president that they watch the live video feed, as Obama claimed he did during the takedown of Osama bin Laden?

According to Panetta's own testimony before the Senate Armed Services Committee on February 7, 2013, he and Dempsey discussed Benghazi with the president for, at most, twenty minutes, and left the White House ten minutes after that.

"We had just picked up the information that something was happening, that there was an apparent attack going on in Benghazi. And I informed the president of that fact, and he at that point directed both myself and General Dempsey to do everything we needed to do to try to protect lives there," Panetta said. "[Obama] basically said, 'Do whatever you need to do to be able to protect our people there.' "

He made no mention of the drone footage.

When Panetta returned to the Pentagon, the Pentagon timeline shows that he convened "a series of meetings" over the next two hours and gave "verbal authorizations" to regional commanders to start preparing to move assets closer to Libya. They also discussed "the potential outbreak of further violence throughout the region, particularly in Tunis, Tripoli, Cairo, and Sana'a"—something you would think they should have been anticipating well before the 9/11 anniversary. In other words, they were more worried about *other* events than responding to the immediate crisis in Benghazi.

As Dempsey later testified, the military response that night was a big lumbering machine, wrapped in layers of bureaucracy. Each emergency response unit required N[otification]+6 hours or N+8 hours before it could move. And even then, most of the units put on alert were told to move to temporary staging areas in other countries, rather than move directly into Libya.

Ever since the attacks, Americans have been asking, why couldn't they scramble F-18 fighter jets from Naval Air Station Sigonella in Sicily? After all, that's where French Rafale fighter jets had based during the NATO air war against Qaddafi. If the French could fly the 430 nautical miles to Libya, hit their targets, and return without incident, couldn't American pilots do the same?

Or why not mobilize F-16s from the 555th Fighter Squadron of the 31st Operation Group at Aviano Air Base in northern Italy and hot-refuel them at Sigonella in the south? They could have made supersonic passes over Benghazi that would have scared the lice off a rat. But, as Dempsey and, later, Admiral Mike Mullen testified, no jets were on strip alert, nor was there single aerial tanker anywhere within a thousand miles of Libya to refuel the jets. And military commanders believed that a supersonic flyover not followed up with actual bombs hitting targets would actually embolden the terrorists.[38]

And why didn't the European Command (EUCOM) Commander's In-Extremis Force (CIF), a specially trained unit of around fifty Special Operations troops, then on counterterrorism training exercises in nearby Croatia, get the go-ahead to gear up and jump into

C-130s, and head to Benghazi, just a two-to-three-hour flight away? The fifty heavily armed JSOC operators, known as C-110, would have made mincemeat of the attackers and were itching to go.

The Pentagon timeline shows that Panetta authorized all of these units—and more—to begin preparations to deploy. He also put a Special Forces response team at Fort Bragg, North Carolina, on alert, with orders to stage in Sigonella the next day.

But there was no sense of a command authority telling the men in charge of the operators on the ground to do it now, make it happen, without fail.

Someone was missing from the fight. It was the commander-in-chief.

Panetta and Dempsey both testified that they never spoke to President Obama again until late the next day. He never once called them, never once followed up, never once sent an underling to see if they needed anything more. President Obama never once inquired as to the welfare of the Americans caught in the jihadis' gunsights. The White House has released no record of him ever visiting the Situation Room and refused to answer my direct question as to whether he did so.

Obama's schedule for that day shows no more meetings after the hasty five o'clock one with Panetta and Dempsey. According to the White House logs, he received no visitors. Nor did he go out. At around 6:30 PM, he spoke with Israeli Prime Minister Benjamin Netanyahu, apparently on unrelated issues, and then retired to prepare for his megamillion fund-raiser in Las Vegas the next afternoon, in all likelihood dialing for dollars. That is why the White House Press Office has deflected all questions about the president's actions during the crisis. Whether it was a sense of fatalism, or a cynical disregard for the lives of the men and women serving their country in hot zones around the world, the president of the United States couldn't be bothered to stay on watch that night.

Secretary of State Hillary Clinton never spoke to Panetta or Dempsey, either, although she later told the Senate Foreign Relations Committee in her only congressional appearance to answer questions about Benghazi that she had been on the phone with CIA Director David Petraeus and his deputy, Mike Morell, several times.

AFRICOM Commander General Carter Ham provided different stories at different times to members of Congress, according to Representative Jason Chaffetz. In an early meeting with Chaffetz three and a half weeks after the attacks, General Ham said the main reason he didn't go "balls to the wall" and get forces to Benghazi was simple: "We were never asked."[39]

Pathetic. But true.

HILLARY'S STAND-DOWN ORDERS

Secretary of State Hillary Clinton has never released a timeline of her own actions that night. But we do know that she played a key role in actually *delaying* the arrival of rescue troops. Clinton insisted that a fifty-man team of U.S. Marines who were part of a FAST (Fleet Antiterrorism Security Team) based at Rota Naval Station in Spain "deplane, change out of their U.S. military uniforms, and put on civilian clothes before flying to Tripoli—a decision that delayed them from launching for approximately 90 minutes."[40]

She didn't want to upset the sensitivities of the Libyans by making it look like the U.S. military was invading their country. There would be no Marines landing "on the shores of Tripoli" on her watch. Just as she had insisted earlier with Lieutenant Colonel Wood's Site Security Team of Special Operators, no military uniforms, insignia, or even boots would be tolerated in her State Department facilities in Libya. "People high up at State resented like hell us being there and doing what we did," Wood said.

House Democrats went to great lengths to get senior military leaders to assert that the ninety-minute delay had no material impact on the fighting in Benghazi, since the FAST arrived anyway in Tripoli well after the fighting was over, at nine the *next night*. Nevertheless, Clinton's order contributed to the overall sense that Washington was in no hurry to send the cavalry to Benghazi.

My investigation has found that this sentiment was felt on the ground and all the way up the command structure. Hillary Clinton despised the military and wanted nothing to do with them. It was an extraordinary attitude for someone who aspired to be Commander in Chief; and the military essentially responded in kind. As the commander of AFRICOM Special Forces, RADM Brian Losey, told Representative Jason Chaffetz when asked if AFRICOM had any sort of contingency plan if something "went awry" at a U.S. embassy, "That would be a State Department question—what goes awry in the embassies."[41]

Hillary Clinton and John Brennan made two additional decisions that night that amounted to an across-the-board stand-down order of key rescue units.

- They refused to convene the Counterterrorism Security Group (CSG), the only structured, experienced, interagency reaction team that could have decided which resources of the government were available for deployment immediately.
- They refused to activate the State Department–led Foreign Emergency Support Team (FEST), an extraordinary operational unit whose sole purpose was to rescue U.S. diplomats under attack and to stand up emergency services within hours using dedicated military, law enforcement, security, medical, and secure communications personnel. They practiced a full-scale exercise twice every year at an overseas location.

Both decisions were clearly based on political opportunity, not operational concerns.

The Counterterrorism Security Group was "the one group that's supposed to know what resources every agency has. They know of multiple options and have the ability to coordinate counterterrorism assets across all the agencies," a high-ranking U.S. government official explained. CBS News investigative reporter Sharyl Attkisson first exposed this extraordinary political decision in November 2012.[42]

Why was the refusal to convene the CSG so important?

"Convening the CSG would have meant there was a terrorist strike under way," said Larry Johnson, a former CIA officer who supported Hillary Clinton in her 2008 presidential bid. "Clearly, they didn't want to make that kind of admission." That decision was made at the White House by John Brennan and by National Security Advisor Tom Donilon, who preferred to handle the crisis in a way that wouldn't sound all sorts of alarm bells.[43]

The refusal to activate the Foreign Emergency Support Team (FEST) also indicated the administration's unwillingness to face facts.

Mark Thompson spent twenty years as a U.S. Marine before joining the State Department to lead their counterterrorism response efforts. In 1998, he was the lead operations officer for the FEST in Nairobi, Kenya, after the al Qaeda bombings of two U.S. embassies in Africa. In 2000, he performed a similar duty with the FEST in Yemen, after al Qaeda attacked the USS *Cole*. He helped rescue hostages in Latin America, Asia, and Africa, and in 2004 set up a special unit in Baghdad to free hostages.

When bad things happened at U.S. diplomatic missions around the world, Mark Thompson was the man to call. But nobody called him on the night of September 11, 2012.

Thompson didn't spend twenty years in the Marines to become a wallflower. He saw what was happening in Benghazi and immediately knew what to do and what resources he had available. Once he learned from the 10:05 PM alert from the State Department Operations Center that Ambassador Stevens had been taken to the safe haven, he alerted his leadership, and recommended that they deploy the FEST. He described what happened next in dramatic congressional testimony:

> I notified the White House of my idea. They indicated that meetings had already taken place that evening that had taken FEST out of the menu of options. I called the office within the State Department that had been represented there, asking them why it had been taken off the table, and was told that it was not the right time and it was not the team that needed to go right then.

He added that Undersecretary Patrick Kennedy's office was the one who told him it was "not the right time" to deploy the FEST. There is no way Kennedy would have made such a decision without consulting with Secretary of State Hillary Clinton.

The FEST included Special Operations troops, FBI agents, DoD pilots and communications officers, and more. It was not run by bureaucrats or subject to the normal Pentagon process of layers and layers of approvals and signoffs. "It's [a process] that can go from a cold start to wheels-up, so to speak, within hours," Thompson said. And they were standing by. Team members were "shocked and amazed that they were not being called on their cell phones, beepers, et cetera, to go," Thompson said.

Not only was the FEST never called, but Hillary Clinton's hand-picked Accountability Review Board never interviewed Mark Thompson, despite his repeated requests to be questioned on the events of that night.[44]

No surprise there.

BREAKOUT

While Obama was meeting with Panetta and Dempsey at the White House, the fighting raged on in Benghazi. Through it all, the DS agents and GRS operators continued to go back into the burning villa to search for the ambassador.

After fifteen minutes of constant shooting, Ty Woods and the other GRS shooters said it was time to go. They were running low on ammo, and the attackers just kept coming, despite the bodies beginning to pile up on the manicured lawn and in the orchard. Soon, they would be completely overrun.

They decided to pull back to the Annex in two stages: first, the DS agents who were in dire need of medical attention, then Ty and the GRS shooters, along with Sean Smith.

So, while they provided cover, the DS agents piled into the armored Land Cruiser that Ubben had parked by the pool. The DS agents weren't familiar with the location of the Annex, so Woods gave them detailed instructions. They were supposed to turn *left* at the front gate and head west toward 17th February Martyrs Brigade territory. Whatever they did, they shouldn't turn *right* out of the main gate. That was jihadiland.

It was 11:15 PM in Benghazi when David Ubben turned on the ignition and gunned the engine and started heading toward the armed mob at the front gate. He was physically exhausted, and could barely speak because of the smoke inhalation. His face was blackened and his eyes were tearing with soot. The GRS shooters laid down supporting fire as he reached the gate, forcing the mob to back off. But for some reason, Ubben turned *right* and headed east on the gravel road, directly into a mass of bodies, all armed.

He slammed the heavy SUV into reverse, then made an emergency U-turn by turning the wheel sharply, throwing the car into gear and standing on the gas. That got him back heading west. But when he got past the main gate a militiaman he believed to be with 17th February signaled him urgently to turn around and head back east. So, he turned around again, slowly this time. It was a mess.

As they roared back into the mob, bullets pinged into the armor plate of the Land Cruiser and impacted in the inch-thick bullet-resistant glass. Spiderweb cracks began to spread in the windshield and side windows. Up ahead, a man eagerly motioned the DS agents to enter his compound but Ubben suspected it was a trap and sped up. He was right. No sooner had he passed the compound than they took sustained bursts of AK-47 fire that nearly penetrated the armored glass and blew out two of their tires. Luckily, the Land Cruiser was equipped with run-flat tires—and they performed. The ARB report noted that "a roadblock was present outside this compound and groups of attackers were seen entering it at about the time this vehicle movement was taking place." That is what the decision makers in Washington could see on the video feed from the Predator drone still overhead.

Ubben accelerated past the crowd, swerving to avoid another group of men. He bypassed a jihadi gun truck parked across the road a bit farther down by jumping up onto the center median. As things got worse, he slipped down into the oncoming lane of traffic, with two cars of shooters in hot pursuit. The pursuers eventually broke off, turning into a warehouse just north of the Annex. One of his colleagues radioed ahead when they were one minute out so the gates would be open when they arrived.

It was 11:30 PM when the Annex gate slammed shut behind them. According to one account the "underside of the vehicle was ablaze."[45]

Ty Woods and the GRS operators made their hot exfil through the front gate at about the time the others arrived at the Annex. They turned left out the front gate, then left again on the north-south road by the 17th February barracks. They made it back in about five minutes. With all U.S. personnel now evacuated from the diplomatic compound, the Predator operator back in the United States was ordered to park his drone over the Annex.

THE "SECRET" BASE

The CIA Annex was supposed to be a secret base—so secret, in fact, that when Congressman Jason Chaffetz, the Utah Republican who chaired the House Oversight and Government Reform subcommittee that was leading the Benghazi investigation, visited Libya shortly after the attacks he was told to never, ever mention its existence.

Top secret, the CIA station chief in Tripoli told him. Classified. Sources and methods. You talk about this and it will end your career.

So, when Charlene Lamb, the deputy assistant secretary of state who became the designated scapegoat for all that went wrong in Benghazi, mentioned the Annex in her testimony at the first public hearing on the debacle, Representative Chaffetz went nuts. Not only did she make reference to it: She displayed a satellite photograph on an easel behind her showing details of the four-villa compound, as well as its location relative to the diplomatic mission compound.

Anyone with minimal map-reading skills could figure out in a jiffy how to find the Annex on the ground.[46]

Chaffetz stopped the hearing on a point of order. He questioned not just the propriety of showing the photo, but said that Charlene Lamb's description of how the Annex team came to the rescue of the entrapped DS officers at the diplomatic compound was "getting into classified issues that deal with sources and methods that would be totally inappropriate in an open forum such as this."

He was eventually overruled when Undersecretary of State Patrick Kennedy, who was also testifying at the hearing, declared that the commercial satellite image, available on Google, was unclassified.[47]

Nevertheless, Chaffetz tried twice more to stop the hearing when he believed classified material was being discussed.

The kerfuffle over the photograph and the very existence of the Annex was just one of many ways that Hillary Clinton's minions were attempting to manipulate Republicans in Congress, prevent them from acquiring accurate information, and when that failed, intimidate them.

It later emerged that Clinton had sent a State Department lawyer to Tripoli to keep tabs on Representative Chaffetz during his fact-finding trip. His primary mission was to ensure that no embassy officials spoke to Representative Chaffetz alone, to give him ground truth of what had actually happened on the night of September 11, 2012, and all the security warnings before. Indeed, instructions to that effect were sent to Gregory Hicks, now the acting chief of mission, before Chaffetz arrived. "Those instructions were to arrange the visit in such a way that Representative Chaffetz and his staff would not have the opportunity to interview myself, John Martinec, and David McFarland alone," Hicks told the committee.[48]

That in itself was unprecedented, as Representative Issa pointed out. "Over the years that you've watched great ambassadors, have you ever failed to see the head of a [congressional] delegation come and get a one-on-one? Isn't that part sort of the ceremony of that relationship and how you treat the head of a congressional delegation [CODEL]?" he asked.

"In every CODEL that I have been involved in, that has been standard," Hicks replied.

Hillary's chief of staff and top lawyer, Cheryl Mills, got a frantic email from the minder she had sent to Tripoli that the CIA chief of station had excluded him from a classified briefing given to Chaffetz because he lacked the necessary security clearances. That gave Chaffetz an opportunity to speak with Hicks, Martinec, and McFarland without the minder's presence. Mills came down on Greg Hicks like a ton of bricks, calling him personally from Washington to learn exactly what he had told Chaffetz and to express her displeasure.

Cheryl Mills was Hillary's enforcer, her one-woman cleanup crew. And she had just begun to go to work. She was going to bury Benghazi so deep that no one would uncover the truth—at least, not until after the November elections. At which point, as her boss famously said, "What difference at this point does it make?"

EVACUATE!

The decision makers in Washington could see the CIA officers on the rooftops of the Annex buildings, as they stared anxiously toward the fires at the diplomatic compound to the northwest.

Those were the first images from the MQ-1 Predator drone once it reached the Annex. The chief of base had ordered a generalized distribution of weapons from the armory, so each of them was carrying an HK 416 assault rifle. Although the images may have been confusing to an untrained eye—the infrared camera the Predator used at night delivered black-and-white imagery, rife with shadows and harsh light—a skilled analyst would be able to interpret them.

Just before midnight—less than a half an hour since the DS agents arrived in their burning Land Cruiser at the Annex—Khalil Harb's men launched their first rocket-propelled grenades against the perimeter wall of the Annex. For the next hour, Ty Woods and the GRS operators took up positions behind sandbags on the roofs, locating the attackers firing RPGs and automatic weapons at them and taking them out one by one. That phase of the attack on the Annex stopped just after one in the morning, although sporadic automatic weapons fire against the Annex continued all night long.

There has been speculation that the attackers only learned the location of the Annex by following the DS agents on their meandering trip back from the diplomatic compound. My Iranian sources tell me that this is simply not true. The Quds Force team had the Annex under surveillance for several weeks at least. They knew how many vehicles were there, how the GRS operators moved, what type of weapons they favored. They knew the layout of the four villas from Google Maps, and could calculate the distance between the villas and the perimeter walls. This became critically important when they moved into the final phase of their plan just before dawn.

In Tripoli, RSO John Martinec alerted the deputy chief of mission, Greg Hicks, that Ansar al-Sharia was claiming responsibility for the attack and was now threatening to attack the embassy in Tripoli. So, Hicks, who was acting chief in Stevens' absence, gave the general evacuation order. In the ensuing cover-up by the administration, the fact that the remaining fifty-five diplomats and intelligence staff in Tripoli were evacuated to the CIA Annex there because of a public threat from Ansar al-Sharia has gone virtually unreported and unnoticed. They moved out at five in the morning in a convoy of armored Suburbans, and hunkered down at the CIA facility for two full days.

Hicks described the chaos of the emergency evacuation in his first public testimony on May 8, 2013:

> Our team responded with amazing discipline and
> courage in Tripoli in organizing our withdrawal. I have
> vivid memories of that. I think the most telling, though,
> was of our communications staff dismantling our com-
> munications equipment to take with us to the annex
> and destroying the classified communications capability.
>
> Our office manager, Amber Pickens, was every-
> where that night, just throwing herself into some task
> that had to be done. First, she was taking a log of what

we were doing. Then she was loading magazines, carrying ammunition to the—carrying our ammunition supply to our vehicles. Then she was smashing hard drives with an axe.

At the same moment Hicks gave the evacuation order—12:30 AM, Libya time—Glen Doherty and his seven-man team were finally wheels up out of Tripoli airport. It had taken time to negotiate the emergency charter of a private jet, plus around $30,000 in cash.

In addition to Doherty and four GRS shooters, the cavalry included two Delta Force operators who were working out of the CIA Annex in Tripoli. They reported to Joint Special Operations Command in Fort Bragg, North Carolina, as opposed to the four remaining members of the SST, now under the command of Lieutenant Colonel S. E. Gibson at the embassy, who reported to AFRICOM Commander General Carter Ham in Stuttgart, Germany.

The very presence of the Delta Force operators in Libya was a closely guarded secret. According to one report, they were part of an eight-man team "on a counterterrorism mission that involved capturing weapons and wanted terrorists from the streets and helping train Libyan forces."[49] My sources say that they were drawn from the C-110 counterterrorism team based in Europe that was scheduled to be assigned to AFRICOM on October 1, 2012, as a dedicated Commander's In-Extremis Force (CIF). They had been dispatched to Libya shortly before the September 11 attacks to prepare for a hostage-rescue simulation in October as part of the Section 1208 Defense Department training of Libyan Special Forces. As this book goes to press, Congress has been unable to determine whether these operators went to Benghazi on their own initiative, following the warrior's ethos of running to the gun, or if they sought to clear their departure with their command.[50]

Doherty and his men came fully armed with weapons and bags of cash. Now help—real help—was on the way.

Their chartered jet touched down at Benghazi's Benina Airport at around one-thirty in the morning. As they started unloading their gear, they were immediately detained by 17th February Martyrs Brigade militiamen under the command of Fathi al-Obeidi. Instead of transporting the team to the Annex, they started a three-hour negotiation over the terms of their assistance.[51]

At two o'clock, Hicks received a call from Secretary of State Hillary Clinton, asking for an update on the search for the ambassador. At that point, no one knew what had happened to him. Hicks remembers telling her that he had just ordered the evacuation of the embassy in Tripoli, so there could be no doubt in her mind that the threats to U.S. diplomatic personnel were ongoing. She said that U.S. Marines were on their way. Two FAST platoons were deploying, one to Tripoli, one to Benghazi.

That's great, Hicks said. When would they arrive?

Tomorrow afternoon, Clinton replied.

"I find it very ironic that when she called Greg Hicks at two AM, she basically said, 'There's nothing we can do, you're on your own.' That to me was a bit chilling," Representative James Lankford told me.

Defense Secretary Leon Panetta later admitted that he had slow-rolled the military response to the attacks. "The basic principle is that you don't deploy forces into harm's way without knowing what's going on, without having some real-time information about what's taking place," he said. "It was really over before, you know, we had the opportunity to really know what was happening."[52]

The chief of base was on an encrypted line with the CIA Operations Center and with his immediate boss in Tripoli, giving minute-by-minute updates throughout the night. Hicks and Martinec were reporting constantly to the State Department Operations Center and to their immediate superiors, Assistant Secretary of State Beth Jones and Charlene Lamb, as Admiral Mullen told the House Oversight Committee in his deposition. The Pentagon was getting situaton reports from the military attaché and the SST commander, LTC Gibson, and had eyes on from the MQ-1 Predator drone that circled over the Annex all during that night.

Exactly what additional intelligence Panetta said he was missing remains unclear.

What is clear, however, was the line of authority. No outside military forces could respond to an attack on a U.S. diplomatic facility without the express authorization of the secretary of state. And Hillary Clinton never gave that authorization. As General Ham and others said, "She never asked" for help.

FIND THE AMBASSADOR

Shortly after Hicks learned from Secretary Clinton at two in the morning that they were on their own, an Arabic-speaking male called from Benghazi using the cell phone Scott Wickland had given Ambassador Stevens in the safe haven. He said he was with "an unresponsive male who matched the physical description of the ambassador" at a local hospital.

At first, they weren't sure what hospital he meant. Once it became clear he was talking about the Benghazi Medical Center, which the Americans now believed was under control of Ansar al-Sharia, they were on their guard. "There was some concern that the call might be a ruse to lure American personnel into a trap," the ARB report stated.

The cellphone video of the U.S. ambassador being dragged out of the diplomatic compound by looters with his shirt and belt unbuttoned shocked Americans when it was posted to YouTube the next day. The sudden shouts of *Allah-o Akbar!* that rang out when someone shined a light on his face and realized he was an American gave rise to speculation that he was raped and dragged through the streets.

But a careful examination of the cellphone video, and subsequent evidence from the doctor who examined Stevens at the Benghazi hospital, suggests otherwise. The young Libyans who discovered the ambassador were not militiamen. They were dressed for the most part in designer jeans and T-shirts and carried no weapons. They clearly stumbled upon him by accident, not as part of the organized attack.

Dr. Ziad Abu Zeid was working the night shift at the Benghazi Medical Center that night. He was used to militiamen dropping off unknown persons during the frequent violence in Benghazi, so at first he didn't pay much attention to the identity of the white male brought in on a gurney shortly past 1:15 AM. He worked on him for nearly forty-five minutes, trying every EMS protocol he knew. He later told a BBC reporter Stevens "might have survived" if he had been pulled out of the building earlier, "because there was no injury to the body, only suffocating from carbon monoxide."[53]

The government in Tripoli was telling Hicks that the ambassador was "in a safe place." However, it soon became clear that the Libyans thought Stevens had evacuated with the others to the Annex. "[W]e keep telling them, no, he is not with us, we do not have him—we do not have him," Hicks said. The GRS team, still stuck at the airport, was morphing into a hostage rescue team. "It looks like . . . we are going to need to send them to try to save the ambassador, who was in a hospital that is, as far as we know, under enemy control," Hicks recalled.

In the end, it was Abdul Basit Haroun—the 17th February Martyrs Brigade commander who had organized the shipments of weapons, including MANPADS, to the Syrian rebels—who formally identified Stevens' body at the hospital early the next morning. That piece of news was contained in an otherwise matter-of-fact obituary written by the Associated Press, which identified Haroun as the "Benghazi security chief."

That closed the circle on the relationship between Ambassador Stevens and the gun-running networks.[54]

INCOMING

Glen Doherty and his team finally left the airport at four-thirty in the morning, after concluding an agreement with Fathi al-Obeidi to provide men and gun trucks to escort them to the Annex. Obeidi personally led the convoy, and later told the Libyan media he had been asked to provide transportation for ten people at the Annex, but was surprised to discover a total of thirty-seven Americans at the compound. (The official U.S. count, including Sean Smith and Stevens, was thirty-five.) The gates slammed shut behind the armored SUVs bringing the fresh team from Tripoli at five in the morning. The 17th February gun trucks and militiamen waited outside.

By this point, the Annex was in clean-out mode. CIA personnel were smashing computers and destroying files. They had been under attack all night long, and the weariness of constant battle showed on their faces. Doherty began searching for his friend Ty Woods. They had been in battle together before. They would die together tonight.

According to the account by fellow special operators Brandon Webb and Jack Murphy, Doherty was told that Woods was on the roof "manning a MK46 machine gun." So, after getting the lay of the land, he climbed the ladder and found Woods along with two other agents. "[T]hey quickly embraced, filled each other in, and retook defensive firing positions."

According to Burton's account, the two former Navy SEALs were reunited inside, just before they heard a fresh round of automatic weapons fire from somewhere just beyond the perimeter wall. Woods "rolled his eyes in an expression of absolute frustration over the audacity of the Libyan terrorists and said, 'I am going to rain down hate among them.'" Together they climbed up onto the roof, along with DS agent David Ubben, to do battle with the jihadis, as a couple of RPGs from nearby positions slammed into the compound's outer walls.

There is no disagreement as to what happened next.

"Incoming!" one of them shouted when they heard the *pop* of the first mortar.

The first round whistled: It was long, sailing over their heads, and exploded out in the street where the 17th February Martyrs Brigade militiamen were lounging. At least one of them was killed on the spot from the blast. The second round was silent until it exploded inside the northwest wall, in between the first two villas. Short.

The jihadis were bracketing their position. It was standard practice among professional mortarmen. One round long, one round short, third round on target. The military referred to the first round as the "registration" round.

Some reports claimed Doherty had spotted the launch site in a vacant lot around eight hundred meters to the northwest, and was lighting it up with a laser targeting device as he had done so often in Afghanistan and Iraq. When the SEALs went into battle with the full power of the United States behind them, a C-130 Spectre gunship circling overhead could then annihilate the attacker with a buzz-saw burst from its 20mm Gatling gun.

But there was no Spectre gunship in the sky over Benghazi, only an unarmed MQ-1 Predator drone, relaying the scene back to the Pentagon and to the White House Situation Room in real time. If Doherty was indeed lighting up the mortar crew, it was more in an attempt to intimidate them, throw them off their game, than to target them. And it didn't work. The mortar men kept dialing in the tubes.

The third round landed on the roof where Ty Woods was firing his MK46 machine gun at the attackers, as they tried to close in on the Annex. Before Glen Doherty could reach his mortally wounded comrade, two more rounds hit the roof, killing him as well. Shrapnel from the round that killed Doherty ripped into David Ubben as he was getting off the ladder to join them on the roof, nearly severing one of his legs. Mark G—, the GRS shooter who had been ordered to remain at the Annex by the chief of base, was also critically wounded in the mortar attacks.

One of the GRS reinforcements from Tripoli saw Ubben go down and scrambled up the ladder. In full combat gear, he strapped the DS agent onto his shoulders and carried him back down to safety, undoubtedly saving his life. (Ubben remained in Walter Reed Army Medical Center recovering from his injuries for over a year.)

And then, just as abruptly as it started, the firing ceased. In the distance, trucks and heavy vehicles of a quasi-government militia called into action by Libyan president Magariaf could be heard approaching. Help—real help—was finally on the way. While it was too late to save any of the four Americans who died that night, the arrival of a large, heavily armed force of pro-government troops saved the remaining thirty-one Americans from certain slaughter, just as spectacular dawn colors were beginning to replace the black sky over Benghazi.

The mortar attack had taken everyone by surprise. "That was a well-coordinated attack," Lieutenant Colonel Andy Wood told me. "It takes a crew of three to five people to man a mortar. That definitely was outside talent. I don't know of anybody in the revolution who shot a mortar."

Lieutenant Colonel Wood explained just how difficult the mortar attack must have been. "They couldn't use a spotter, because there was no high point, no height advantage, and there were walls all round. That's why [the CIA] picked that compound the way they did. It was the highest thing around.

"I've shot mortars, and they're hard," he went on. "It's a perishable skill. I bet they took that mortar out, paced it out from where they were going to fire it, and went out in the desert to practice at the exact same angle, exact same distance. They probably fired it a number of times, then moved it into town to fire that night. Those guys were experienced mortarmen."

Qaddafi's army, such as it was, was not known for its mortar or artillery skills. But the men of the Quds Force had cut their teeth during the eight horrible years of the Iran-Iraq War, where fierce artillery and mortar battles alternated with human-wave assaults.

Former CIA operations officer John Maguire agreed. "I am convinced the mortar team were not some bunch of Libyans we used to call 'the Flintstones.' That's a mortarman who's been firing mortars for a long time. That's the way the Quds Force fired on us from across the river in Baghdad. When they fired from that side of the river, people were hauling ass for cover because the Quds Force mortarmen knew how to shoot."

Maguire believed the attackers were intending to assault the Annex, but had been prevented from getting close by the GRS shooters on the roof. "They needed a standoff position. And they succeeded in silencing the Annex with just three (fatal) rounds."

The Quds Force tradecraft was evident to another Baghdad veteran as well, Blackwater founder Erik Prince. The final attack was carried out by "a very, very professional mortar team, which I believe to be Iranian Quds Force, an Iranian special operations team. To get on target with three or four mortar rounds in an area . . . you don't do that. That takes some skill."[55]

Accounts differ of how many jihadis died that night. Estimates range from sixty to over one hundred, versus four dead Americans. Those results are a tribute to the training, professionalism, and discipline of the U.S. Special Forces.

15

THE COVER-UP

The cover-up, the lying, and the dissembling began even before Woods and Doherty were dead.

It was 10 PM in Washington—4 AM in Benghazi—when Hillary Clinton phoned the White House. She had been speaking off and on during the afternoon and early evening with National Security Advisor Tom Donilon, but this was her first call to the president. She wanted to make sure that they had their story straight. She read him a draft of the statement she was about to release, which blamed the attacks on an Internet video no one in Benghazi had seen.

Here is the complete statement, which she released that evening:

> I condemn in the strongest terms the attack on our
> mission in Benghazi today. As we work to secure our
> personnel and facilities, we have confirmed that one of
> our State Department officers was killed. We are heart-
> broken by this terrible loss. Our thoughts and prayers
> are with his family and those who have suffered in this
> attack.

This evening, I called Libyan President Magariaf to coordinate additional support to protect Americans in Libya. President Magariaf expressed his condemnation and condolences and pledged his government's full cooperation.

Some have sought to justify this vicious behavior as a response to inflammatory material posted on the Internet. The United States deplores any intentional effort to denigrate the religious beliefs of others. Our commitment to religious tolerance goes back to the very beginning of our nation. But let me be clear: There is never any justification for violent acts of this kind.

In light of the events of today, the United States government is working with partner countries around the world to protect our personnel, our missions, and American citizens worldwide.[1]

If you read that statement carefully, you can see that Hillary was hedging her bets. She didn't yet mention the Internet video by name, but the allusion was clear. If the story floated with the media, they could go with it.

Obama also carefully hedged his bets when he broke into his schedule the next morning at 10:43 to deliver remarks to the press in the Rose Garden. He used the word "attack" and "attackers" repeatedly, but he never identified the attackers as "terrorists." In fact, as Republicans on the Senate Intelligence Committee pointed out, "the President passed up at least ten opportunities to clearly identify the Benghazi attacks as terrorist attacks." Instead, he used weasel words like "killed in an attack, "outrageous and shocking attack," "killers who attacked," "this type of senseless violence," and more.[2]

Like Hillary, he attempted from the get-go to pin the blame on religious bigotry *emanating from the United States.* "Since our founding, the United States has been a nation that respects all faiths," Obama said in his first public statement on the attacks. "We reject all efforts to denigrate the religious beliefs of others."[3]

ORIGINAL SIN

It was clear from the beginning that the administration was refusing to label the events in Benghazi a terrorist attack, since that would call into question their policy in Libya and throughout the Muslim world. So, they fell back on a well-worn tactic of the liberal intelligentsia: blame America first. Both Hillary Clinton and President Obama were betting that liberal reporters would go along with the fiction that an Internet video, made with $5 million raised by a "right-wing" pastor in Florida, was the real cause of the Benghazi attacks.

Hillary had already identified Pastor Terry Jones as a great campaign prop. When he first announced he would burn Korans two years earlier, during the 2010 midterm elections, she blasted him in a speech before the Council on Foreign Relations. "It's regrettable that a pastor in Gainesville, Florida, with a church of no more than fifty people can make this outrageous and distressful, disgraceful plan and get, you know, the world's attention," she said.[4]

For Hillary, it was a twofer. Not only would the story take the heat off the administration for what had been going on in Benghazi, but if Republicans didn't condemn the video as self-righteously as she did, she would tar them as bigots as well. Heck, she'd even accuse them of being responsible for Chris Stevens' death, since they had the temerity to "slash" the State Department budget. This, in fact, became a much-repeated Democrat talking point in the dozens of public hearings Congress eventually held.

That's just the way Hillary and Obama worked. It wasn't a conspiracy. It wasn't planned. For them, it was standard operating procedure for the blood sport of politics.

But there was a problem: The story just wasn't true.

At 12:07 AM, the State Department Operations Center put out a cable stating that Ansar al-Sharia had claimed responsibility for the Benghazi attack, based on the reporting they had been receiving from Libya that night. In addition, the Islamist group was now threatening to strike the U.S. Embassy in Tripoli, as well. The cable made no mention of an Internet video or a protest and was widely distributed throughout the government, including the White House, the CIA, the Pentagon, the FBI, as well as to top executives at the Department of State.

The next day, September 12, Acting Assistant Secretary of State Beth Jones, the top career diplomat for the region, sent an email to Hillary Clinton's spokesperson, Victoria Nuland, and to the top leadership at the State Department, to relay what she had learned about the perpetrators.

"I spoke to the Libyan Ambassador and emphasized the importance of Libyan leaders continuing to make strong statements. When he said his government suspected that former Qadhafi regime elements carried out the attacks, I told him that the group that conducted the attacks, Ansar al-Sharia, is affiliated with Islamic terrorists." This was one of many documents the State Department refused to release. The only reason it has come to light is thanks to Representative James Lankford, who read excerpts from the still-classified cable in an exchange with Greg Hicks during the May 8, 2013, House Oversight Committee hearing.

The CIA was warning in the days just before the attacks that they were picking up "a lot of chatter" among known al Qaeda operatives in Benghazi that suggested they were getting ready to attack the Special Mission Compound. "We learned that the CIA informed their folks at the Annex, hey, there's been a lot of chatter, we want you on high alert," said Representative Lynn Westmoreland, who interviewed the five surviving GRS operators based at the Annex. "They even invited the RSOs to spend the night [of 9/11] with them at the Annex."

Fox News intelligence correspondent Catherine Herridge added that the CIA had actually "posted a notice" at the Annex telling everyone to be on high alert on the 9/11 anniversary, because of the likelihood of attacks on Western targets. "So when [the GRS operator] saw Ambassador Stevens show up in Benghazi on September 10 with essentially no additional security, they testified that they figured there must have been some reason, something important that brought him to Benghazi at that time."[5]

The day after the attacks, the CIA chief of base in Benghazi reported to Washington that they had intercepted communications during the attacks showing that the attack was carried out by known terrorist elements. That, too, was swept aside in crafting the talking points.[6]

The evidence showing that Benghazi was a coordinated, well-planned terrorist attack was extensive, multisourced, and undisputed at the State Department and within the intelligence community. That made fabricating the story of the protest-turned-violent all the more brazen. It was a complete and utter invention that Hillary and Obama concocted during their ten o'clock phone call on the night of the attack.

Because it was a lie, it needed to be covered up.

SPIN DOCTOR

Jay Carney was a former *Time* reporter who got promoted for bashing George W. Bush. I met him briefly while working out of the *Time* Washington, D.C., bureau at the beginning of the Clinton administration. He was a political animal already, working the White House press beat. I can't remember a word of criticism for the Clinton White House ever passing his lips at the weekly editorial meetings I attended. He was a creature of Washington, loyal to his masters, whether he was wearing his media hat, or later, his hat as spokesman for President Obama.

The record shows that Carney played a key role in putting out the cover story that the Benghazi attack had begun as a demonstration over an Internet video, which got out of hand and degenerated into a riot.

In fact, Carney put out that story several times before the much-scrutinized CIA talking points eventually provided to then–U.S. permanent representative to the United Nations, Susan Rice, for use in the Sunday TV talk shows the first weekend after the attack.

On September 12, 2012, just as the bodies evacuated from Benghazi were arriving at a U.S. air base in Germany, Carney held his "gaggle" with reporters on board Air Force One en route to Las Vegas. The death of four Americans serving their country overseas in a terrorist attack clearly was not a sufficient reason to cancel the president's long-scheduled fund-raiser.

He revealed that Obama had sat down earlier in the day with CBS *60 Minutes* reporter Steve Kroft and talked at length about the Internet video as it related to the attack on the embassy in Cairo. In a highly unusual move, Carney read from a White House transcript of the interview, which wouldn't air for several weeks. In it, Obama lashed out at his political opponent. "There are times when we set politics aside, and one of those is when we've got a direct threat to American personnel who are overseas." He went on to accuse Romney of getting his facts wrong. "Governor Romney seems to have a tendency to shoot first and aim later," Obama said.[7]

But Obama was talking about the attack on the U.S. Embassy *in Cairo*, not the Benghazi attacks. He blasted Romney's condemnation of the statement put out by the U.S. Embassy in Cairo, which apologized for "the ongoing efforts by misguided individuals to hurt the religious feelings of Muslims," a reference to the Internet video. Romney thought that was a bit rich. "It's disgraceful that the Obama administration's first response was not to condemn attacks on our diplomatic missions, but to sympathize with those who waged the attacks," Romney said.[8]

The White House was carefully baiting the trap, and Romney fell right in.

The *Washington Post* awarded the Republican nominee "three Pinocchios" for "misplaced" comments about the attacks, and the rest of the media piled on.[9]

At his briefing on September 14, Carney felt the wind at his back. Asked what happened in Benghazi, he said unequivocally, "We have no information to suggest that it was a preplanned attack. The unrest we've seen around the region has been in reaction to a video that Muslims, many Muslims find offensive. And while the violence is reprehensible and unjustified, it is not a reaction to the 9/11 anniversary that we know of, or to U.S. policy."

ABC News reporter Jake Tapper wasn't buying it.

TAPPER: But the group around the Benghazi post was well armed. It was a well-coordinated attack. Do you think it was a spontaneous protest against a movie?

MR. CARNEY: Look, this is obviously under investigation, and I don't have—

TAPPER: But your operating assumption is that that was in response to the video, in Benghazi? I just want to clear that up. That's the framework? That's the operating assumption?

MR. CARNEY: Look, it's not an assumption—

TAPPER: Because there are administration officials who don't—who dispute that, who say that it looks like this was something other than a protest.

MR. CARNEY: I think there has been news reports on this, Jake, even in the press, which some of it has been speculative. What I'm telling you is this is under investigation. *The unrest around the region has been in response to this video.* [Author's emphasis added.] We do not, at this moment, have information to suggest or to tell you that would indicate that any of this unrest was preplanned.[10]

THE TALKING POINTS

Carney began his White House press briefing that morning at 11:42, as the CIA was still drafting the talking points. The whole idea of releasing unclassified talking points on the attacks came out of a coffee that morning between CIA Director David H. Petraeus and Representative Dutch Ruppersberger, the ranking Democrat on the House Permanent Select Intelligence Committee.

The CIA's Near East Division and the Office of Terrorism Analysis (OTA) conducted a "FLASH coordination" of the talking points over lunch. One concern, of course, was that information they released could impact future prosecutions, so they gave the talking points a legal read and found that only the first point posed a potential problem, because CIA reporting showed that it simply wasn't true. Here's how it read in this early version:

> We believe based on currently available information that
> the attacks in Benghazi were spontaneously inspired by
> the protests at the US Embassy in Cairo and evolved

into a direct assault against the US Consulate and
subsequently its annex. This assessment may change as
additional information is collected and analyzed. . . .

The CIA felt they should add a line on the warning they had sent
to the U.S. Embassy in Cairo about the call for protests by a Salafist
group, to provide "some insulation" to the lie about Benghazi. Other
bullet points mentioned that Ansar al-Sharia initially had claimed
responsibility for the attacks, and that the agency had been reporting
regularly on "the threat of extremists linked to al-Qaeda in Benghazi
and eastern Libya." They also noted that the "wide availability of
weapons and experienced fighters in Libya almost certainly contributed to the lethality of the attacks."

When the agency finally transmitted the talking points to the State
Department and the White House that afternoon, they went nuts.
Spokesperson Victoria Nuland insisted that any mention of Ansar
al-Sharia be taken out. She also wanted all mention of the CIA warning removed, because it "could be abused by Members [of Congress] to
beat the State Department for not paying attention to Agency warnings, so why do we want to feed that either?" Hillary's deputy chief of
staff, Jake Sullivan, played the role of enforcer during the final editing
process on Saturday morning.

CIA Director David Petraeus was so disgusted when he saw the
watered-down version that he nearly pulled the plug on the whole
thing. "No mention of the cable to Cairo, either?" he wrote to his deputy, Michael Morell, who had negotiated the final copy with top White
House and State Department aides. "Frankly, I'd just as soon not use
these. NSS's call, to be sure; however, this is certainly not what Vice
Chairman Ruppersberger was hoping to get for [unclassified] use." [11]

Like so many things about Benghazi, the White House only released the talking points emails kicking and screaming many months

later. They initially made them available for members of Congress to consult in a Sensitive Compartmented Information Facility (SCIF) on Capitol Hill, but insisted that they were not allowed to take notes. Accusations of stonewalling and footdragging from the five Republican committee chairs investigating Benghazi in their "Interim Report" in April 2013, went unheeded. In the end, it took threats from Senate Republicans to block the confirmation of John Brennan as CIA chief to get the White House to release the one-hundred-page email chain.[12]

Calls from the Sunday talk shows started coming in late Friday afternoon and on Saturday, but Hillary Clinton refused to go. She had always hated the Sunday talk shows. She may also have been worried that one of the hosts would actually ask her a real question about what she knew and when she knew it. So, at White House request, Susan Rice became the designated water carrier.

In nearly identical statements on five morning TV shows, Rice claimed that the attack on Benghazi was a spontaneous protest in response to a "hateful video" and that the administration did "not have any information that leads us to conclude that this was premeditated or preplanned."

Her mention of the "hateful video" is significant, since the video did not appear in any draft of the CIA talking points the administration claimed were the sole basis of her information. It was never mentioned even in the back-and-forth email discussion. So, clearly, Susan Rice had been briefed by White House officials Denis McDonough and Tom Donilon, Morell, or by Hillary Clinton's advisors, who together were coordinating the cover-up. The notion that she was presenting the "best evidence" given her by the intelligence community, as the White House claimed, was yet another lie.

At the first congressional hearing on the attacks, convened by House Oversight and Governmental Reform Committee chairman Representative Darrell Issa on October 10, 2012, Undersecretary of State Patrick Kennedy played coy over the video protest claim. "There were reports that we received saying that there were protests. And I will not go any farther than that. And then things evolved. Period." When pressed by Issa why he wouldn't go any further, Kennedy said he couldn't discuss the "reports" in open session.[13]

Journalist Stephen Hayes wrote the best analysis of the talking points and believes the reference to the protests most likely grew out of a real-time communications intercept between one of the participants in the attacks and a superior. "In that intercept, the participant mentioned that he'd seen the protests in Cairo and then gave an update on the attack in Benghazi," Hayes told me. "As I understand it, there was no causal link between the two in that original communication. So, contrary to administration claims, no one ever said: *We saw the Cairo protest and then decided to launch an attack*. But that mention of Cairo was used to explain Benghazi. I can't say with certainty that it was done on purpose, but given everything else we know about how they misused and abused the intelligence to tell the story they wanted, it's not a stretch."

Even more significant was the fact that the CIA had already received a reporting memo from the chief of base disputing claims of any protest, "calling the attacks what they were, and offering a detailed report on the participants," Hayes added.[14]

As this book went to print, it emerged that Deputy CIA Director Mike Morell "downplayed or dismissed reporting from his own people in Libya" as he deleted key information from the final version of the talking points used by Susan Rice, according to Fox News reporter Catherine Herridge. Among the information he discounted were situation reports from the CIA station chief in Tripoli and his deputy in Benghazi that included "raw intelligence reporting" clearly indicating there was no protest that sparked the attacks, and the testimony of survivors on the ground who gave their accounts of the attacks in a secure video-teleconference three days later. When CIA personnel in Libya learned the agency was drafting talking points that would blame the attacks on an Internet video, they felt so strongly they emailed Morell directly to insist there had been no protest.

Since retiring from the agency, Morell has taken several high-profile assignments for the administration, and gone to work for Beacon Global Strategies, a company cofounded by Andrew Shapiro, the former State Department official in charge of MANPADS collection and Philippe Reines, described by the *New York Times Magazine* as Hillary Clinton's "principal gatekeeper." Reines went to the State Department with Clinton in 2009 as her spokesman, and was promoted to become deputy secretary of state. Whenever his boss came under scrutiny for her conduct during and after Benghazi, Reines was sent out on the attack.[15]

Morell's apparent suppression of evidence led to fresh calls to Speaker Boehner to appoint a select committee to investigate Benghazi, from family members of the fallen and from ninety national security professionals, including more than two dozen retired U.S. flag officers.[16] Morell was called back by the House intelligence committee to give public testimony on his actions on April 2, 2014. Under intense grilling for over three hours, he insisted that he gave greater weight to analysts in Washington, D.C., who based their assessment on press reports than to his own station chief in Tripoli or to the GRS officers who tried to rescue Chris Stevens in Benghazi. It was a stunning admission verging on professional incompetence. Morell apparently preferred falling on his sword than admitting that he had skewed the intelligence for the political benefit of President Obama and Secretary of State Clinton. "These allegations are false," he said.

Few members of Congress were convinced by his performance. "Morell lied," Senator Lindsay Graham told me that afternoon. "The administration created a story for their political benefit at the expense of the dead."

Hillary and Obama were desperate to cover up their failed policies in Libya and throughout the region. They didn't want reporters or

Congress nosing around into who knew what and when they knew it about CIA warnings, lax security, gun running in Benghazi, or the missing MANPADS. They certainly didn't want information about the Iranian connection to come out just before the election, since the president was still hoping that Valerie Jarrett could work her back channel to the Supreme Leader in Tehran to convince the Iranians to come to the table in time for the foreign policy debate with Romney at the end of October.[17]

And it was personal. Representative Adam Kinzinger (R-IL) revealed at a town hall meeting that just two days after the attacks, Hillary "screamed at a member of Congress who dared suggest that this was a terrorist attack" during a closed-door briefing.[18]

I believe they latched on to the tiny Salafist protest in Cairo because it fit their worldview that America was "hateful" and was misrepresenting Islam. They wanted Muslims to understand that the United States "is not at war with Islam," as Obama stated repeatedly. They even had the State Department spend $70,000 of taxpayer money to produce a sixty-second television spot for Pakistani television, explaining that America had absolutely nothing to do with producing the "hateful video" and thoroughly condemned it.[19]

At the ramp ceremony at Andrews Air Force Base to welcome home the bodies of the four dead Americans on September 14, Hillary Clinton put on her face of deepest scorn. The attacks, she said, were caused by "an awful Internet video that we had *nothing* to do with. It is hard for the American people to make sense of that, because it *is* senseless. And it is totally unacceptable." She privately then promised Charles Woods, father of Navy SEAL Ty Woods, "We will make sure that the person who made that film is arrested and prosecuted." She made a similar vow to the mother of Sean Smith. CBS reporter Sharyl Attkisson, who covered the Benghazi scandal over the protests of her bosses, pointed out the obvious: "Why would we be out to get the people who made the video more than the people who carried out the attacks?"[20] Attkisson left CBS in frustration on March 10, 2014, over their refusal to air more of her Benghazi coverage.

Hillary Clinton knew how to use government against people she identified as her enemies. The Egyptian Coptic Christian who made the amateur video she falsely claimed provoked the Benghazi attacks was arrested at midnight by five Los Angeles County sheriff's deputies and thrown in jail without bond for a parole violation. He was ultimately sentenced in federal court to an additional year in prison and only released in late September 2013. As Rich Lowry of the *National Review* pointed out, Nakoula Basseley Nakoula's real "crime" was that "[h]e exercised his First Amendment rights."

The truth, of course, is that the violent hatred Islamists turned on the United States in Benghazi had little to do with public statements of our leaders, even less an Internet video. It grew out of an Islamist worldview that wanted the West out of the Muslim world altogether, and was nurtured by every drone strike on every village in Pakistan and Yemen. In every battle, they danced in the blood of the infidels.

Former Tripoli Military Council commander Abdelhakim Belhaj explained their thinking the day after the Benghazi attacks in an interview with a local radio station, according to a transcript posted on the station's website (translation courtesy of Walid Shoebat):

> I am indeed a member of the [Libyan Islamic Fighting Group]; I'm proud of it; I will not deny that. For this, I have spent the best years of my life in the prisons, under torture and I'm still a member of this group. I will not give up on its edicts. Everyone must know this in the West, before the East.
>
> I've committed jihad operations in all parts of the globe from Morocco to Yemen to Somalia to Algeria, even to Afghanistan and Pakistan. I give great respect to

[al Qaeda leader] Sheikh Ayman al-Zawahri and to the first teacher, Abu Musaab al-Zarkawi. I was saddened when Hassan al-Qaidi in Pakistan was martyred. Me and my brothers . . . have formed a committee in Tripoli and we will not allow any foreigner, no matter who he is, to dictate what we do.

. . . There is no place for America, Great Britain, and all the West in Libya. Libya is a nation of Islam and jihad. The light of Islam will shine forth from it despite the noses of everyone. The weapons are here and the mujahideen from every corner of the earth are here with us and we have all the weaponry—that was prohibited before—with us now. We will not hesitate to use it against anyone who touches the land of Libya and that is the end of this discussion.[21]

This is the Libya that Obama and Hillary brought into being.

This is the Libya that murdered Chris Stevens, Sean Smith, Ty Woods, and Glen Doherty.

CONSEQUENCES

When Greg Hicks saw Susan Rice talk about a protest at the diplomatic compound in Benghazi, he said, "My jaw dropped."

The Internet movie was "a nonevent," he told Congress. He also tried to tell his hierarchy in the State Department, but got chewed out by his immediate boss, Assistant Secretary of State Beth Jones, and subsequently demoted because he wouldn't buy into the cover-up.

Just before Rice appeared on CBS's *Meet the Press*, Libyan President Mohamed Magariaf told host Bob Schieffer in no uncertain terms that the terrorist attack in Benghazi was planned and premeditated.

"It was planned by foreigners who arrived in Libya a few months before," he revealed. That dovetails with the information that I have presented in this book.

Magariaf had gone to Benghazi two days after the attacks to lay a wreath in commemoration of Chris Stevens and his colleagues, a trip that Greg Hicks said involved "great personal and political risk." During that trip, he took to heart the plea from Assistant Secretary of State Beth Jones, who initially asked him to deny and rebut statements from other Libyan government officials blaming the attacks on former Qaddafi regime elements. Hicks said that Magariaf's performance in Benghazi "was a gift for us from a policy perspective."

But when Magariaf heard Susan Rice talk about demonstrations and an Internet video, he was furious. "He was insulted in front of his own people, in front of the world. His credibility was reduced. His ability to lead his own country was damaged," Hicks said. "A friend of mine who ate dinner with him in New York during the UN season told me that he was still steamed about the talk shows two weeks later."[22]

Susan Rice had just called the Libyan president a liar—or worse, a fool. Hicks believed that the insult prompted Magariaf to slow-roll approvals for an FBI forensics team to go to Benghazi. "It was a long slog of seventeen days to get the FBI team to Benghazi, working with various ministries to get, ultimately, agreement to support that visit," Hicks said. When they finally reached the burned-out diplomatic compound in Benghazi on October 4, journalists and looters had already pored through the wreckage. And the main Quds Force team was long gone.

Those who remained embedded with the "kidnapped" Red Crescent workers—got the last laugh. They were miraculously released on October 6 in Benghazi and appeared for a group photo with the Libyan Red Crescent, then were put on a plane to Tripoli and flown back immediately from there to Tehran.

Home free. Mission accomplished. Americans gone.

16

AFTERMATH

The Benghazi attacks sent ripple effects all across the region.

Dedicated adversaries, such as the Islamic Republic of Iran, took their victory to the bank. They understood that America was weak and that they could confront us just about wherever and whenever they pleased, with nothing to fear. As I argued in a commentary the day after Benghazi, weakness invites attack. "The evil ones must understand that they cannot torture or murder Americans with impunity," I wrote. "As long as they continue to perceive America as weak, you can be sure they will attack us again when it suits their interest."[1]

Supposed friends, such as the Muslim Brotherhood president of Egypt Mohamed Morsi, took advantage of the administration's misguided support for Islamist regimes by assuming dictatorial powers and attempting to rapidly dismantle Egypt's secular institutions. He, too, thought he could get away with murder.

Al Qaeda, fresh for another victory against America, was on the march and well-armed with weapons from Qaddafi's stockpiles that the Americans had been unable or unwilling to prevent them from acquiring. One of the al Qaeda lieutenants who reportedly took part in the Benghazi attacks, Mokhtar bel Mokhtar, launched a sophisticated assault on the Amenas natural gas plant in Algeria four months later, taking Westerners hostage and blowing up parts of the facility. He was so well armed and well organized that it took several days for Algerian Special Forces troops to dislodge him. And, even then, he managed to escape.

In neighboring Mali, al Qaeda seized control of the northern half of the country and was threatening the capital, prompting a unilateral French military intervention that began on January 17, 2013. Later that week, the British government issued a warning to British citizens in Benghazi "to leave immediately" because of a "specific and imminent threat to Westerners." Britain's new ambassador to Libya, Michael Aron, told the *Daily Telegraph* that there was a "security vacuum" in Libya that al Qaeda was exploiting, especially in Benghazi and Derna, where they had set up training camps and operational cells.[2]

Despite this evidence, which was well-known to U.S. officials at virtually all levels of the national security and policy establishment, the Obama administration continued to claim that it had defeated al Qaeda.

THE SUSPECTS

When this book goes to press, I intend to turn over to the FBI the information that I have gathered on the instrumental role played in the attacks by Quds Force officers and Khalil Harb, the top Hezbollah commander. I will describe to them the communications equipment used by the Quds Force, and how they intercepted VHF radios used by the CIA and by the Diplomatic Security personnel in Benghazi. And I will lay out what I have learned about the money trail from IRGC-controlled accounts, especially in Malaysia. A source involved in the investigation tells me that the FBI has none of this information at this point.

Information has come out, however, on several Libyan and Egyptian suspects who allegedly took part in the attacks. Some of these individuals were part of the core group recruited and paid by the Quds Force. Others were simply part of the pickup team that joined in during subsequent waves of the attack.

Hamouda Sufian bin Qumu, the former Guantánamo detainee befriended by Chris Stevens once he was transferred to a Libyan jail, was prime among the suspected organizers. He was the founder and commander of Ansar al-Sharia, the group that publicly claimed responsibility for the attacks. Fox News intelligence correspondent Catherine Herridge reported that her sources confirmed that bin Qumu was in Benghazi on the night of the attacks along with Mohammad Jamal al-Kashif. Both men were believed to have trained terrorists involved in the attack, and to have been in direct connect with al Qaeda senior leadership in Pakistan.[3]

When Representative Cynthia Lummis (R-WY) asked Ambassador Tom Pickering at a House Oversight Committee hearing whether "a former GITMO detainee who knew Stevens claimed responsibility" for the attacks, he demurred. "I believe you're getting into classified intelligence." He was corrected by committee chair Darrell Issa. "That is not true. That is not classified. . . . They made a very public claim." They were referring, of course, to Sufian bin Qumu of Ansar al-Sharia.[4]

Mohammad Jamal was briefly detained by Egyptian authorities when Morsi was still president, but was eventually released. One year after the Benghazi attacks, the State Department designated Jamal and his network as an international terrorist organization, "citing letters he exchanged with al Qaeda leader Ayman al-Zawahri, where Jamal asked for money and explained the scope of his training camps, which included Libya and the Sinai."[5]

Mohammad Jamal al-Kashif and his henchman, a U.S.-trained former Egyptian army officer named Tarek Taha Abu al-Azm, would appear to be the linchpins connecting the pickup team involved in the second wave of attacks on the diplomatic compound, and the al Qaeda central leadership. The FBI certainly needs to be interrogating them, if they haven't already.[6]

Abdelhakim Belhaj, the LIFG founder and former commander of the Tripoli Military Council, was key in Islamizing the anti-Qaddafi insurrection. He was a founder of the 17th February Martyrs Brigade, which was supposed to be guarding the U.S. diplomatic compound. His statement about expelling the West from Libya, which I cited in the last chapter, speaks volumes.[7]

One of his fellow LIFG members was arrested in Pakistan in March 2013 and deported to Libya. Faraj al-Shalabi (full name Faraj Husayn Hasan al-Shalabi al-Urfi) claimed he had been injured during the Benghazi attacks, but told his FBI interrogators he had nothing to do with the attacks.

"They said that I was a resident of the eastern region [of Libya] and the attack on the U.S. Consulate was carried out there," he said in a published interview. "Moreover, because I was in Abu-Selim Prison, they thought that I knew some of them [people in photos], as some of the photos were of people that were believed to be prisoners in Abu-Selim Prison. I told them that I do not know any of them, because they were young, while the people who were with me in the prison were older."

Shalabi was one of six people whose photograph was posted to the FBI website as a suspect in the attacks. Despite his strong al Qaeda pedigree—he fled to Tora Bora with Osama bin Laden in October 2001—he was released by the Libyan authorities on June 13, 2013, for lack of evidence.

Shalabi said that the FBI was confused in its questioning. "At times they say that it happened against the backdrop of a popular attack over the film that offends the prophet, peace be upon him, and at other times they say it was planned. If it was planned, they do not know who did it."[8]

Another key suspect was Ahmed Abu Khattala, a top Ansar al-Sharia leader. A Libyan Interior Ministry official told Reuters in October 2012 that "a photograph was taken of Abu Khattala at the consulate at the time of the September attacks, but there was not enough evidence to arrest him."

His name continued to appear in the Libyan and international media in connection with the attacks, and the FBI was seeking to interview him, but apparently couldn't find him. The irony was that Abu Khattala was not hard to get: he spoke to just about every journalist in town, denying any involvement in the attacks. "I am in Benghazi, have a job and live my life normally. I have not been accused by any party with any allegations . . . I am not a fugitive or in hiding," he told an Associated Press reporter.[9]

Just one week before President Obama leaked that the Justice Department had identified suspects in a sealed indictment—including, apparently, Abu Khattala—CNN correspondent Arwa Damon met with the supposed fugitive Libyan for two hours at a Benghazi hotel. Abu Khattala's involvement in the jihadi movement and his willingness to use brutal force were so notorious during the revolution that he became a thinly veiled character in French thriller writer Gerard de Villiers's spy novel *Les Fous de Benghazi*.[10]

U.S. influence in the region had fallen to such a low point that the government of Tunisia, which the Obama White House had helped come to power after Islamists ousted President Ben Ali from power, kept a team of FBI investigators on ice for five weeks waiting to interview a suspect. Ali Ani al-Harzi came to the attention of the FBI after he had posted a live update on the attacks on a social media site, and was seen as the first big break in the case. He initially fled to Turkey, was arrested, then was sent home to Tunisia, where officials shielded him from the FBI.[11]

"Keep in mind Tunisia is what they call a 'Millennium Challenge country.' Last year, we gave them $320 million in foreign aid," said Representative Frank Wolf. It took reminders of that aid from Senator Lindsey Graham and others to the government of Tunisia to convince the Tunisians to invite the FBI team back in to interview the suspect in December 2012. "They had him for three hours, three hours," then the Tunisians released him. "We can't find him and you saw on YouTube them celebrating with him. Here was somebody that was directly involved," Representative Wolf said.[12]

The same thing happened in Egypt. The authorities had arrested Mohammad Jamal, but they wouldn't make him available to the FBI. Representative Wolf traveled to Cairo in February 2013 with a letter for then-president Morsi, asking him to make Jamal available. U.S. Ambassador Anne Patterson insisted that Wolf not give the letter to Morsi when the two were scheduled to meet. "She said she would get it to him. Whether it got to him or not, I'm not sure," Wolf said.

Ambassador Patterson was so tight with President Morsi that in the massive popular demonstrations demanding his ouster in June 2013, Egyptians in Tahrir Square carried giant banners in English with the slogans: "Obama and Patterson support terrorism in Egypt," and "Obama Supports Terrorism." Others, using obscenities, called for her to be recalled by the United States.[13]

"DR. MORSI SENT US"

There is an extensive body of evidence suggesting that Egyptian Muslim Brotherhood operatives close to President Morsi also took part in the Benghazi attacks.

A cellphone video from the night of the attacks captures two individuals, speaking a distinctly Egyptian dialect of Arabic, calling out urgently to militiamen attacking the diplomatic compound who have trained their guns on them. "Don't shoot!" they shout. "Dr. Morsi sent us." [14]

My Iranian sources believed that Iranian intelligence may have engineered that cellphone video to frame Morsi and the Egyptian Muslim Brotherhood for the attack. This seemed possible after the public falling-out between Morsi and the Iranian leadership at the Non-Aligned Summit in Tehran on August 30, fewer than two weeks before the attacks. Morsi angrily denounced Iran's chief ally in the region, President Assad, and said it was his "ethical duty" to support the Syrian opposition. That prompted a walkout by the Syrian delegation and much public indigestion in Tehran. Whether that flare-up was just a public display meant to mislead analysts is another possibility. [15]

But there is much more.

An official letter purporting to summarize the early phase of the Libyan government investigation into the Benghazi attacks was leaked in late June 2013 to the Kuwaiti paper *Al Ra'i*. In a letter to Libya's interior minister dated September 15, 2012, Mahmoud Ibrahim Sharif, director of national security in Libya, reported that his services had unearthed "an Egyptian [terror] cell" that had been involved in the planning and execution of the attack. "[C]onfessions derived from some of those arrested at the scene" led them to a neighborhood in Tripoli where they were able to arrest six members of Ansar al-Sharia, "all of them Egyptians." Some of those arrested mentioned a direct connection to Egyptian president Morsi and to senior Muslim Brotherhood figures in Egypt. [16]

The Libyan authorities announced the first arrests just two days after the attack. But so far, the Libyan government has declined to confirm the authenticity of this document. One glaring contradiction leaps out: No one was arrested "at the scene" of the attacks, since the government in Tripoli had limited influence in Benghazi. Given that the document appeared just as a popular revolution, backed by the military, was brewing against Morsi in Egypt, could it have been produced by Egyptian intelligence? Raymond Ibrahim, Cynthia Farhat, and Walid Shoebat—all of whom are native Arabic speakers who focus on translating and analyzing public-source documents that have appeared in the Arabic media—all believe that it is authentic. The video and the intelligence document, taken together, "may just constitute a 'smoking gun' " of Egyptian Muslim Brotherhood involvement, Shoebat wrote. Just days after it was leaked, two of the Egyptian Islamists mentioned in the purported Libyan document—Sheikh Hazim Abu Ismail and Safwat Hegazy—were arrested by the new government in Egypt.[17]

Ambassador Tom Pickering provided direct confirmation of the Muslim Brotherhood role in the Benghazi attacks during his congressional testimony, when prodded by Representative Cynthia Lummis. "Our report indicates that one Egyptian organization which is named in the report was possibly involved. I think that's in the unclassified— I hope it is," he said.

The only Egyptian group mentioned in the unclassified version of the ARB report was the Imprisoned Sheikh Omar Abdul-Rahman Brigades. Their alleged involvement in the attack has given rise to suggestions that they were hoping to kidnap Ambassador Stevens and exchange him for the Blind Sheikh. Walid Shoebat believes it more likely that Pickering was referring to the Mohammad Jamal network.[18] (As I explain in the afterword, no hard evidence of a kidnapping plot has as yet come to light.)

One month after Morsi was ousted and placed under arrest, the Egyptian police arrested the Muslim Brotherhood's top strategist, Khaiat El-Shater. While raiding his home, the police found the passport of Mohsen al-Azazi, who had been identified as one of those present at the Special Mission Compound on the night of the attacks. Clearly, El-Shater had been sheltering Azazi, perhaps helping him to lie low.

A prominent Egyptian television personality explained to his audience on July 30 why Azazi was so important:

> Ambassador Stevens was killed in Benghazi, and you know who killed him, the U.S. administration knows who killed him. . . . The assassin is now present at Rabia Al-Adawiya [mosque protest]. His name is, do you know it or you would like me to inform you? He's affiliated with Al Qaeda in Libya, his name is Mohsen Al-Azazi. His passport was found in the house of Khairat El-Shater.[19]

Azazi provides a hard link between the Benghazi attackers and the Muslim Brotherhood in Egypt. TV commentator Ahmed Moussa went on to berate the Obama administration and Ambassador Anne Patterson for supporting Ambassador Stevens' killer and the heads of the clandestine apparatus of the Muslim Brotherhood in Egypt. He concluded: "Your administration aids terrorism."

The Obama White House "declined to comment" when I asked whether they had been in communication with the new Egyptian government over Azazi.

THE MISSING MANPADS (CONTINUED)

Admiral Mullen made an extraordinary admission in his sworn deposition with the House Oversight Committee staff that has gotten little attention. Asked about the lack of a military response, Mullen acknowledged that military readiness under Obama was so degraded they had no "on-call" assets available anywhere. Then he gave up the story:

> And frankly I've heard, quote, "Well, why didn't you just fly a fighter jet over and try and scare them with the noise or something?" end quote. Well, given the number of surface-to-air missiles that have disappeared from Qadhafi's arsenal, I would not have approved sending an aircraft, a single aircraft over Benghazi under those circumstances.[20]

General Carter Ham expressed the same concern.

> And as I look back on the events of that night and say—and think in my own mind would air have made a difference? And in my military judgment, I believe the answer is no. It was a very uncertain situation in an environment which we know we had an unknown surface-to-air threat with the proliferation particularly of shoulder-fired surface-to-air missiles, many of which remain unaccounted for.[21]

Libya's missing MANPADS had a nasty way of turning up all over the region in the hands of dedicated terrorists, including in Benghazi, where prominent jihadis, such as Abdel-Basit Haroun, were calling on former militia leaders to supposedly donate them to the cause of the anti-Assad rebels in Syria.

America's allies were thrust in the unpleasant position of having to mop up.

On October 24, 2012, Israeli F-16s swooped low over the Yarmouk military complex outside Khartoum, Sudan, and destroyed a warehouse where forty shipping containers containing surface-to-air missiles and other munitions had recently arrived from Iran. Sudan was a well-known hub for al Qaeda arms shipments as well as Iranian shipments to Hamas and Palestinian Islamic Jihad, both in Gaza. Israel neither confirmed nor denied hitting the facility. Satellite imagery analyzed by the Satellite Sentinel Project in the United States showed six fifty-two-foot-wide craters in the spot where the shipping containers had been just days earlier.[22]

That same day, a top Russian military commander said Russia had "reliable evidence" that the rebels in Syria had acquired surface-to-air missiles, "including U.S.-made Stingers." The statement by Army Headquarters General Nikolai Makarov caused a stir, given his rank and the extent of the Russian military and intelligence presence on the ground in Syria. In Washington, Hillary Clinton's spokesperson, Victoria Nuland, had been trying for weeks to tamp down rumors that MANPADS from Libya had reached Syria. "We have already made it clear how much we are concerned about the treatment of this type of weapons around the world. We work with governments trying to withdraw MANPADS from circulation," she said on October 16.[23]

In early November, Israel launched eight days of air strikes against targets in Gaza. Initial reporting suggested this was to prepare a ground incursion, but my sources told me that, from the start, Israel was going after weapons depots where they believed Hamas had stockpiled surface-to-air missiles acquired from arms brokers in Libya and from Iran.

Later that month, an Egyptian border patrol intercepted two more weapons shipment from Libya bound for Gaza. One convoy carried 108 Grad rockets, apparently bound for Hamas. The second was much larger, and include 185 crates with RPGs, antitank rockets, explosives, land mines—and 150 surface-to-air missiles.[24]

As President Morsi asserted greater authority over the military and border patrol, interdictions slowed. But the arms smuggling did not.[25]

Once Morsi was removed from power in July 2013, the Egyptian military stepped up its efforts to patrol the Sinai, where Morsi's friends, Mohammad Zawahiri and Mohammad Jamal, had established training camps for al Qaeda guerillas.

Egyptian leader General Abdul Fattah al-Sisi told a delegation of retired U.S. military officers, led by Major General Paul Vallely, that his military desperately needed spare parts for its AH-64 Apache helicopters so it could continue to monitor the vast Sinai Desert and track the estimated 10,000 guerilla fighters who were training there. "The Apache is the best surveillance platform they own," General Vallely said. But President Obama had blocked military supplies to Egypt to protest Morsi's removal from office, including parts for the Apaches. Despite the limited availability of the helicopters, al-Sisi's men managed to intercept several weapons convoys from Libya heading into the Sinai carrying "sophisticated weapons, including missiles" that autumn.[26]

The vast Sinai Desert borders Israel in the south at the famous Red Sea resort towns of Sharm el-Sheikh (in Egypt) and Eilat (in Israel). The threat of al Qaeda teams in the Sinai armed with Russian or U.S.-made surface-to-air missiles became so acute in August 2013 that Israeli Army Chief of Staff Lieutenant General Benny Gantz ordered the Eilat airport closed on August 8, 2013, and all flights diverted to a nearby military airport. The Israelis had intelligence suggesting an imminent terrorist attack using one of Libya's missing MANPADS.[27]

The U.S. media establishment has downplayed the arms smuggling out of Libya, just as they tried to pooh-pooh a slew of YouTube videos posted by Syrian jihadis to show off the surface-to-air missiles they were getting from their "brothers" in Libya.

An early video shows Syrian rebels with Russian-made SA-16 and SA-24 missiles, some of them training versions. It includes footage of a

Syrian MI-8 military helicopter shot down by rebels on November 16, 2012.[28]

The *New York Times* claimed it had confirmed the "first" Syrian rebel shoot-down of a Syrian military aircraft using a surface-to-air missile on November 27, 2012—ten weeks after the Benghazi attacks.[29]

The U.S. strategic forecasting group STRATFOR cited Russian television footage showing Syrian rebels shooting down a government aircraft again on February 5, 2013.[30] In May, a Syrian rebel group called Alasala Watanmia uploaded more expansive footage where they showed off a whole arsenal of Russian-origin MANPADS, including a fully operational 9K338 SA-24 IGLA-S, a 9K32M Strela-2M (SA-7b Grail), a Chinese FN-6, and an 9K310 Igla-1 (SA-16 Gimlet).[31] More of the Libyan MANPADS, fitted with the Egyptian gripstocks and CIA-supplied batteries, turned up in the hands of jihadi fighters in Mali.[32]

The Israeli government told United Nations investigators in November 2013 that MANPADS from Libya had been smuggled to terrorists in Gaza, who attempted to down an Israeli military helicopter. Just two months later, terrorists opposed to the interim government of General al-Sisi in Egypt successfully shot down an Egyptian military helicopter patrolling the Sinai.[33]

The latest United Nations report, released in March 2014, was even more stark than the first two. Noting that "the vast majority of Libyan [weapons] stockpiles are under the control of non-state actors," the UN panel determined that Libya had become "a primary international source of illicit weapons trafficking." Libyan weapons were "likely to enhance the capacity of terrorist groups in areas such as Egypt, Mali, Nigeria, the Syrian Arab Republic, Tunisia and the Gaza Strip," the report concluded.[34]

One day, without fail, an al Qaeda team armed with missiles the CIA saw leaving Libya but failed to interdict will shoot down a U.S. passenger aircraft. Or perhaps the Taliban will shoot down a U.S.

military C-5 aircraft with hundreds of U.S. troops on board, using a Stinger delivered by Qatar. Is that what it will take for a full-blown congressional and media investigation into what the CIA knew was going on in Libya?

OCTOBER SURPRISE

On October 1, 2012, President Obama dispatched Valerie Jarrett to Doha, Qatar, on an ultimate mission to meet with Ali Akbar Velayati, the former Iranian foreign minister who was now the chief foreign policy advisor to the Supreme Leader. The hope was that Jarrett could leverage her family ties to Velayati to convince the Iranians to make a public announcement before the U.S. presidential election that they were prepared to negotiate a nuclear deal with the Western powers.

Reza Kahlili, who spied for the United States for over a decade inside Iran's Revolutionary Guards Corps, maintains top-level sources inside the Iranian regime and broke the story the next day. His sources told him that "one Obama representative, the woman, who had met Velayati before, urged Velayati to announce a halt, even if it is only for a week or two, to uranium enrichment prior to the U.S. election."[35]

The *New York Times* put their own reporters on it and confirmed the basic outlines of the story. "The United States and Iran have agreed in principle for the first time to one-on-one negotiations over Iran's nuclear program, according to Obama administration officials, setting the stage for what could be a last-ditch diplomatic effort to avert a military strike on Iran."

The White House issued a nondenial denial that no "final agreement" had been reached, which the story had never claimed. "It's not true that the United States and Iran have agreed to one-on-one talks or any meeting after the American elections," White House spokesman

Tommy Vietor said. But they were "open to such talks" and "had said from the outset that we would be prepared to meet bilaterally."[36]

Israeli reporters and government officials also confirmed the story, providing the flight number of the Lufthansa flight Jarrett traveled on after connecting in Frankfurt, Germany, to Tehran.

The English-language website of the Tel Aviv daily *Yediot Aharonot* reported that the U.S.-Iran "talks are not only planned but have been going on for months and are being led by presidential advisor Valerie Jarrett."

Jarrett's involvement caught people's attention, and the leaks from Tehran caused panic in the White House—even today. *Commentary* magazine noted that "by putting someone with no background on security issues in charge of this track, Obama may be signaling that the president's goal here is not an Iranian surrender of nuclear capability, but rather a political compromise that may not eliminate the threat of an Islamist bomb sometime down the road."[37]

I asked the White House press office for comment on many questions for this book, but the only one they responded to—repeatedly—was the Valerie Jarrett back channel. They denied the story, said it was false, "lies," then came back with more denials when I sent emails on completely different subjects that they refused to address, such as the president's schedule on the night of September 11.

This attempt at pulling off an "October surprise" turned out to be a flop, because the Obama White House put an amateur in charge of the operation.

However, it may have contributed to opening the door for the official back-channel talks the administration later admitted intensified over the summer of 2013, despite repeated State Department denials to Congress that any such talks were under way. The administration ultimately made sweeping concessions to Iran's nuclear program in Geneva in November 2013.[38]

STONEWALLING CONGRESS

In Washington, the lying and the cover-up intensified.

Congress held multiple hearings, and the administration stonewalled every one, withholding documents, preventing witnesses from testifying, and providing inaccurate and sometimes false information.

Whistle-blowers still in government jobs—such as Greg Hicks, the State Department diplomat who so ably leapt into the breach when his ambassador was under attack—were threatened, demoted, and publicly tarnished.

CIA Director David Petraeus precipitously lost his job on November 8, 2012—just two days after Obama was reelected. In his letter to CIA staff, he blamed his lack of judgment for conducting an extramarital affair.

But the timing of his resignation raised many eyebrows, especially once it became clear that Petraeus had been so unhappy with the editing of the talking points by Hillary Clinton's staff that he considered not releasing them. Further questions were raised when it became apparent that the FBI—and possibly the White House—had known about Petraeus's affair with biographer Paula Broadwell months earlier, and Petraeus knew it, because the FBI had told him about their investigation. This prompted conservative columnist Charles Krauthammer to speculate that Petraeus soft-pedaled what he knew about Benghazi in his initial closed-door briefing to Congress on September 13 in exchange for keeping his job. "[H]e understood that his job, his reputation, his legacy, his whole celebrated life was in the hands of the administration. And he expected they would protect him by keeping it quiet," Krauthammer said. Petraeus himself has refused to comment.[39]

CBS reporter Sharyl Attkisson, who knew Petraeus fairly well, also believed he had been fired over Benghazi. "You'd have to be brain dead not to wonder," she said.[40]

If the White House knew about the FBI investigation and felt it was so damaging, why didn't they force Petraeus to resign earlier? The answer is obvious: because then the press would have begun to dig, and the whole Benghazi cover story, which the president felt was critical to his reelection campaign, would have started unraveling.

When Congress did begin digging deeper, they sought access to Petraeus's after-action report, based on the secret trip he made to Libya to debrief the chief of station and the chief of base in Benghazi, but the CIA "flatly refused" to make it available. Senator Richard Burr (R-NC) asked former White House counterterrorism chief John Brennan in February 2013 if he would pledge to provide all relevant documents to the intelligence committee as the new CIA director. Brennan said he would try to "reach an accommodation" with the committee, but would not commit to full cooperation. "[O]ur Founding Fathers did sort of separate the branches of government: judicial, legislative, and executive. And so I want to be mindful of that separation, but at the same time, meet your legitimate interests," he said.[41]

With Brennan now at CIA, in charge of the cover-up, Congress became increasingly frustrated and sought to interview the CIA case officers, analysts, and GRS operators at the Annex. Several congressional committees were involved in this effort. Some wanted to get a more complete picture of the attack itself. But others started to investigate what the Annex was actually tasked to do.

Almost immediately, they ran into another brick wall. The CIA officers told them they were being required to take repeated polygraphs to determine if they were speaking to the press or to Congress. Some accounts said they were getting polygraphed monthly. The contractors said they were forced to sign a special nondisclosure agreement (NDA) in addition to the blanket form they had signed earlier when accepting employment with the Global Response Staff.

"A CIA polygraph in itself is a daunting experience," former CIA polygrapher Kevin Shipp told me. "If it's given once a month, that is intimidation. I never saw that frequency given to CIA employees ever, not once. If the polygraph is used for intimidation, it's illegal."

The same goes for the nondisclosure agreements. "I've issued thousands of them," Shipp said. "They are a very powerful tool because they stipulate criminal penalties for unauthorized disclosures. So the people who sign them are not just afraid for their careers, but afraid of going to jail." He added that it was highly unusual to make contractors sign a second one, regardless of whether they were leaving CIA employment. "It just never happens." The fact that some of the GRS contractors are still with the agency made it even more unusual.[42]

John Brennan asserted publicly there were no polygraphs, no pressure, no NDAs, and that he was actually encouraging CIA employees to tell their stories. But Congress wasn't buying it. Representative Trey Gowdy (R-SC), an active member of Issa's investigative committee, told Fox News host Greta Van Susteren that the CIA had gone to great lengths to hide the survivors, "including changing names, creating aliases." He explained:

> Stop and think what things are most calculated to get
> at the truth? Talk to people with first-hand knowledge.
> What creates the appearance and perhaps the real-
> ity of a cover-up? Not letting us talk with people who
> have the most amount of information, dispersing them
> around the country and changing their names.[43]

Washington, D.C., attorney Mark Zaid, who represented three members of the elite security team based at the Annex, said he had been unable to confirm that the CIA had ordered extra polygraphs for anyone involved in Benghazi. "None of my clients was polygraphed," he told me. "But they were asked by the government to sign new secrecy agreements, which is completely unusual and out of the norm."[44]

It wasn't just Republicans who felt that they were being stonewalled and lied to. Senator Dianne Feinstein, the California Democrat who chaired the Senate Select Committee on Intelligence, "rolled her eyes" at the CIA flacks sent to brief the committee behind closed doors on the Annex mission statement and its activities. "Is that it? This doesn't do anything for U.S. interests," she said, according to a government official who was present at the briefing.

"The CIA and the State people who were briefing her were telling lies, bullsh–t lies," the official told me. "Later, out in the hallway, they were high-fiving each other like they had put something over on her." In the end, Senator Feinstein circled the wagons and said it was "nonsense" that Benghazi should keep Hillary from the presidency.[45]

However, Feinstein wouldn't paper over the damage Benghazi and the administration's policy of promoting the Muslim Brotherhood had done to our nation's security. Ironically, she made that admission to Candy Crowley, the CNN anchor who so conveniently helped Obama in his foreign policy debate with Mitt Romney by "reminding" him that he had called Benghazi a "terrorist attack" the very next day, which, of course, wasn't true.

Here is a remarkable passage from Crowley's interview with Senator Feinstein and her Republican counterpart in the House, Representative Mike Rogers, from CNN's *State of the Union* on December 1, 2013. Once again, it shows Crowley's true colors.

CROWLEY: The big question that's always asked, are we safer now than we were a year ago, two years ago? In general?

FEINSTEIN: I don't think so. I think terror is up worldwide, the statistics indicate that, the fatalities are way up. The numbers are way up. There are new bombs, very big bombs, trucks being reinforced for those bombs. There are bombs that go through magnetometers. The bomb maker is still alive. There are more groups that ever and there's huge malevolence out there.

CROWLEY: So Congressman Rogers, *I have to say, that is not the answer I expected* [emphasis mine]. I expected to hear, oh, we're safer. Do you agree?

ROGERS: Oh, I absolutely agree that we're not safer today for the same very reasons.[46]

NO ACCOUNTABILITY FOR HILLARY

Hillary Clinton gave several closed-door briefings to members of Congress in the days immediately following the attacks, but refused to appear at public hearings. She said she first wanted to see the results of the Accountability Review Board she had appointed to investigate what went wrong in Benghazi, who was responsible, and what needed to change to make sure it "never happened again."

She appointed two old friends to co-chair the board, Ambassador Thomas Pickering and Admiral Mike Mullen, the former chairman of the Joint Chiefs of Staff. Undersecretary of State for Management Patrick Kennedy chose the staff from among State Department loyalists. According to Kevin Maxwell, one of the four State Department employees who were recommended for sanctions by the ARB, "Admiral Mullen stated, said on three separate occasions, 'This has to stop at the assistant secretary level.' He said it on three separate occasions. . . . The 'this' is the accountability, the process, the blame."[47] Mullen told committee staff he couldn't remember having made that remark.

Both Mullen and Pickering acknowledged publicly that they never interviewed the one person who had statutory responsibility for the diplomatic compound in Benghazi, Secretary of State Hillary Clinton. In an astonishing revelation buried deep in his interview with congressional staff, Mullen stated: "We never found any evidence whatsoever that she was involved in the day-to-day security decisions with respect to Benghazi. . . ." They also exempted from blame Undersecretary Kennedy, even though he personally signed off on renting the Benghazi complex knowing that it fell far short of the department's security requirements.[48]

Kennedy told the ARB that his decision was to "occupy the real estate," and that all security decisions were the responsibility of underlings at the Bureau of Diplomatic Security. And yet, in testimony before the House Foreign Affairs Committee, Kennedy said, "We review the threat level every day." In the case of threats against another embassy in the region (Damascus, Syria), he testified that he personally made the determination to keep the post open. But when it came to Benghazi, the ARB gave him a pass. "We did not ask him what conversations he had with the Secretary of State on Benghazi," Ambassador Pickering said.[49]

Before releasing the report to Congress and the public in early December 2012, Mullen gave it to Hillary Clinton and Cheryl Mills, her chief enforcer. Asked why he gave her a heads-up, he said, "It was her report, we were working for her." So much for impartiality and accountability.

When Hillary's supporters tout her record as secretary of state, they point to the number of miles she traveled ("956,733 to be exact"). They say she advanced "women's rights" around the world (in particular, the right to government-paid abortion) and knew how to party with the best of them. Photographs of Hillary knocking back a beer from the bottle in Cartagena, Colombia, or dancing in South Africa with Nelson Mandela have made headlines.

But in the end, even her supporters fret that she was a lightweight and point to "a dilemma for Mrs. Clinton in the effort to define her tenure as Secretary of State," according to the *New York Times*.[50] The book you are now reading will be appearing at about the same time as a memoir Hillary will be co-authoring with two former assistants from the State Department that will portray her as a world-class diplomat and statesman. If Benghazi gets mentioned at all, she will state that she "assumed full responsibility" for the attacks, and then blame the hired help for having let her down. Fifty-one percent of Americans polled in March 2014 weren't buying it, and said they didn't believe her story that she "never saw requests for more security" from her own diplomats in Libya before the Benghazi attacks.[51]

"What difference in the end does it make?" as she famously exclaimed during her one day of testimony on the Benghazi debacle on January 23, 2013.

She sacrificed diplomats and special warriors to cover up the illegal arms walking in Benghazi, put out a fake cover story about an Internet video, then pretended indignantly that her underlings were at fault.

Benghazi revealed Hillary Clinton's true character and showed precisely why she does not have the qualities required to become commander-in-chief.

Appendix I: Questions & Answers

It is important to dispel myths and conspiracy theories, because the facts themselves, as I have tried to demonstrate, are troubling enough.

Here are some of the main questions about Benghazi that I believe have now been answered through multiple congressional investigations, as well as my own.

ASKED AND ANSWERED

Q: Was there a demonstration in front of the Benghazi diplomatic mission caused by an Internet video?

A: No, there is no evidence there was a demonstration. Secretary of State Hillary Clinton was informed at 6:02 PM that it was a terrorist attack, but put out an invented story about an Internet video four hours later, possibly confusing what was happening in Benghazi with reports from Cairo. Former Deputy CIA Director Mike Morell, who was instrumental in "sanitizing" the talking points to eliminate all mention of a terror attack, absurdly claimed in his public testimony before Congress on April 2, 2014, that he and CIA analysts in Washington, D.C., gave more credence to after-the-fact press reports of a protest turned violent than to cables and emails from CIA officers on the ground in Libya. After leaving the CIA, Morell went to work for a consulting group founded and owned by Hillary Clinton protégés. His testimony was widely viewed as an unprofessional attempt at currying favor with his former political bosses.

Q: Why did Chris Stevens go to Benghazi in the first place?

A: According to Greg Hicks, his second in command in Tripoli, he went on orders from his principals in Washington. The ostensible reason was to allow Hillary Clinton to stage a photo op later that year when officially opening the Benghazi consulate, to showcase her success as secretary of state.

Q: Was Stevens involved in weapons smuggling from Benghazi to the Syrian rebels?

A: Only on the periphery. He was aware of the weapons smuggling, and may even have tried to shut down unauthorized arms shipments. My sources say that he was supporting the administration effort to buy back missing surface-to-air missiles. The State Department has refused to release Stevens' classified cables or to make all of them available to Congress, so we don't yet know the full extent of his involvement with weapons being smuggled from Libya to Islamist rebels in Syria.

Q: What was the CIA Annex actually doing?

A: During the revolution, the CIA Annex was running guns to the Libyan rebels, with authorization from the Obama White House and notification to Congress. As of June 2012, most of the clandestine staff involved in that operation left for southern Turkey and Syria. From then on, the main mission of the CIA/NSA team at the Annex was to collect intelligence on suspected terrorists, monitor their communications, and assist the MANPADS collection effort.

Q: Did the United States ship Stingers to the Libya rebels?

A: Not directly. The CIA used the Qatari Special Forces as a cut-out for these shipments, and later tried to buy back the missing missiles, with limited success.

Q: Did the United States ship MANPADS from Qaddafi's arsenal to the Syrian rebels?

A: Unclear. But CIA officers certainly observed those shipments and were aware of them. They were informed of the "ant trade" by the UN Experts Panel on Libya and by their own sources. In some cases, they had Predator drone coverage of smugglers in the desert and did nothing to interdict them. CIA Director David Petraeus was personally informed of the MANPADS smuggling effort through Agadez (Niger) by former CIA officers. While he was still at the White House, John Brennan intervened with Petraeus's staff to make sure that the CIA took no action against the MANPADS deliveries to jihadi groups. Why he did so is a key question that the administration has refused to address.

Q: Were MANPADS being stockpiled at the diplomatic compound or the Annex?

A: I have found no evidence at this point to suggest MANPADS were being stockpiled at either facility.

Q: Was the attack a hostage-taking gone wrong?

A: Egyptian opponents of ousted president Mohamed Morsi have alleged that Muslim Brotherhood operatives close to Morsi wanted to kidnap Ambassador Stevens and exchange him for the Blind Sheikh, serving a life prison sentence in the United States. While that may have motivated some of the individuals who took part in the attack—especially the pickup crowd, once the Quds Force team had torched the buildings—I have seen no evidence there was any serious attempt to reach Ambassador Stevens in the safe haven let alone kidnap him by *any* of the terrorists.

Q: Who actually carried out the attacks?

A: My investigation determined that the attacks were a military operation planned, financed, and orchestrated by the Quds Force of Iran's Revolutionary Guards Corps. The Iranians

and a Hezbollah associate recruited and paid local terrorist brigade commanders from Ansar al-Sharia and other
groups to take part in the attacks. Quds Force specialists
supervised and may have personally carried out the deadly
mortar attack that killed Ty Woods and Glen Doherty at
the Annex.

In other words, Benghazi was a state-sponsored terrorist attack by the Islamic Republic of Iran.

Q: Was Ambassador Stevens raped and dragged through the streets?

A: No. Initial reports suggested that he had been sexually mo-
lested, and they were further fueled by Hillary Clinton's false
ten o'clock statement that the attack was a "demonstration"
that had gotten out of hand. But a careful examination of the
video of him being pulled out of the charred building, and eyewitness reports from the doctor who examined him and from
others who saw the body in Libya, give no suggestion that he
was sexually assaulted or otherwise abused.

Q: Did someone in the White House issue a "stand-down" order
that prevented a military response?

A: Not in those terms. The White House did worse: The president of the United States never issued a *stand-up* order, which
would have started by convening the Counterterrorism Security Group (CSG). The president never told his advisors that he
was 100 percent engaged, that he was personally following the
crisis, let alone leading the response. That lack of leadership
accounts for most of the foot-dragging that night.

There are three notable exceptions.

- AFRICOM turned down a request from Colonel Gibson in
  Tripoli to send four remaining SST special operations troops
  in the early morning hours of September 12 to Benghazi,
  judging that they were desperately needed in Tripoli as the

Tripoli embassy was forced to evacuate because of terrorist threats. Military commanders interviewed by the House Armed Services Committee all concurred—as did Gibson himself—that the order was justified.

- Hillary Clinton never gave the order to mobilize the State Department's Foreign Emergency Support Team (FEST), which could have secured the compound and the Annex and prevented the loss of classified data.
- The Pentagon refused to allow C-110, a fifty-man team of Special Operations troops on a training mission in Croatia, to gear up and fly immediately to Benghazi, just a few hours away, where they could have provided substantial reinforcements to the Annex before the final attack. Who ordered them to stand down remains classified information.

Q: Could anything have been done that would have prevented the murder of Ambassador Stevens and Sean Smith?

A: Yes. As I describe in chapter 14, if the CIA chief of base at the Annex had allowed Ty Woods and the GRS shooters to depart immediately for the compound, as they tried to do, there's a strong likelihood they could have saved the ambassador and Sean Smith.

QUESTIONS LEFT UNANSWERED

Representative Frank Wolf represents the suburban Virginia congressional district that includes CIA headquarters. He has said publicly that he has met with CIA officers in local supermarkets who have given him information about Benghazi that conflicts with the administration's narrative, and has lined up, at last count, 190 co-sponsors of a bill, H. Res. 36, calling on House Speaker Boehner to empanel a select committee with broad subpoena powers to definitively answer all the ambiguities of the events surrounding the Benghazi attacks.

Representative Wolf has asked a series of twelve questions, many of them political, he believes the administration needs to answer and that require a select committee. They include:

- Why did the administration order Benghazi survivors not to talk to Congress for months on end?
- Who in the White House knew what was going on in the CIA Annex?
- What happened in Washington on the night of the attack and in the days to follow?
- Which foreign consulates did Chris Stevens call for help, and how did they respond?
- Why were Ty Woods and the GRS contractors told to "stand down" by the CIA chief of base? [1]

Representative Devin Nunes (R-CA), who chairs the House Permanent Select Intelligence Committee subcommittee investigating Benghazi, identified nine questions he believes need to be answered, including:

- Why was the FEST not deployed?
- Did the GRS operators at the Annex call for air support? If so, why was it not provided?
- Which terrorist group "had the access and training needed to carry out" the mortar attack on the Annex?
- A full accounting of CIA and State Department operations in Benghazi, including what foreigners they met with over the year preceding the attacks.
- Why has the United States failed to capture or kill any of the perpetrators?
- Were any U.S. government officials engaged in intimidating potential witnesses to prevent them from testifying to Congress? [2]

Appendix II: United Nations Inventory of Weapons Found on the *Letfallah II*

ARMS AND AMMUNITION SEIZED ON BOARD THE *LETFALLAH II*

Type of items	Quantity
Weapon systems	
Kalashnikov-type assault rifles	23
FN FALs	14
Dragunov sniper rifles	3
BKT [*general purpose machine gun*]	13
12.7mm heavy machine gun	1
14.5mm heavy machine guns—twin-barrel	2
RPG	24
Antitank recoilless rifles	4
	(+ 4 bases)
120mm mortar	1
82mm mortar	1
60mm mortar	1
Ammunition	
Rocket 130mm	11
Tank rounds 115 and 125mm	6
SA-24 Igla-S	2

SA-7b	10
	(+ 1 gripstock and 6 batteries)
Antitank missiles	23
	(Including 2 MILAN, 5 KONKURS-M,
	3 METIS-M, 1 MALUTKA)
7.62x39, 7.62x51, 7,62x54mmR	378,274
12.7mm	165,960
14.5mm	22,450
23mm	6,285
57mm rockets	758
68mm rockets	201
Different types of artillery rounds	18
PG rockets	1,640
Other rockets	9
Recoilless rifle rounds 73mm	483
Recoilless rifle rounds 106mm HEAT	24
107mm Katyusha	9
Different types of mortar rounds	528
Defensive and offensive hand grenades	786
Different types of rifle grenades	319

Mines and explosives

Antitank mines	8
Semtex H	40kg
400g TNT blocks	4

Other items

Sights, magazines, cleaning kits, spare parts for weapons, military uniforms, helmets, gas masks, communication equipment (71 radios), propulsive charges for rockets and fuses.

Source: United Nations Security Council, final report of the Panel of Experts established pursuant to Security Council resolution 1973 (2011) concerning Libya, March 9, 2013, S/2013/99; 77; citing information received from "Lebanese authorities."

Notes

PROLOGUE

1 Rebecca Shabad, "Inhofe rips 'outrageous lie' on Benghazi," *The Hill*, Feb. 3, 2014.
2 U.S. Senate Select Committee on Intelligence, "Review of the Terrorist Attacks on U.S. Facilities in Benghazi, Libya, September 11–12, 2012," Jan. 15, 2014. Hereafter, Jan. 2014 SSCI report.

1. FROM TERRORIST TO FRIEND

1 To get a sense of the high tension of these engagements, readers can listen to the audio exchange between AWACS flight controllers and two F-14 pilots, and much more, at the Reader's Guide page for this book. http://www.kentimmerman.com/darkforces.htm. The Reader's Guide contains many additional features not found in the print edition of this book, including documents cited in this book, unique photographs, hyperlinks to source articles, and more.
2 Reagan's interchange with a much younger Helen Thomas can be watched via the Reader's Guide.
3 Interview with Thomas Fortune Fay, an attorney representing U.S. victims of the La Belle bombing, Dec. 6, 2006. German attorney Andreas Schulz told a Berlin court that Qaddafi admitted to ordering the La Belle bombing in a conversation with German Chancellor Gerhard Schroeder's chief foreign policy advisor, Michael Steiner. See "Qaddafi admitted 1986 Berlin bombing—claim," Agence France-Presse, May 15, 2001.
4 I was based in France at the time, where the refusal by French Prime Minister Jacques Chirac—who pretended to be a friend of the United States—to

grant overflight rights to the U.S. aviators became a public spectacle. Chirac's action forced the F-111s launched from the 48th Tactical Fighter Wing at Lakenheath Air Base in England to take a long detour around Spain and Portugal, adding 1,300 nautical miles, additional refueling, and three hours to the flight in each direction. Pilot fatigue contributed to missed targets and stray bombs, which killed several hundred civilians.

5 Kenneth R. Timmerman, "The Semtex Scandal," *Middle East Defense News/MEDNEWS*, April 16, 1990.

6 Al-Hayat (London), October 20, 1995, cited in Gary C. Gambill, "The Libyan Islamic Fighting Group," *Jamestown Terrorism Monitor*, volume 3, issue 6, March 23, 2005.

7 "The Shayler Affair: The Spooks, the Colonel and the Jailed Whistle-blower," *Observer*, August 9, 1998.

8 George Tenet, *At the Center of the Storm* (New York: HarperCollins, 2007), 288.

9 Ron Suskind, "The Tyrant Who Came in from the Cold," *Washington Monthly*, October 2006.

10 Glen Owen, "Ex-Spy is BP's Lawrence of Arabia," *Daily Mail*, Sept. 6, 2009.

11 Interview with senior British diplomat, Tripoli, March 4, 2004.

12 Interview with senior British diplomat, Tripoli. See also Michael Smith, "Blair may take credit, but it was all down to an MI6 spy in a Bedouin tent," *Daily Telegraph*, Dec. 22, 2003.

13 Qaddafi's translator and aide de camp, Muftah Misoori, described these events to me in Sirte, March 3, 2004.

14 Suskind, "Tyrant."

15 Tenet, *At the Center of the Storm*, 289.

16 Interview in Sirte, March 4, 2004.

17 Abdelaziz Barrouhi, "Le dauphin de Kadhafi sous surveillance," *Jeune Afrique*, Feb. 10, 2009.

18 "For the Urgent Personal Attention of Musa Kusa," letter from Mark Allen, Special Intelligence Service, June 19, 2003. This letter was among a trove of secret intelligence documents, some of them bearing the signatures of Mark Allen and Steve Kappes, discovered in the office of Musa Kusa in Tripoli after the fall of the Qaddafi regime in September 2011 by researchers from Human Rights Watch. See Human Rights Watch, "Delivered into Enemy Hands: U.S.-led Abuse and Rendition of Oppo-

nents in Qaddafi's Libya," September 2012, 187. See also Human Rights Watch, "US/UK: Documents Reveal Libya Rendition Details," September 9, 2011.

19 William J. Broad, David E. Sanger, and Raymond Bonner, "A Tale of Nuclear Proliferation: How Pakistani Built His Network," *New York Times*, Feb. 12, 2004. See also Raymond Bonner, "Dubai Company Denies Tie to Nuclear Spread," *New York Times*, Feb. 13, 2004.

20 Human Rights Watch, "Delivered," 187.

21 Judith Miller, "How Gadhafi Lost His Groove," *Wall Street Journal*, May 16, 2006.

22 Interview with Representative Curt Weldon, Sirte, March 3, 2004. Weldon heard this story from Ukranian Foreign Minister Constantine Grushenko in Kiev shortly after Kuchma had returned from Tripoli.

23 Interviews with DeSutter and top aides, February–April 2004.

24 House Committee on International Relations, "Weapons of Mass Destruction, Terrorism, Human Rights, and the Future of U.S.-Libyan Relations," Testimony of Assistant Secretary of State for Verification and Compliance, Paula DeSutter, March 10, 2004.

25 Interview with former CIA officer, Ishmael Jones, Feb. 3, 2009. "Jones" wrote a scathing critique of the CIA's risk-averse management and top-heavy bureaucracy. Ishmael Jones, *The Human Factor: Inside the CIA's Dysfunctional Intelligence Culture* (New York: Encounter Books, 2008), which referred mockingly to Kappes as "Mr. Suspenders."

26 Interview with retired CIA station chief who worked with Kappes, Nov. 15, 2006.

27 Interview with Duane (Dewey) Clarridge, Sept. 19, 2007.

28 Human Rights Watch, "Delivered," 177.

29 Leaked Stratfor email, Jamie Smith to Fred Burton, Sept. 14, 2011, http://english.al-akhbar.com/gi-files/966063insight-libya-info-about-opposition-leadership-ly0700.

2. THE MAKING OF AN AMBASSADOR

1 U.S. Consulate Jerusalem, "Aid Agencies Consider Assistance to Address Humanitarian Crisis," March 16, 2006, http://wikileaks.org/cable/2006/03/06JERUSALEM1009.html.

2 U.S. Embassy Tripoli, "Qadhafi Hosts Sarkozy Following Departure of Bulgarian Nurses," July 27, 2007, http://wikileaks.org/cable/2007/07/07TRIPOLI 641.html.

3 U.S. Embassy Tripoli, "Libyan Reaction to AQ-LIFG Merger," Nov. 7, 2007, http://wikileaks.org/cable/2007/11/07TRIPOLI945.html.

4 U.S. Embassy Tripoli, "Libya: Follow-Up on Access to Returned GTMO Detainees," SECRET, Dec. 13, 2007, http://wikileaks.org/cable/2007/12/07TRIP OLI1039.html.

5 U.S. Embassy Tripoli, "GoL Further Protests Planned Transfer of Libyan Detainee at GTMO to U.K.," Dec. 13, 2007, http://wikileaks.org/cable/2007 /12/07TRIPOLI1040.html.

6 U.S. Embassy Tripoli, "U.S. Companies Win $2 Billion Worth of Infrastructure Contracts as Reward for Political Relationship," Dec. 18, 2007, http://wikileaks.org/cable/2007/12/07TRIPOLI1053.html.

7 United States Department of State and the Broadcasting Board of Governors Office of Inspector General, "Report of Inspection, Embassy Tripoli, Libya," Report Number ISP-I-09-01A, December 2008, 6.

8 U.S. Embassy Tripoli, "Libya's National Security Council: Experiencing Growing Pains," Dec. 27, 2007, http://wikileaks.org/cable/2007/12/07 TRIPOLI1063.html.

9 U.S. Embassy Tripoli, "Libyan Reaction to AQ-LIFG Merger."

10 Interview with Senator Joe Biden, Sirte, March 2004.

11 U.S. Embassy Tripoli, "Follow up on Human Rights Activist Fathi El-Jahmi," March 4, 2008, http://wikileaks.org/cable/2008/03/08TRIPOLI183 .html.

12 U.S. Embassy Tripoli, "Fathi el-Jahmi's Son Asks Embassy to Stop Visiting Him," June 27, 2008, http://wikileaks.org/cable/2008/06/08TRIPOLI506 .html.

13 U.S. Embassy Tripoli, "Embassy Visits Human Rights Activist Fathi el-Jahmi," Feb. 21, 2008, http://wikileaks.org/cable/2008/02/08TRIPOLI142.html.

14 U.S. Embassy Tripoli, "Fathi el-Jahmi and Family Decide He Should Travel to Switzerland for Medical Treatment," Jan. 22, 2009, http://wikileaks.org /cable/2009/01/09TRIPOLI47.html.

15 Jack Murphy and Brandon Webb, *Benghazi: The Definitive Report* (New York: HarperCollins 2013), Kindle edition. Murphy, an ex-Ranger, par-

ticipated in Objective Massey as well as 2005 raids in Mosul "that killed or captured Libyan foreign fighters."

16 U.S. Embassy Tripoli, "Extremism in Eastern Libya," SECRET, Feb. 15, 2008, http://wikileaks.org/cable/2008/02/08TRIPOLI120.html.

17 United States Department of State and the Broadcasting Board of Governors Office of Inspector General, "Report of Inspection, Embassy Tripoli, Libya," 11.

18 U.S. Embassy Tripoli, "Die Hard in Derna," June 2, 2008, Confidential/NOFORN, http://wikileaks.org/cable/2008/06/08TRIPOLI430.html.

3. "A NEW BEGINNING"

1 Daryl Kimball, "Fact Sheet: MANPADS at a Glance," Arms Control Association, March 2013.

2 Dean Nelson and Alim Remtullah, "Wikileaks Afghanistan: Taliban used heat-seeker missiles against Nato helicopter," *Daily Telegraph* (London), July 26, 2010, cited in http://en.wikipedia.org/wiki/List_of_aviation_accidents_and _incidents_in_the_War_in_Afghanistan#cite_note-148.

3 Afghan War Logs for May 30, 2007, leaked by Private Bradley Manning; compiled and analyzed by the author from http://wikileaks.org/wiki/Afghan _War_Diary,_2004-2010/. The shoot-down is described in report 2A34FD1C -F601-40C8-8483-C8A6A64F818D. Follow-on attacks using MANPADS are logged in reports 9602FBFA-8614-4CE7-B9D9-188B4D4E68B5 and E9E91309-9BCF-45C3-9000-8308D34D1F09.

4 Afghan War Logs., ibid., report identifier E9E91309-9BCF-45C3-9000-8308D34D1F09.

5 Tom Coghlan, "Taliban in First Heat-Seeking Missile Attack," *Daily Telegraph* (London), July 28, 2007. While a NATO spokesman refused to comment on the attack, the July 22 Afghan War log entry for the attack later included a reference to Coghlan's article.

6 Afghan War Logs for September 16, 2005, report identifier 77BE60E5-2219-0B3F-9F4E8AB7A08D885D.

7 Andrew C. McCarthy, *The Grand Jihad* (New York: Encounter Books, 2010), 240.

8 "Barack Takes a Bow," *Washington Times*, April 9, 2009.

9 Michael D. Shear and Kevin Sullivan, "Obama Portrays Another Side of U.S.," *Washington Post*, April 7, 2009.

10 Kenneth R. Timmerman, "Obama Egypt Strategy Could Place US at Risk," *Newsmax Media*, Jan. 31, 2011.

11 The White House, "Press Briefing by Press Secretary Robert Gibbs, Speechwriter Ben Rhodes, and Deputy National Security Advisor for Strategic Communications Denis McDonough," June 3, 2009.

4. TEHRAN SUMMER, ARAB SPRING

1 This account is based on contemporaneous interviews with top Mousavi advisors, Green Movement leaders, and former Iranian regime intelligence officers. I reported on these events as they occurred at http://www.iran .org/index2009.html and at *Newsmax Media*: http://www.newsmax.com /Archives/KenTimmerman/90/2009/6.

2 CNN set up a website where viewers could post amateur videos. http:// www.ireport.com/docs/DOC-273227.

3 Kenneth R. Timmerman, "Key Iranian Dissident Riled at Obama's Approach," *Newsmax Media*, June 23, 2009.

4 See http://www.cnn.com/2009/POLITICS/06/24/iran.obama.letter/.

5 Obama's approach of direct talks was not new. The Bush administration held twenty-eight high-level talks with the Iranian regime that mostly went nowhere. Obama was proposing to send the very same State Department official to talk to the Iranians in Geneva in October 2009 who had conducted the last set of failed talks with the Iranians in July 2008. See Kenneth R. Timmerman, "Obama in Wonderland," *Washington Times*, May 20, 2009.

6 Robert Kagan, "Obama, Siding with the Regime," *Washington Post*, June 17, 2009.

7 Kenneth R. Timmerman, "Clandestine Iran-U.S. Huddles Seek Mutual Aid," *Newsmax Media*, April 30, 2009.

8 Interview with the family of Ayoub Adnan Ayoub, Karakhosh, Iraq, Feb. 16, 2011. See also Kenneth R. Timmerman, "Muslim Terrorists Murder 58 Iraqi Christians in Church," *Newsmax Media,* Nov. 2, 2010, and "Were Iraqi Security Forces Involved in Baghdad Church Massacre?" *Newsmax Media*, March 1, 2011.

9 I went on numerous reporting and mission trips to Jordan, Iraq, and Lebanon between 2007 and 2011 to interview Christian refugees. Those stories are available at kentimmerman.com/articles.htm. For the quotes in these paragraphs, see Kenneth R. Timmerman, "Congress to Press Obama on Religious Persecution," *Newsmax Media,* Jan. 21, 2011.

10 "Report: Obama's Muslim Advisers Block Middle Eastern Christians' Access to the White House," Breitbart.com, Oct. 25, 2011. Every year since taking office, President Obama has continued the tradition, first established by President George W. Bush, of an annual Iftar dinner at the White House to mark the end of Ramadan, the Muslim month of fasting. Guests at the August 13, 2010, Iftar dinner, for example, included Saleha Abedin, the vice dean of the Muslim Brotherhood–associated Dar El-Hekma college in Saudi Arabia and mother of Huma Abedin; Salam al Marayati, of MPAC; Hasan Chandoo, Obama's Pakistani companion from Columbia University days; Imam Mohamed Haj Magid of the All Dulles Area Muslim Society; Ingrid Mattson, president of the Islamic Society of North America; and many more. See "Expected Attendees at the White House Iftar Dinner," White House Press Office, http://www.whitehouse.gov/the-press-office/2010/08/13/expected -attendees-white-house-iftar-dinner.

11 U.S. Department of State, Daily Briefing for Jan. 4, 2011.

12 The White House, "Previewing the Upcoming State Visit of President Hu of China," Briefing by Treasury Secretary Tim Geithner, National Security Advisor Tom Donilon, and Press Secretary Robert Gibbs, Jan. 14, 2011. A search of the White House Press Office website finds only a dozen or so mentions of "Tunisia" before this date, none of them related to the disturbances.

13 Josh Gerstein, "Has Obama been riding Mubarak over rights?" *Politico,* Jan. 28, 2011.

14 Craig Whitlock, "U.S. reexamining its relationship with Muslim Brotherhood opposition group," *Washington Post,* Feb. 2, 2011.

15 Robert Spencer, "Barack Obama and the Muslim Brotherhood," *FrontPage Magazine,* Feb. 2, 2011, http://www.aina.org/news/20110203104344.htm.

16 George Soros, "Why Obama has to get Egypt right," *Washington Post,* Feb. 3, 2011.

17 "Qaradawi Guards Ban Google's Wael Ghonim from Cairo Stage," *Middle East Online,* Feb. 18, 2011, http://www.middle-east-online.com/english/?id= 44456.

18 U.S. Department of State, "Congressional Budget Justification—Summary Tables for Fiscal Year 2012." See also Marjorie Ann Brown, "United Nations System Funding," Congressional Research Service, Jan. 15, 2013, http://www.fas.org/sgp/crs/row/RL33611.pdf.

5. QADDAFI, THE ENEMY

1 Paul Schemm, "Battle at army base broke Gadhafi hold in Benghazi," Associated Press, Feb. 25, 2011.

2 "Deadly 'Day of Rage' in Libya," Al Jazeera, Feb. 18, 2011.

3 Mariam Fam and Ola Galal, "Libya Violence Deepens as Protestors Claim Control of Second-Largest City," Bloomberg News, Feb. 21, 2011.

4 http://en.wikipedia.org/wiki/Zenga_Zenga.

5 Interview with Amnesty International researchers who visited Abu Selim prison, April 29, 2004.

6 U.S. Embassy Tripoli, "Qadhafi Development Foundation on Negotiations for Release of LIFG Members," March 4, 2008, http://wikileaks.org/cable/2008/03/08TRIPOLI182.html.

7 Ben Wedeman, "In eastern Libya, citizens buoyant and cautious as they await Gadhafi's move," CNN news blog, Feb. 21, 2011.

8 Alistair MacDonald, "Cameron Doesn't Rule Out Military Force for Libya," *Wall Street Journal*, March 1, 2011.

9 Ryan Lizza, "The Consequentialist: How the Arab Spring Remade Obama's Foreign Policy," *New Yorker*, May 2, 2011.

10 Robert M. Gates, *Duty: Memoirs of a Secretary at War* (New York: Knopf, 2014).

11 Lizza, "The Consequentialist."

12 Email from U.S. defense contractor, Benghazi, Libya, March 18, 2011.

13 Michael Waller, blog post, Oct. 7, 2008, http://jmw.typepad.com/political_warfare/2008/10/video-shows-qaddafi-praising-obama-as-a-brother-and-fellow-muslim.html.

14 U.S. Embassy Tripoli, "Musa Kusa Seeks Cooperation on Africa, al-Qaeda—and POTUS Meeting with al-Qadhafi," May 3, 2009, http://wikileaks.org/cable/2009/05/09TRIPOLI362.html. For the compensation fund demands, see U.S. Embassy Tripoli, "Lipstick on a Pig: Libya Renames Claims Compensation Fund," May 21, 2009, http://wikileaks.org/cable/2009/05/09TRIPOLI413.html.

15 U.S. Embassy Tripoli, "FM Kusa Nonchalant on Return of Libyan GTMO Detainees," June 22, 2009, http://wikileaks.org/cable/2009/06/09TRIPOLI 494.html.

16 U.S. Embassy Tripoli, "The Cast of the Jamahiriya: Background Information for NEA AA/S Feltman's Visit to Tripoli," July 21, 2009, http://wiki leaks.org/cable/2009/07/09TRIPOLI590.html.

17 U.S. Embassy Tripoli, "Libyan Reaction To Claims Settlement Agreement Positive, High Expectations For U.S. Influence on Al-Qadhafi," Aug. 28, 2008, http://www.cablegatesearch.net/cable.php?id=08TRIPOLI666&q=sarkozy.

18 U.S. Embassy Tripoli, "Qadhafi Personally Welcomes Lockerbie Bomber," Aug. 23, 2009, http://wikileaks.org/cable/2009/08/09TRIPOLI686.html. See also Mark Isakoff, "U.S., British officials benefited from thaw in Libya relations," NBC News, Feb. 28, 2011.

19 U.S. Embassy Tripoli, "Qadhafi in Rome: The 'Man of History' Visits the Eternal City," June 18, 2009, http://wikileaks.org/cable/2009/06/09TRIPOLI492 .html.

20 Clare Morgana Gillis, "Contempt for Gadhafi unites diverse rebels," *USA Today*, March 30, 2011.

21 Praveen Swami, Nick Squires, and Duncan Gardham, "Libyan rebel commander admits his fighters have al-Qaeda links," *Daily Telegraph* (London), March 25, 2011.

22 Memorandum for Commander, United States Southern Command, "Update Recommendation to Transfer to Control of Another Country for Continued Detention (TRCD) for Guantánamo Detainee, ISN: US9LY-0557DP," April 22, 2005 (JTF GTMO Detainee Assessment of Abu Sufian Ibrahim Ahmed Hamouda Bin Qumu).

23 U.S. Embassy Tripoli, "Libya: Family Visit for Returned Gitmo Detainee Confirmed," Jan. 3, 2008, 08TRIPOLI14. See also 07TRIPOLI1039, the Dec. 13, 2007, cable cited in chapter 2; still classified cable 07TRIPOLI 1060, Dec. 25, 2007, describes the Christmas Day prison visit; and Holly Watt, "WikiLeaks: Guantánamo detainee is now Libyan rebel leader," *Daily Telegraph* (London), April 26, 2011. Most U.S. government documents refer to him as "Hamouda."

24 "Al Qaeda in Libya," ibid., 17; original source: *Al-Ahwal* (Tripoli), May 5, 2012, http://www.facebook. com/photo.php?fbid=371532919560380&set=a.3 53682918012047.78426.353433611370 311&type=1&theater.

25 Lauranne Provenzano, "Liasons dangereuses entre AQMI et les insurgés libyens," *Jeune Afrique*, March 30, 2011. For an English language version, see http://www.alarabiya.net/articles/2011/03/25/143013.html.

26 Australian Broadcasting Corporation, March 7, 2011, http://www.abc.net.au /am/content/2011/s3156632.htm.

27 Email exchange with the author, July 3, 2013.

28 *La Lettre du Continent*, n°531, Dec. 20, 2007, cited in Zobel Behalal and Solene Margerit, *Le développement piégé: les transferts d'armes et le développement au Tchad (2005–2010)*, CCFD-Terre Solidaire, Paris, January 2012, 37.

29 Interview with former advisor to French President Sarkozy, Oct. 20, 2013.

30 Charles Levinson and Margaret Coker, "Inside a Flawed Spy Machine as Gadhafi's Rule Crumbled," *Wall Street Journal*, September 2, 2011.

31 The bin Laden profile, titled "This Man Wants You Dead," appeared in late June 1998 in *Reader's Digest*, just three weeks before bin Laden and al Qaeda burst onto the international scene with the twin truck bombing attacks against the U.S. embassies in Kenya and Tanzania, http://kentimmerman .com/news/rd_OBL.htm.

32 A September 1987 cable from the U.S. embassy in London recounted the Younis Khalis caper; U.S. intelligence officers involved in the effort to track down the missing missiles told me two months later that Younis Khalis had been "ambushed," no doubt trying to cover for his venality. See "Afghanistan: how the Stingers were 'confiscated,'" *Middle East Defense News (MEDNEWS)*, Nov. 2, 1987.

33 U.S. Embassy Abu Dhabi, "Iraq and Iran Key Topics in MbZ Meeting with Admiral William J. Fallon," Sept 25, 2007, http://wikileaks.org/cable /2007/09/07ABUDHABI1595.html.

34 Department of State, Directorate of Defense Trade Controls, "Proposed Charging Letter," March 21, 2013, http://www.pmddtc.state.gov/compliance /consent_agreements/pdf/Raytheon_PCL_13.pdf.

35 Mark Hosenball, "Exclusive: Obama authorizes secret help for Libya rebels," Reuters, March 30, 2011.

36 Jonathan Allen and Meredith Shiner, "Hillary Clinton, Robert Gates win no love on Capitol Hill," *Politico*, March 30, 2011.

37 Interview with Representative Pete Hoekstra, June 24, 2013.

38 Interview with Senator James Inhofe, Sept. 25, 2013.

39 Interview with Ambassador Ali Aujali, July 9, 2013.

6. INTO THE DANGER ZONE

1 Corbett Daly, "Clinton on Qaddafi: 'We came, we saw, he died,'" CBS News, Oct. 21, 2011.

2 Mario Montoyo, "Mission to a Revolution," *State*, November 2011, http://www.state.gov/documents/organization/178204.pdf. Montoyo named the ten DS agents who accompanied Stevens to Benghazi as Brian Haggerty, Kent Anderson, Josh Vincent, Chris Deedy, James Mcanelly, Jason Bierly, Ken Davis, Mike Ranger, and Keith Carter. Montoyo himself was the tenth.

3 March 2014 UN Panel of Experts report, S2014/106, 92.

4 Department of State, Accountability and Review Board Report, December 2012, 14. Hereafter, ARB report.

5 House Oversight and Government Reform Committee, prepared testimony of Deputy Assistant Secretary of State Charlene Lamb, Oct. 10, 2012, 2–3.

6 Levinson and Coker, "Inside a Flawed Spy Machine."

7 "Feltman meets with Libyan rebels," Associated Press, May 23, 2011.

8 Murphy and Webb, *Benghazi: The Definitive Report*.

9 Yazan al-Saadi, "US Mercenary 'Took Part' in Qaddafi Killing; Sent to Assist Syrian opposition," *Al Akhbar*, March 19, 2012, http://english.al-akhbar.com/contenat/blackwater-veteran-took-part-Qaddafi-killing-asked-us-help-syrian-opposition. The online version has links to the original emails.

10 Christian Aniort, "Carcassonne: Temoignange de l'Espouse de l'Epouse de Pierre Marziali, tué en Libye," *La Depeche du Midi*, March 19, 2011; Kareem Fahim and Maia de la Baume, "Head of French Company Is Killed in Libyan City," *New York Times*, May 12, 2011.

11 Communications with a U.S. defense contractor in Benghazi, March 19, 2011, Sept. 11, 2013.

12 James Risen, Mark Mazzetti, and Michael S. Schmidt, "U.S.-Approved Arms for Libya Rebels Fell Into Jihadis' Hands," *New York Times*, Dec. 5, 2012.

13 Ibid.

14 Interview with Simon Henderson, July 3, 2013.

15 Chris Stephens, Robert Tuttle, and Caroline Alexander, "Qatar may win big if Libyan rebels prevail," *BusinessWeek*, July 14, 2011.

16 Mussab Al-Khairalla, "Qatari plane supplies ammunition to Libyan rebels," Reuters, Aug. 6, 2011.

17 U.S. Embassy Doha, "Synchronizing USG Engagement with Qatar on Critical Infrastructure Protection," SECRET, May 20, 2009, http://wikileaks.org/cable/2009/05/09DOHA329.html.

18 U.S. Embassy Doha, to Secretary of State Hillary Clinton, "ADM Olson Discusses Military Relationship with Senior Qataris During Doha Visit," March 9, 2009, http://wikileaks.org/cable/2009/03/09DOHA204.html.

19 Boeing Corporation, "Boeing Delivers Qatar's 2nd C-17 Globemaster III," Sept. 10, 2009, http://boeing.mediaroom.com/index.php?s=20295&item=832.

20 Ashraf Abdul Wahab, "Gunmen at Tripoli Airport prevent Qatar Airways plane landing," *Libya Herald*, Aug. 4, 2013.

21 United Nations Security Council, *Final report of the Panel of Experts established pursuant to Security Council resolution 1973 (2011) concerning Libya*, March 20, 2012, S/2012/163;§94–96. Hereafter the March 2012 UN Panel of Experts report. The panel was reauthorized and made a second "final report" the following year, and a third in March 2014.

22 Ibid., 69–70: "Rebuttal from the State of Qatar."

23 Ibid., §98–100.

24 Ibid., §121–26.

25 The White House, "Report to Congress on United States Activities in Libya," June 15, 2011.

26 Bruno Waterfield and Richard Spencer, "Libya: Col Qaddafi still has quarter of chemical weapons stockpile," *Daily Telegraph* (London), April 28, 2011.

27 Graeme Smith, "General's family drives wedge of suspicion into Libya's rebellion," *Globe and Mail*, Aug. 2, 2011.

28 Ali Shuaib, "Libya says ex-deputy PM suspect in general's killing," Reuters, Nov. 28. 2011.

29 Rami al-Shaheibi, "Officer accuses fellow rebels in Libya killing," Associated Press, July 29, 2011.

30 Adrian Blomfield, Damien McElroy, and Ruth Sherlock, "Islamists blamed for killing General Abdel Fattah Younis as Libya's rebels face up to enemy within," *Daily Telegraph*, July 30, 2011.

31 Leela Jacinto, "Shifting loyalties among Libya's Islamists," *France 24*, Aug. 10, 2011.

32 "Al-Qaeda in Libya: A Profile," Library of Congress, Federal Research Division, and U.S. Department of Defense, Combating Terrorism Technical Support Office, Irregular Warfare Support Program, August 2012, 8.

33 http://libyaagainstsuperpowermedia.com/2012/08/05/libya-the-libyan-last-report-01082012/.

34 Interview with John Maguire, Nov. 5, 2013.

7. HOUSE OF WAR

1 John T. Bennett, "Hillary Clinton pressed on negotiating with Islamic militants in Afghanistan," *The Hill*, Oct. 27, 2011.

2 The White House, Remarks by Assistant to the President for Homeland Security and Counterterrorism John Brennan at CSIS, May 26, 2010.

3 "Belgistan? Sharia Showdown Looms In Brussels," Christian Broadcasting Network, March 15, 2012.

4 Steven Nelson, "Muslim Brotherhood a Secular Organization, Says Clapper," *Daily Caller*, Feb. 10, 2011.

5 Comments by Dr. Sebastian Gorka, Sept 6, 2013, Westminster Institute, Washington, D.C., author's transcript.

6 The White House, *Empowering Local Partners to Prevent Violent Extremism in the United States*, August 2011.

7 http://www.whitehouse.gov/blog/2013/02/05/working-counter-online-radicalization-violence-united-states.

8 Quintam Wiktorowicz, "The New Global Threat: Transnational Salafis and Jihadi," *Middle East Policy* 8, No. 4, December 2001, pp. 29–30.

9 Comments by Dr. Patrick Sookhdeo, Sept. 6, 2013, Westminster Institute, Washington, D.C., author's transcript. Sookhdeo and Gorka have expanded on these concepts in their contributions to a remarkable short book of essays: Katharine C. Gorka and Patrick Sookhdeo (eds.), *Fighting the Ideological War: Winning Strategies from Communism to Islamism* (McLean, VA: Isaac, 2012).

10 Josh Gerstein, "Report: Leon Panetta revealed classified SEAL unit info," *Politico*, June 5, 2013. Panetta's apparent security breach was described in detail in a draft inspector general report on the administration's cooperation

with the *Zero Dark Thirty* film project, but reference to it was deleted when the final report was released a few weeks later. So was a line mentioning that Undersecretary of Defense for Intelligence Michael Vickers, one of the key officials helping the filmmakers, had disclosed to them the identity of the Neptune Spear mission planner, which also was classified. See http://abc news.go.com/blogs/headlines/2013/06/defense-report-omits-panettas-seal -team-6-disclosure/.

11 Brigadier General Jeffrey Colt, "Memorandum for Commander, United States Central Command, Investigation Findings and Recommendations (Crash of CH-47 Aircraft in Wardak Province, Afghanistan on 6 August 2011)," Department of the Army, Sept. 9, 2011, SECRET; Declassified. Hereafter, Colt report.

12 Colt report, 6.

13 Ibid., 56: Exhibit 20.

14 Communication from a senior Special Forces officer, Oct. 2, 2013.

15 *Charles and Mary Ann Strange et al v. Joseph Robinette Biden et al.*, 13-cv-974, Complaint 10; see also https://www.youtube.com/watch?v=G968rxOjiMs& feature=player_embedded.

8. WEAPONS, WEAPONS, EVERYWHERE

1 Scott Stewart, "The Continuing Threat of Libyan Missiles," *Stratfor Security Weekly*, May 2, 2012, http://www.stratfor.com/weekly/continuing-threat -libyan-missiles.

2 Martha Raddatz, "Hillary Clinton Visits Libya to Meet Rebel Leaders," ABC News, Oct. 18, 2011.

3 Author's correspondence with DynCorps spokesperson Ashley Burke, Oct. 10, 2013.

4 C. J. Chivers and Rod Nordland, "Missiles Missing from Libyan Stockpile," New York Times News Service, Sept. 8, 2011.

5 Reza Kahlili, "Almost 5,000 SAM-7 missiles missing in Libya," blog post, Oct. 2, 2011, http://atimetobetray.com/blog/almost-5000-sam-7-missiles -missing-in-libya/.

6 Con Coughlin, "Will a Middle East awash with weapons be Qaddafi's final legacy?" *Daily Telegraph*, Oct. 23, 2011.

7 Chivers and Nordland, "Missiles Missing from Libyan Stockpile."

8 United States Senator John McCain, press release, April 22, 2011; see also Peter Kenyon, "McCain on Libyan Rebels: They are my heroes," *National Public Radio*, April 22, 2011.

9 http://gawker.com/5852485/heres-the-clearest-video-yet-of-Qaddafis -capture.

10 Raghida Dergham, "The West is hijacking Arab revolutions to the benefit of the Islamists," *Al Arabiya*, Oct. 30, 2011.

11 Michael Boyle, "Obama leading from behind on Libya," *Guardian* (London), Aug. 27, 2011, http://www.theguardian.com/commentisfree/cifamerica /2011/aug/27/obama-libya-leadership-nato.

12 Neil Munro, "Sharia: Obama-encouraged Libyan transitional council approves polygamy, bans banking interest," *Daily Caller*, Oct. 24, 2011. See also Aymenn Jawad Al-Tamimi, "Libya headed toward Islamism."

13 "Libyan Jew Blocked from Tripoli Synagogue," *Guardian*, Oct. 4, 2011. See also Aymenn Jawad Al-Tamimi, "Libya headed toward Islamism."

14 Ruth Sherlock, "Leading Libyan Islamist met Free Syrian Army opposition group," *Daily Telegraph*, Nov. 27, 2011. The *Daily Telegraph* account had him arrested as he was leaving Tripoli. A chronology of 230 security incidents compiled by the head of the U.S. Embassy's Regional Security Office, Eric Nordstrom, had him arrested as he was returning to Tripoli on Nov. 24, ostensibly from a trip to Qatar.

9. THE THREAT MATRIX

1 House Oversight and Government Reform Committee (HOGR), "Letter to President Obama from Reps. Darrell Issa and Jason Chaffetz," Oct. 19, 2012, 8. Hereafter, Issa and Chaffetz to Obama.

2 (SBU) Action Memorandum from AS Jeffrey Feltman for Undersecretary for Management Patrick F. Kennedy, "Future Operations in Benghazi, Libya," December 27, 2011; included in Issa and Chaffetz to Obama.

3 U.S. Department of State Foreign Affairs Manual Volume 12, Diplomatic Security, 12 FAM 315.1; included in Issa and Chaffetz to Obama.

4 House Oversight and Government Reform Committee, "Interview of Admiral Michael Mullen," June 19, 2013, 112.

5 Ibid., 131.

6 Email exchange, Eric. A. Nordstrom to Shawn P. Crowley et al., "Re: DS coverage for Benghazi," Feb. 11, 2012. Included in Issa and Chaffetz to Obama.

7 Email exchange, Shawn P. Crowley to Patrick A. Tillow et al., "DS coverage for Benghazi," Feb. 11, 2012.

8 Email exchange, Eric A. Nordstrom to James P. Bacigalupo, "Re: DS Coverage for Benghazi," Feb. 12, 2012.

9 U.S. Embassy Tripoli, "Possibility of Demonstrations on February 17, 2012," http://libya.usembassy.gov/service/information-for-travelers/warden -messages2.html.

10 U.S. Embassy Tripoli, Memo from Deputy Chief of Mission Joan Polaschik to Deputy Assistant Secretary of State Charlene Lamb, Feb. 28, 2012, quoted in Elise Labott, "US Embassy in Libya sought extension of 'essential' security team," CNN security blogs, Oct. 8, 2012. The CNN article mistakenly reports that the SST was approved through August. In fact, it was scheduled to depart on April 5, 2012. Polaschik convinced Lamb to approve a 120-day extention through August 5. See the email exchanges on the SST quoted in House Oversight and Government Reform Committee, "Benghazi Attacks: Investigative Update Interim Report on the Accountability Review Board," Sept. 16, 2013, 57–58.

11 Ibid., 58.

12 House Oversight and Government Reform Committee, "The Security Failures of Benghazi," October 10, 2012, 87 (Lamb), and 88 (Nordstrom). See also U.S. Embassy Tripoli, "Request for DS TDY and FTE Support," March 28, 2012 (12TRIPOLI 130), in which Eric Nordstrom details the steps he was taking "[a]s recommended by the Department" to "transition from emergency to normalized security operations." Included in Issa and Chaffetz to Obama.

13 Jan. 2014 SSCI report, "Additional Views of Vice Chairman Chambliss and Senators Burr, Risch, Coats, Rubio, and Coburn," 9, citing the committee transcript of the member and staff interview of Eric Nordstrom, June 27, 2013, 24–25.

14 Eric Nordstrom, "Security Incidents Since June 2011," 23. The incidents that follow have been paraphrased from Nordstrom's report.

15 Interview with Lieutenant Colonel Andrew Wood, Utah National Guard, Sept. 23, 2013.

16 Austin Long, "Partners or Proxies? U.S. and Host Nation Cooperation in Counter-terrorism Operations," *CTC Sentinel*, Nov. 30, 2011, 1, http://www.ctc .usma.edu/posts/partners-or-proxies-u-s-and-host-nation-cooperation-in -counterterrorism-operations.

17 Ambassador Gene Cretz signed off on the counterterrorism training program. Part of it got funded through Section 1207 of the NDAA, a broadly worded authority that funds programs to "defeat terrorist organizations and their ability to gain recruits. The programs and projects focused on denying support and sanctuary to terrorists through strategically targeted development assistance and public diplomacy, while increasing the reach of good governance and rule of law." See Congressional Research Service, "Department of Defense: Section 1209 and Section 1203(b) Report to Congress on Foreign-Assistance Related Programs for Fiscal Years 2008, 2009, and 2010," April 2012, 5. DoD allocated $6.2 million to Special Operations Command in FY 2012 to build capacity of two Libyan special operations forces companies under a slightly changed statute (1207) that went into effect in 2012 that was "jointly administered and funded by the Department of Defense (DOD) and the State Department." See Nina Serafino, "Global Security Contingency Fund: Summary and Issue Overview," Congressional Research Service, Jan. 22, 2013, CRS Report R42641. At the beginning, however, the funding came from JSOC under 1208.

18 Assistant Secretary of State Anthony Shapiro, "Addressing the Challenge of MANPADS Proliferation," Stimson Center, Washington, D.C., Feb. 2, 2012. Quotes from the Q&A after his presentation were transcribed by the author from the event. See the online Reader's Guide for more.

19 Senate Armed Services Committee, "Worldwide Threats to U.S. National Security," Feb. 16, 2012, "Testimony of Director of National Intelligence Lt. Gen. James Clapper."

20 Ibid., "Testimony of Defense Intelligence Agency Director Lt. Gen. Ron Burgess."

21 "Bodies of Hurlburt airmen returned home," PNJ.com, Feb. 21, 2012.

22 Bill Warner blog post, "Al-Shabaab in Somalia Publish Article on 'Somali-Memo' Website About American U-28A Plane Going Down," Feb. 22,

2012, http://www.billwarnerpi.com/2012/02/al-shabaab-in-somalia-publish
-article.html.

23 Email exchange with Colonel Dick Brauer, Oct. 8, 2013; see also Brian
Everstine, "Fatah Djibouti U-28A crash laid to crew error," *Air Force Times*,
Oct. 24, 2012.

24 Noor Ali Farah, "Is Al-Shabaab Using Heat-Seeking Missiles?" *Somali-
Report*, March 4, 2012.

25 http://www.strategypage.com/htmw/htada/20120322.aspx.

26 Source interview, Sept. 26, 2013.

27 House Committee on Foreign Affairs, "LRA, Boko Haram, al-Shabaab,
AQIM and Other Sources of Instability in Africa," April 25, 2012, U.S. Gov-
ernment Printing Office, Serial 112-142, "Statement of Daniel Benjamin,
Ambassador-at-Large, Coordinator for Counterterrorism, Department of
State," 29–30.

28 See, *inter alia*, January 2014 SSCI report, 40; and "Interim Progress Report
for the Members of the House Republican Conference on the Events Surrounding the September 11, 2012 Terrorist Attacks in Benghazi," from the
Chairmen of the Committees on Armed Services, Foreign Affairs, Judi-
ciary, Oversight and Government Reform, and Permanent Select Committee on Intelligence, April 23, 2013, 11, 38.

29 House Committee on Foreign Affairs, "LRA, Boko Haram, al-Shabaab,
AQIM and Other Sources of Instability in Africa," "Statement of
Amanda J. Dory, Deputy Assistant Secretary of Defense for African Af-
fairs," 43.

30 Ibid., "Response to questions submitted for the record to Deputy Assistant Secretary of State Donald Yamamoto by Representative Ileana Ros-
Lehtinen," 76.

31 Ibid.

32 The White House, Statement by the President on the White House Organization for Homeland Security and Counterterrorism, May 26, 2009, Online
Reader's Guide Extra.

33 Senate Appropriations Committee, "State, Foreign Operations, and Related
Programs Appropriations for Fiscal Year 2013," Feb. 28, 2012, U.S. Govern-
ment Printing Office, Hearing 112-669, written replies to questions from
Senator Lindsay Graham.

34 March 2014 UN Panel of Experts report, 86, for the period Jan. 1, 2013, through April 30, 2013.

35 House Appropriations Subcommittee on State, Foreign Operations, and Related Programs, February 29, 2012, Statement of Secretary of State Hillary Rodham Clinton.

10. THE ARMS PIPELINE TO SYRIA

1 U.S. Embassy Tripoli, Remarks by Ambassador Gene Cretz, "Inta Liby Hoor" Award Ceremony, Tripoli, Libya, Feb. 26, 2012.

2 UN Panel of Experts report, March 9, 2013, paragraphs 171–82.

3 Official Lebanese Army photographs, Selaata, Lebanon, April 28, 2012, Online Reader's Guide Extra.

4 March 2014 UN Experts report; 89. See also,"Syria arms ship impounded, crew held for questioning," *Daily Star*, April 28, 2012.

5 United Nations General Assembly/Security Council, Document A/66/803-S/2012/316, "Letter dated 11 May 2012 from the Permanent Representative of Turkey to the United Nations addressed to the Secretary-General and the President of the Security Council." The Turkish ambassador was responding to a Syrian government letter to the secretary general and the president of the Security Council dated May 10, 2012, alleging that "the case of the ship Lutfullah 2 [*sic*], which was intercepted by the Lebanese Army, proves that Libya and Turkey are cooperating with other States to send murderous weapons to terrorist groups, in order to wreak more carnage and destruction."

6 March 2014 UN Panel of Experts report, 89.

7 *Al Akhbar* daily in Arabic, May 4, 2012. A similar account appeared the same day in the more balanced *Al-Jumhuriya* daily, which mentioned 100 Stingers and 30 French-made MILAN antitank missiles, whereas the Lebanese authorities said they had found just 2 MILANs and 19 Russian-made antitank missiles of various sorts.

8 "N. Lebanon man arrested for selling arms to Syrian rebels," *Daily Star*, Sept. 9, 2013.

9 March 2013 UN Experts report, §180.

10 Interview with U.S. official in Libya, September 2013.

11 Interview with Selim Raad, July 9, 2013.

12 "From Libya to the world: Egypt makes its biggest arms bust ever," *Russia Today*, May 20, 2012; "Egyptian intercepts Libyan weapons heading for Gaza," *Libya Herald*, July 1, 2012.

13 "Iran funding smuggling of Libyan Arms into Sinai—Source," *Al Sharq al Awsat*, May 9, 2012.

14 Interviews and email exchanges with John Maguire, October–November 2013. On the Qayed brothers, see Walid Shoebat blog post, "Libya Leaks: More Secret Documents Reveal Obama's Failure in Libya," Nov. 3, 2012; Human Rights Watch, "Delivered into Enemy Hands: U.S.-led Abuse and Rendition of Opponents in Qaddafi's Libya," 26.

15 View the photo and other supporting documents at http://kentimmerman.com/darkforces.htm.

16 Jack Murphy and Brandon Webb, *Benghazi: The Definitive Report*.

17 U.S. Department of Treasury, "Treasury Targets Key Al-Qa'ida Funding and Support Network Using Iran as a Critical Transit Point," July 28, 2011.

18 U.S. Department of State, "Briefing on New Rewards for Justice Reward Offer," Dec. 22, 2011.

11. OBAMA'S AMBASSADOR

1 Issa and Chaffetz to Obama, 99: "Report of Security Incidents since June 2011" from the Tripoli Embassy Regional Security Office.

2 Adam Entous, Siobhan Gorman, and Margaret Coker, "CIA Takes Heat for Role in Libya," *Wall Street Journal*, Nov. 1, 2012.

3 Interview with Lieutenant Colonel Andy Wood, Sept. 23, 2013, U.S. Embassy, Tripoli, Press Release, May 21, 2012, http://libya.usembassy.gov/ne_052112.html.

4 Interview, Sept. 23, 2013. Wood went over this in detail in his congressional testimony as well. "General Ham made it very clear to Ambassador Cretz that he could have the SST as long as he needed them," he told the House Oversight and Governmental Affairs Committee on Oct. 10, 2012. "It was also made clear to Joan Polaschik, who took over as charge d'affaires in between Ambassador Cretz and Ambassador Stevens. He came personally and told her that." HOGR hearing transcript, 67.

5 Eric Nordstrom, "Security Incidents Since June 2011," 44. See also Joe Sterling, "In Libya, militias 'running the show,' analyst says," CNN, Sept. 13, 2012.

6 Issa and Chaffetz to Obama, 5, citing transcribed interview of Assistant Regional Security Officer David Oliveira, Oct. 9, 2012.

7 Conversation with Representative James Lankford (R-OK-5), Oct. 9, 2013.

8 Email chain between Ambassador Chris Stevens and John Moretti, June 7, 2012, 3:34 AM, Subject: MSD/Tripoli. Cited in *Interim Report*, 8.

9 Douglas Jehl, "C.I.A. Officers Played Role in Sheik Visas," *New York Times*, July 22, 1993.

10 Dugald McConnell and Brian Todd, "Egyptian lawmaker met U.S. officials despite affiliation with terrorist group," CNN, June 23, 2012.

11 "Shariah Guardians Reviving Islamic Revolution in Libya," *PressTV*, June 8, 2012. Watch this extraordinary two-minute report at http://kentimmerman .com/darkforces.htm.

12 Hannah Draper blog, "The Slow Move East." This particular post, dated June 10, 2012, is titled, "Hello from Sunny Tripoli!," http://hannahdraper .blogspot.com/2012/06/hello-from-sunny-tripoli.html.

13 Hannah Draper blog, "Tripoli: the good, the bad, and the sandy," Aug. 14, 2012, http://hannahdraper.blogspot.com/2012/08/tripoli-good-bad-and-sandy.html.

14 Eric Schmitt, "C.I.A. Said to Aid in Steering Arms to Syrian Opposition," *New York Times*, June 21, 2012.

15 Cited in ibid.

16 Paul Kane, "CIA Says Pelosi Was Briefed on Use of 'Enhanced Interrogations,' " *Washington Post*, May 7, 2009.

17 Alfred Cumming, "Statutory procedures Under Which Congress is to be Informed of U.S. intelligence activities, Including Covert Actions," Congressional Research Service memorandum, Jan. 18, 2006, p. 4.

18 Interview with former senior CIA operations officer, Oct. 18, 2013.

19 Michael Gordon and Mark Landler, "Backstage Glimpses of Clinton as Dogged Diplomat, Win or Lose," *New York Times*, Feb. 2, 2013.

20 U.S. Embassy Tripoli, 12TRIPOLI 622, June 25, 2012. The June 22 cable has not been released by the State Department and is cited in Issa and Chaffetz to Obama, 5.

12. PRELUDE TO MURDER

1 Hannah Draper blog, "Elections," July 5, 2012; http://hannahdraper .blogspot.com/2012/07/elections.html.

2 Ben Geman, "Sen. McCain Observes Libyan elections, heralds 'historic day,' " *The Hill*, July 7, 2012. Hannah Draper updated her blog on July 14, 2012, with a brief account of escorting McCain in a post titled, "Lift Your Head High, You're a Free Libyan," http://hannahdraper.blogspot.com/2012/07 /lift-your-head-high-youre-free-libyan.html.

3 George Grant, "Elections Analysis: So who are they and what do they actually stand for?," *Libya Herald*, June 30, 2012.

4 Eric Nordstrom, "Security Incidents Since June 2011," cited in Issa and Chaffetz to Obama.

5 Senator John McCain, "Remarks by Senator John McCain on the Attack on the U.S. Consulate in Benghazi on the Floor of the U.S. Senate," Sept. 12, 2012, http://www.mccain.senate.gov/public/index.cfm/floor-statements?ID =bbec5934-ed6b-d64d-7238-c9e72d66778d.

6 Issa and Chaffetz to Obama, 5.

7 "Under current arrangements, Post's thirty-four U.S. security personnel (16 SST, 11 MSD, 1 WAE TDY, 1 RSO, 2 ARSO, and 3 TDY ARSO) will draw down to twenty-seven security personnel on 7/13. On 8/05, post will reduce U.S. security personnel to 4 MSD trainers, 1 RSO, 2 ARSOs, and 3 TDY ARSOs, with a further reduction to seven U.S. security personnel on 8/13, which includes four MSD trainers not generally supporting transportation security, VIP visits, or RSO programs." U.S. EMBASSY TRIPOLI, "Tripoli: Request for extension of TDY security personnel," 12TRIPOLI 690, July 9, 2012. Included in Issa and Chaffetz to Obama.

8 Gregory Hicks, "Benghazi and the Smearing of Chris Stevens," *Wall Street Journal*, Jan. 22, 2014.

9 Eric Nordstrom, "Security Incidents Since June 2011."

10 Oct. 10, 2012, HOGR hearing, 108.

11 Confidential source interview, October 2013. Deputy Assistant Secretary of State Charlene Lamb testified at the Oct. 10, 2012, HOGR hearing that there were seven members of the "rapid reaction force" stationed at the Annex (69). Former Diplomatic Security officer Fred Kaplan wrote in his

account, *Under Fire*, that there were ten GRS "shooters" (107). The Senate Select Committee on Intelligence reported there were nine "security officers," most likely adding Glen Doherty and the GRS colleague who had flown in from Tripoli that night to the seven security personnel mentioned by Charlene Lamb. See U.S. Senate Select Committee on Intelligence, "Review of the Terrorist Attacks on U.S. Facilities in Benghazi, Libya, September 11–12, 2012," Jan. 15, 2014, 20. Hereafter, Jan. 2014 SSCI report.

12 Communications with AEI analyst Will Fulton. See Ali Alfoneh, "Iran's Secret Network: Major General Qassem Suleimani's Inner Circle," American Enterprise Institute, March 3, 2011; and Will Fulton, "The IRGC Command Network: Formal Structures and Informal Influence," American Enterprise Institute, July 18, 2013.

13 "Libya militia interrogates Iran Red Crescent team," Agence France-Presse, Aug. 1, 2012.

14 Sergeant Morgan Jones and Damien Lewis, *Embassy House* (New York: Simon & Schuster, 2013), Kindle edition. Davies wrote his account under a pseudonym. Simon & Schuster withdrew the book in November 2013, less than one month after publication, after allegations that Davies had exaggerated his own role on the night of the attack.

15 For photos and more, visit the Online Reader's Guide, http://kentimmerman .com/darkforces.htm.

16 Edward Klein, *The Amateur* (Washington, DC: Regnery, 2012), 85–86.

17 Helene Cooper and Mark Landlert, "U.S. Officials Say Iran Has Agreed to Nuclear Talks," *New York Times*, Oct. 21, 2012. This followed several stories by former CIA undercover agent Reza Kahlili, author of *A Time to Betray*, at WND.com.

18 Reza Kahlili, "October Surprise? Obama Secret Iran Deal Cut," Oct. 4, 2012.

19 Source interview, May 23, 2013.

20 U.S. Embassy Tripoli, "Request to Add LES Ambassador Protective Detail Bodyguard Positions in US Embassy Tripoli," Aug. 2, 2012 (12 Tripoli 944). Included in the appendix of the Oct. 10, 2012, HOGR hearing transcript, 115–116.

21 U.S. Embassy Tripoli, "The Guns of August: Security in Eastern Libya," Aug. 8, 2012 (12 Tripoli 952), included in Issa and Chaffetz to Obama, 37–38.

22 Interview with Lieutenant Colonel Andy Wood, Sept. 23, 2013. The attempted car-jacking was mentioned briefly in both the Oct. 10, 2012, and the May 8, 2013, HOGR hearings.

23 Library of Congress, "Al Qaeda in Libya: A Profile," 1.

24 Susan Katz Keating and Richard Miniter, "Intelligence Warnings on Benghazi Were Loud And Clear," *Investor's Business Daily*, Nov. 12, 2013.

25 "Decision *In Re Terrorist Attacks on September 11, 2001, 03 MDL 1570 (GBD),*" December 2011, § 116; http://www.law.com/jsp/decision.jsp?id=120 2536770916. The case is better known as *Havlish et al v. Islamic Republic of Iran et al*. More documents from the case are available at http://iran911case .com/http://iran911case.com. My affidavit laying out thirty years of Iran's support for Sunni Muslim terrorist groups was presented to the court as part of the plaintiff's evidence, http://information.iran911case.com/Exhibit_2.pdf.

26 Bill Roggio, "US targets al Qaeda operatives with links to Iran, Pakistan," *Long War Journal*, June 6, 2008; Kenneth R. Timmerman, "U.S. Treasury: Al-Qaida Worked Closely With Iran," *Newsmax Media*, January 19, 2009; Joby Warrick, "US accuses Iran of aiding al Qaeda," *Washington Post*, July 28, 2011.

27 Muaad al-Maqtari, "Saudi Arabia Accuses Iran of Supporting Ansar al-Sharia in Yemen," *Yemen Times*, April 30, 2012.

28 One of my sources provided pictures of this training from the Quds Force academy that I published in *Countdown to Crisis: The Coming Nuclear Showdown with Iran* (New York: Crown Forum, 2005), http://kentimmerman .com/countdown.htm.

29 Communication with source, Nov. 19, 2013. Khalil Harb was eventually blacklisted by the U.S. Treasury Department in August 2013, although not in connection with the Benghazi attacks. See Department of Treasury, "Treasury Sanctions Hizballah Leadership," Aug. 22, 2013, http://www .treasury.gov/press-center/press-releases/Pages/jl2147.aspx.

30 "Erdoğan now kissed his forehead," *Gerçek Gündem*, Aug. 26, 2012, http:// arsiv.gercekgundem.com/?p=484567.

31 Yazan al-Saadfi, "U.S. Mercenary 'Took Part' in Qaddafi Killing; Sent to Assist Syrian opposition," *Al Akhbar*, March 19, 2012.

32 Mariam Karouny, "Exclusive—Libyan fighters join Syrian revolt against Assad," Reuters, Aug. 14, 2012. For a more detailed profile of al-Harati, see Mary Fitzgerald, "The Syrian Rebels' Libyan Weapon," *Foreign Policy*, Aug. 9, 2012.

33 Online Reader's Guide Extra, http://kentimmerman.com/darkforces.htm.

34 Obaida Elwani Facebook post, Aug. 7, 2012. Online Reader's Guide Extra.

35 C. J. Chivers, "In Syria, Potential Threat to Government Air Power Emerges," *New York Times*, "At War" blog, Aug. 7, 2012. See also Richard Engel, "Turning point? Syrian rebels obtain missiles," *NBC Nightly News*, July 31, 2012.

36 C. J. Chivers, Eric Schmitt, and Mark Mazzetti, "In Turnabout, Syria Rebels Get Libyan Weapons," *New York Times*, June 21, 2013. At the time he made these comments to a *New York Times* reporter, Buktatef had just been named Libyan ambassador to Uganda. He was still commander of the 17th February Brigade at the time of the Benghazi attacks.

37 ArmyRecognition.com, Oct. 6, 2012, http://www.armyrecognition.com /syrian_syria_conflict_revolution/figting_continues_in_several_cities_in _syria_as_rebels_allegedly_downed_syrian_warplane_0610122.html.

38 Jones and Lewis, *Embassy House*, chapters 6–7.

39 House Armed Services Committee, "DOD Force Posture in Anticipation of September 11, 2012," Transcript of Testimony by General Martin Dempsey, Chairman of the Joint Chiefs of Staff, Oct. 10, 2013, Top Secret (redacted), 14.

40 Eric Nordstrom, the Tripoli-based RSO until July 26, 2012, said he was "awestruck" when asked by the State Department's ARB why he never recommended putting a .50-caliber machine gun on the roof of the Benghazi compound. "I said, if we are to the point where we have to have machine gun nests at a diplomatic institution, isn't the larger question, what are we doing? Why do we have staff there?" May 8, 2013, HOGR hearing, 65.

41 Burton and Katz, *Under Fire*, 63–64.

42 U.S. Embassy Tripoli, August 16, 2012 (12 TRIPOLI 55), cited in Jan. 2014 SSCI report, 15–16.

43 House Foreign Affairs Committee, "Benghazi: Where Is the State Department Accountability," Sept. 18, 2013, U.S. Government Printing Office Serial No. 113-93, 55.

44 "Benghazi Weekly Report: Special Eid al-Fitr Edition," Aug 20, 2012 (12TRIPOLI 1020), included in Issa and Chaffetz to Obama, 55.

45 Cited in Jan. 2014 SSCI report, 16.

46 http://travel.state.gov/travel/cis_pa_tw/tw/tw_5762.html.

47 Richard A. Oppel, Jr., "11 Are Killed as U.S. Copter Goes Down in Afghanistan," *New York Times*, Aug. 16, 2012. An earlier account, from

the Associated Press, reported seven American deaths. See Fox News, "7 Americans among dead in crash of US military helicopter caught in Afghan firefight," Aug. 16, 2012.

48 October 2013 communication with a senior active-duty U.S. Special Forces officer. Under pressure from Congress, the administration canceled the final tranche of a $1.1 billion deal to purchase sixty-five Mi-17 helicopters for the Afghan Army on November 14, 2013. See Radio Voice of Russia, "US cancels purchase of Russian Mi-17 helicopters," Nov. 14, 2013.

49 Lee Foran, "American Killed in Libya Was on Intel Mission to Track Weapons," ABC News, Sept. 13, 2012.

50 Interview with Lieutenant Colonel Andy Wood, Sept. 23, 2013. Greg Hicks also described Stevens' dilemma over renewing the SST without a Status of Forces Agreement. See Hicks, "Benghazi and the Smearing of Chris Stevens."

51 Nancy A. Youssef, "Ambassador Stevens twice said no to military offers of more security, U.S. officials say," McClatchy Newspapers, May 15, 2013.

13. BAITING THE TRAP

1 Sheeran Frenkel, "Syrian rebels squabble over weapons as biggest shipload arrives from Libya; Turkey," *Times* (London), September 14, 2012, 23.

2 Michael Kelley, "There's a reason why all of the reports about Benghazi are so confusing," *Business Insider*, Nov. 3, 2012.

3 Jessica Donati, Ghaith Shennib, and Firas Bosalum, "The adventures of a Libyan weapons dealer in Syria," Reuters, June 18, 2013.

4 Harald Doornbos and Jenan Moussa, "Comrades in Arms: How Libya sends weapons to Syria's rebels," *Foreign Policy*, July 10, 2013.

5 UN Experts Panel report, March 2013, ¶ 184–88.

6 Interview with John Maguire, Nov. 5, 2013.

7 Associated Press, "J. Christopher Stevens: Obituary," *Boston Globe*, September 12, 2012.

8 For Colonel Adams' earlier public appearance, see Stimson Center, Remarks by Assistant Secretary of State Anthony Shapiro, "Addressing the Challenge of MANPADS Proliferation," Stimson Center, Washington, D.C., Feb. 2, 2012.

9 Jack Murphy and Brandon Webb, "Breaking: The Benghazi Diary, A Hero Ambassador's Final Thoughts," June 24, 2013, *SOFREP*, http://sofrep .com/22460/ambassador-chris-stevens-benghazi-diary/#ixzz2ler7kCqw.

10 HOGR hearing transcript, May 8, 2013, 90.

11 Ibid., 40.

12 Ibid., 76.

13 Associated Press, "US State Department blasts CNN report on Christopher Stevens' diary," Sept. 22, 2012.

14 U.S. Embassy Tripoli, "Benghazi Weekly Report—September 4, 2012" (12 Tripoli 1078), Issa and Chafettz to Obama, 38.

15 The security lapses were described in a draft letter from the U.S. Mission, Benghazi, to Mohammad Obeidi, the head of the Libyan Ministry of Foreign Affairs office in Benghazi, Sept. 11, 2012. A crumpled printout of the letter was found on the floor of the burned-out compound several weeks later by journalists and included in Issa to Clinton, Nov. 1, 2012.

16 Many of the details of Stevens' arrival, movements, and conversations on Sept. 10–11, 2012, are described in the State Department's Accountability Review Board report. I have supplemented these with descriptions provided by eyewitnesses to various congressional committees; committee testimony; documents released by the State Department and the Department of Defense; other documents discovered in the burned-out Special Mission Compound by journalists; and Fred Burton and Samuel M. Katz's account in *Under Fire*.

17 Stevens' conversation with the Blue Mountain guards is also mentioned in Burton and Katz, *Under Fire*, 93.

18 Email from Alec Henderson to John B. Martinec, "RE: Benghazi QRF agreement" (Sep. 9, 2012, 11:31 PM), cited in "Interim Progress Report," 7.

19 See Diana West, "Benghazi: The Turkish Timeline," blog post, Dec. 13, 2012, http://www.dianawest.net/Home/tabid/36/EntryId/2345/Benghazi -The-Turkish-Timeline.aspx.

20 Senate Foreign Relations Committee, "Benghazi: The Attacks and the Lessons Learned," Jan. 23, 2013, 34.

21 Interview with Larry Johnson, Sept. 22, 2013.

22 In an email exchange with me, Akin said he would no longer respond to questions from reporters by phone or by email.

23 U.S. Embassy Tripoli, "Benghazi Weekly Report, September 11, 2012," Issa and Chaffetz to Obama, 56–57. The State Department has not released *any* of the many classified cables that deal with the weapons collection effort, even though they have talked about it publicly when it suited their political needs. See Stimson Center remarks, Assistant Secretary of State Anthony Shapiro.

24 The meeting between Wisam bin Hamed and al-Gharabi was described in "Benghazi Weekly Report, September 11, 2012," 12TRIPOLI 1098. The other cites are from the Library of Congress report, "Al Qaeda in Libya," 1, 12, 15, 36, etc. Henderson showed Stevens a draft of the new QRF agreement earlier that day. A copy of it was found by journalists in the charred ruins of the Special Mission Compound. See, *inter alia*, http://benghazipost .blogspot.fr/2012/10/qrf-security-agreement-benghazi.html.

14. THE ATTACKS

1 The White House, "Readout of the President's Meeting with Senior Administration Officials on Our Preparedness and Security Posture on the Eleventh Anniversary of September 11th," Sept. 10, 2012.

2 Senate Armed Services Committee, "Attack on U.S. Facilities in Benghazi, Libya," Feb. 7, 2013, Exchange with Senator James Inhofe (R-OK), 22.

3 "Prominent Shumoukh Member Calls To Burn Down U.S. Embassy In Egypt With All Staff Inside In Order To Pressure U.S. To Release Blind Sheikh," *Global Jihad News*, MEMRI, Sept. 7, 2012.

4 Raymond Ibrahim blog post, "New Evidence Ties Morsi top Zawahri and to al Qaeda," Sept. 20, 2013, http://www.raymondibrahim.com/from-the -arab-world/new-evidence-ties-ousted-morsi-government-to-al-qaeda.

5 Nic Robertson, "Exclusive: Al Qaeda leader's brother offers peace plan," CNN, Sept. 11, 2012, Online Reader's Guide Extra.

6 Raymond Ibrahim, "Jihadis Threaten to Burn U.S. Embassy in Cairo," *PJ Tatler*, Sept. 10, 2012, http://pjmedia.com/tatler/2012/09/10/jihadis-threaten -to-burn-u-s-embassy-in-cairo.

7 Andrew C. McCarthy, "'Blame It On the Video' Was a Fraud for the Cairo Rioting, Too," *National Review Online*, May 13, 2013. I will come back to the talking points in the next chapter. See also an excellent analysis by Stephen Hayes, "The Benghazi Scandal Grows," *Weekly Standard*, May 20, 2013.

8 Several Marines posted her order on their blogs. The State Department quickly denied it although from all accounts, the Marines remained disarmed. See http://freebeacon.com/reports-marines-not-permitted-live-ammo.

9 Discussion with RADM Richard Landolt, February 2014.

10 Issa to Clinton, Nov. 1, 2012.

11 Ibid. See also Harold Doornbos, Jenan Moussa, "Troubling Surveillance Before Benghazi Attack," *Foreign Policy*, Nov. 1, 2012, where the letters first appeared.

12 Sean Flynn, "Murder of an Idealist," *GQ*, December 2012.

13 Department of State, background conference call with senior State Department officials, Oct. 9, 2012, Online Reader's Guide Extra.

14 Burton and Katz, *Under Fire*, 93. See also the final entry of Stevens' diary.

15 Thomas Joscelyn, "Al Qaeda-linked jihadists helped incite 9/11 Cairo protest," *Long War Journal*, Oct. 26, 2012.

16 Interview with Lieutenant Colonel Andy Wood, Oct. 15, 2013.

17 Jones and Lewis, *Embassy House*, 169.

18 Nathan Max, "San Diegan slain in Libya attack 'lived to serve,' " *San Diego Union Tribune*, Sept. 17, 2012.

19 Alex "The Mittani" Gianturco, "RIP Vile Rat," online tribute, Sept. 12, 2012, http://themittani.com/news/rip-vile-rat.

20 The online gaming community eventually raised $125,000 to support Smith's family. For more tributes from his online gaming friends, see http://www.wired.com/dangerroom/2012/09/vilerat/, http://community.eveonline.com/news/dev-blogs/73406, and http://www.somethingawful.com/news/sean-smith-vilerat.

21 Cited in "Farewell to Vilerat," http://www.somethingawful.com/news/sean-smith-vilerat.

22 This chronology, unless otherwise noted, follows that of the State Department's Accountability Review Board. They identify Alec Henderson as "TDY RSO," Scott Wickland as "ARSO-1," and David Ubben as "ARSO-2." So far, the identities of the two African-American DS agents the ambassador brought with him from Tripoli have not been released. The ARB refers to them as "ARSO-3" and "ARSO-4."

23 Chris Stephens, "US consulate attack in Benghazi: A challenge to official version of events," *Guardian*, Sept. 9, 2013.

24 Burton and Katz, *Under Fire*, 102.

25 The initial eyewitness accounts from the Libyan guards—both the 17th February militiamen and the Blue Mountain guards—were wildly contradictory and unreliable. By the time they met with reporters they had already been aggressively interviewed by the FBI, who accused them of failing the "fight to the death" to defend the compound. See Shashank Bengali, "Libyan Guards Speak Out on Attack That Killed U.S. Ambassador," *Los Angeles Times*, Oct. 13, 2012. In separate interviews with *Time*, the 17th February guards claimed they brought up reinforcements, one of many statements of bravura in this article that simply wasn't true. See Steven Sotloff, "The Other 9/11: Libyan Guards Recount What Happened in Benghazi," *Time*, Oct. 21, 2012.

26 Conversation with Representative James Lankford, Feb. 5, 2014. Representative Lankford said he was pressing the administration to release the surveillance video and the drone footage "so the American people can see it."

27 Representative Lynn Westmoreland, interview with CNN, "New Day," Nov. 18, 2013.

28 "Radical Yemen Cleric: No Fatwa Needed to Kill Americans," Voice of America, Nov. 7, 2010.

29 Gretel C. Kovach, Debbi Baker, and Nathan Max, "Two SEAL vets from SD killed in Libya," *San Diego Union Tribune*, Sept 13, 2012.

30 Rowan Scarborough, "U.S. commandos fought in Benghazi," *Washington Times*, Oct. 31, 2013.

31 None of the official timelines for the events of this night match. In the CIA's initial timeline, Ty Woods and his team departed the Annex at 10:02 PM. The State Department's ARB report said they left at 10:05 PM. In a much later timeline developed by the House Permanent Select Committee on Intelligence, they departed at 10:07 PM—and arrived simultaneously! See David Ignatius, "In Benghazi timeline, CIA errors but no evidence of conspiracy," *Washington Post*, Nov. 1, 2012; and "HPSCI Update on Benghazi," Nov. 14, 2013. The Senate Intelligence Committee cites surveillance video from the Annex showing that they departed at 10:03 PM, but then adds wiggle room that they departed "approximately 20–25 minutes after the first call came into the Annex," Jan. 2014 SSCI report, 4.

32 EMTs discovered the cyanide effect after a nightclub fire in West Warwick, Rhode Island, in 2003. See Michael E. Murphy, "Hydrogen Cyanide in Structure Fires," July 2010, New York State Association of Fire Chiefs on-

line magazine, http://inletemergencyservices.files.wordpress.com/2010/07/hydrogencyanide1.pdf.

33 "Glasses of tea": Burton and Katz, *Under Fire*, 162. The fiction of the forty militiamen was repeated in Steven Sotloff's account in *Time*, cited above, where one of the 17th February militiamen claimed they had "jogged alongside" the cars of the arriving GRS team. That simply didn't happen.

34 Interview with RADM Richard Landolt, Feb. 4, 2014. As this book went to press, no congressional committee had yet asked Landolt or Leidig to testify.

35 House Armed Services Committee, Subcommittee on Oversight and Investigations, Redacted Transcript of Executive Session (Top Secret), "AFRICOM and SOCAFRICA and the Terrorist Attacks in Benghazi," 35. Hereafter, General Ham testimony.

36 Ibid., 38.

37 http://www.youtube.com/watch?v=bhipJw8QB_0.

38 Interview with RADM Richard Landolt.

39 House Armed Services Committee, Subcommittee on Oversight and Investigations, Redacted Transcript of Executive Session (Top Secret) Briefing with Major General Darryl Roberson, Vice Director, Operations, Joint Staff, May 21, 2013, 83.

40 Jennifer Griffin and Adam Houseley, "Facts and questions about what happened in Benghazi," Fox News, Jan. 22, 2013.

41 House Armed Services Committee, Subcommittee on Oversight and Investigations, Redacted Transcript of Executive Session (Top Secret), "AFRICOM and SOCAFRICA and the Terrorist Attacks in Benghazi," transcript of briefing by RADM Brian Losey, 149.

42 Sharyl Attkisson, "Sources: Key task force not convened during Benghazi consulate attack," CBS News, Nov. 2, 2012.

43 Interview with Larry Johnson.

44 HOGR hearing on Benghazi, May 8, 2013; testimony of Mark Thompson.

45 Burton and Katz, *Under Fire*, 198.

46 Here is the photo that she displayed: http://cryptome.org/2012-info/cia-benghazi/cia-benghazi.htm.

47 HOGR hearing, Oct. 10, 2012, 32.

48 Hicks' statement was in a prehearing deposition before committee staff. He reaffirmed it when it was read back to him by Representative Trey Gowdy at the May 8, 2013, HOGR hearing, 106.

49 Scarborough, "U.S. commandos fought in Benghazi."

50 The presence of the two JSOC operators was first alluded to by General Martin Dempsey in congressional testimony, and was highlighted in questioning by staff for Representative Darrell Issa in the committee's transcribed interview with Admiral Mike Mullen, the co-chair of the State Department's Accountability Review Board. See House Oversight and Government Reform Committee, Mullen transcript, June 19, 2013, 89–90. They are also mentioned in an administration email previewing a DoD press release timed to coincide with the May 8, 2013, HOGR hearing at which Greg Hicks first testified. It's extraordinary that the administration was still only just releasing this information fully eight months after the attacks. See HOGR, May 8, 2013, hearing, 119.

51 "Série d'arrestations en Libye après la mort de l'ambassadeur américain," *Le Monde*, Sept. 13, 2012. See also Joe Herring and Dr. Mark Christian, "Where is Fathi al-Obeidi?" *American Thinker*, Nov. 9, 2013. On the negotiations at the airport, see Burton and Katz, *Under Fire*, 228.

52 Stephen Dinan and Shaun Waterman, "Panetta: Benghazi intelligence too sketchy to send troops," *Washington Times*, Oct. 25, 2012.

53 *BBC Newshour*, "Benghazi doctor who treated US Ambassador to Libya," undated, http://audioboo.fm/boos/957631-benghazi-doctor-who-treated-us-ambassador-to-libya.

54 Associated Press, obituary of Ambassador J. Christopher Stevens, Boston Globe, Sept 12, 2012, http://www.legacy.com/obituaries/bostonglobe/obituary.aspx?pid=159852423.

55 Spencer Ackerkman, "Blackwater founder: My company could have prevented Benghazi deaths," *Guardian*, Nov. 22, 2013.

15. THE COVER-UP

1 http://www.state.gov/secretary/rm/2012/09/197628.htm.

2 Jan. 2014 SSIC report, Additional Views, 5.

3 http://www.whitehouse.gov/the-press-office/2012/09/12/remarks-president-deaths-us-embassy-staff-libya.

4 Lucy Madison, "Hillary Clinton, Joe Lieberman Denounce Florida Pastor's Planned Quran Burning Event," CBS News, September 8, 2010.

5 Megyn Kelly, "Representative Westmoreland says there is 'no excuse' for inadequate Benghazi security," Fox News *(The Kelly File)*, Nov. 27, 2013.

6 Kimberly Dozier and Matt Apuzzo, "CIA station chief in Libya linked Benghazi attack to militants within 24 hours," Associated Press, Oct. 19, 2012.

7 The White House, Press Gaggle by Press Secretary Jay Carney en route Las Vegas, NV, Sept. 12, 2012.

8 Sarah Huisenga and Matt Vasilogambros, "Romney Blasts Obama on Cairo Embassy Statement of 'Sympathy,' " *National Journal*, Sept. 11, 2012.

9 http://www.washingtonpost.com/blogs/fact-checker/post/the-romney-cam paigns-repeated-errors-on-the-cairo-embassy-statement/2012/09/13/978 a6be6-fdf0-11e1-b153-218509a954e1_blog.html.

10 The White House, Briefing by Press Secretary Jay Carney, Sept. 14, 2012.

11 Email from Petraeus to Morell, Sept. 15, 2012, 2:27 PM; 95 of the talking points emails as posted by the White House.

12 For an analysis of the emails, see Stephen F. Hayes, "Benghazi Emails Directly Contradict White House Claims," *Weekly Standard*, May 16, 2013. The full text is available as an Online Reader's Guide Extra at http://kentim merman.com/darkforces.htm.

13 HOGR hearing, Oct. 10, 2012, 72.

14 Communication with Stephen Hayes, Dec. 4, 2013.

15 Catherine Herridge, "Did CIA official suppress Benghazi narrative?" Fox News, Feb. 14, 2014.

16 "Benghazi letter to Speaker Boehner," March 3, 2014, http://judicialwatch .org/document-archive/next-benghazi-letter. The author was one of the signatories.

17 The White House vigorously denies that Valerie Jarrett had any back-channel dealings with Tehran, and repeatedly emailed me to call these reports "rumors" and "lies."

18 Video of Kinzinger's comments at the Town Hall meeting: http://www .youtube.com/watch?v=4xwoAaW5DJk.

19 http://www.youtube.com/watch?v=6akGlF6g-Zw.

20 Rich Lowry, "The Benghazi Patsy," National Review Online, May 9, 2013; "Sharyl Attkisson on Benghazi Attack Investigation," *C-SPAN (Washington Journal)*, May 12, 2013, http://www.c-spanvideo.org/program/Shary.

C-SPAN played the ramp ceremony video of Hillary Clinton's remarks in the first few minutes of Attkisson's interview.

21 Abdelhakim Belhadj, Samir Shalwi Program, Sept. 12, 2012, *Radio Derna*, http://www.zangetna.com/t57930-topic.

22 HOGR hearing, May 8, 2013, 70.

16. AFTERMATH

1 "Murder of U.S. Ambassador Shows U.S. Weakness; Republican Candidate Calls for Muscular Response," *Daily Caller*, Sept. 12, 2012, http://dailycaller.com/2012/09/13/libya-killilngs-we-need-a-more-muscular-response/.

2 Duncan Gardham, Rowena Mason, and Richard Spencer, "Britons and other Westerners told to leave Benghazi after 'imminent' terror threat," *Daily Telegraph*, Jan. 24, 2013.

3 Catherine Herridge and Pamela Browne, "Former Guantánamo detainee was on ground in Benghazi during terror attack, source says," Fox News, Oct. 25, 2013.

4 HOGR hearing, Sept 19, 2013. This was one of several questions I had suggested the committee ask Pickering and Mullen in an op-ed that appeared that morning in the *Washington Times*.

5 Herridge, "Former Guantánamo detainee."

6 Walid Shoebat, "State Dept acknowledges Egyptian Terror Group in Libya," Oct. 8, 2013, http://shoebat.com/2013/10/08/state-dept-acknowledges-egyptian-terror-group-libya/. Thomas Joscelyn notes that the United Nations has also listed Jamal as an al Qaeda–affiliated terrorist, http://www.longwarjournal.org/archives/2013/10/un_adds_benghazi_sus.php.

7 For more on his background, see Walid Shoebat, "Libyan Leaks: More Secret Documents Reveal Obama's Failure in Libya," Nov. 3, 2012, http://shoebat.com/2012/11/03/libyan-leaks-obamas-gift-to-al-qaeda-libya-itself/.

8 Bill Gertz, "Benghazi Attack Suspect Walks," *Washington Free Beacon*, June 27, 2013.

9 For a good summary of the press reporting on Abu Khattala, including links to original sources, see Walid Shoebat, "CNN interviews 'lead suspect' in Benghazi attacks," July 310, 2013, http://shoebat.com/2013/07/31/cnn-interviews-lead-suspect-in-benghazi-attacks.

10 De Villiers called him "Abu Bukhattala," and accused the fictional character of personally assassinated General Abdel Fattah Yunis in July 2011, a key turning point in the insurrection that put the Islamists, led by Abdelhakim Belhaj, firmly in charge of the rebel military.

11 Eli Lake, "Benghazi suspect held in Tunisia," *Daily Beast*, Oct. 23, 2012. See also Mark Hosenball, "U.S. investigators to get access to Benghazi suspect," Reuters, Nov. 3, 2012.

12 Committee on Benghazi, Remarks by Representative Frank Wolf. Transcript: http://www.aim.org/benghazi/rep-frank-wolf-ccb/. The Tunisian authorities released Harzi in January 2013.

13 Jason Howerton, "15 Anti-Obama Photos from Tahrir Square Protests That You Probably Haven't Seen," *The Blaze*, July 3, 2013.

14 "Farahat: video shows Egyptians in Benghazi, Dr. Morsi sent us," *FrontPage*, May 31, 2013, http://frontpagemag.com/2013/cynthia-farahat /benghazi-terrorists-dr-morsi-sent-us.

15 "Morsi criticises Syria at Tehran meeting," Al Jazeera, Aug. 30, 2013.

16 U.S.-Egyptian researcher Raymond Ibrahim, who calls his blog "Islam translated," unearthed the original document and posted a preliminary translation; two weeks later, he said that Arab-language media further confirmed it. Raymond Ibrahim, "Involved in U.S. Consulate Attack," June 26, 2013, http://www.raymondibrahim.com/from-the-arab-world/libyan-intelligence -muslim-brotherhood-morsi-involved-in-u-s-consulate-attack. For a complete translation see Jerome Corsi, "Libyan official ties Morsi to Benghazi attack," *WorldNetDaily*, July 10, 2013.

17 Walid Shoebat, Ben Barrack, and Keith Davies, "Ironclad: Egypt Involved in Benghazi Attacks," June 30, 2013, http://shoebat.com/2013/06/30/benghazi -turning-a-blind-eye-for-the-blind-sheikh/. Also see Raymond Ibrahim, "Leading Egyptian Islamists Arrested, Protesters Celebrate," July 2, 2013, http://www.raymondibrahim.com/from-the-arab-world/leading-egyptian -islamist-arrested-protesters-celebrate.

18 House Oversight and Government Reform Committee, "Review of the Benghazi Attack and Unanswered Questions," Sept. 19, 2013. See also http://shoe bat.com/2013/09/25/benghazi-arb-chair-egypt-involved-benghazi-attacks.

19 Cited by Egyptian blogger Cynthia Farahat, "Is Muslim Brotherhood working together with Amb. Chris Stevens' assassin?" Aug. 12, 2013, http://

cynthiafarahat.com/2013/08/12/is-muslim-brotherhood-working-together-with-amb-chris-stevens-assassin/.

20 House Government Reform Committee, Mullen interview, 62.

21 House Armed Services Committee, General Ham testimony, 46.

22 Associated Press, "Satellite pictures suggest Sudanese weapons factory hit by air strike," *Guardian*, Oct. 27, 2012.

23 "Syrian Rebels Have US Stinger Missiles—Russian General," *Ria-Novosti*, Oct. 24, 2012.

24 On Jan. 4, 2013, local Bedouins tipped off the Egyptian military to an arms convoy heading toward Gaza through the northern Sinai. They seized a handful of short-range rockets, allegedly U.S.-made. See Chana Ya'ar, "Egypt Stops Libyan Arms Shipment to Gaza," *Israel National News*, Nov. 28, 2012.

25 Cnaan Liphshiz, "Egypt intercepts Gaza-bound missile shipment," *Jewish Telegraph Agency*, Jan. 4, 2013.

26 Westminster Institute, "Delegation of Former U.S. Military Leaders, Counterterrorism Experts, Reports on Findings from Sept. 27–30 Visit to Egypt," National Press Club, Oct. 1, 2013; "Egypt intercepts convoys from Libya with 'sophisticated weapons' for insurgents in Sinai," *World Tribune*, Oct. 27, 2013.

27 "Report Sinai militias have SAMs closes Israeli tourist airport," *World Tribune*, Aug. 9, 2013.

28 Photos and videos available via the Online Reader's Guide, http://kentimmerman.com/darkforces.htm.

29 http://atwar.blogs.nytimes.com/2012/11/27/videos-from-syria-appear-to-show-first-confirmed-hit-of-aircraft-by-surface-to-air-missile/?partner=rss&_r=0.

30 http://www.stratfor.com/sample/situation-report/syria-rebels-down-military-aircraft-target-iranian-plane-report-says.

31 http://rogueadventurer.com/2013/05/31/9k338-igla-s-and-other-manpads-in-syria/—more-1547. See also the earlier *New York Times* "At War" blog post from November 2012: http://atwar.blogs.nytimes.com/2012/11/13/possible-score-for-syrian-rebels-pictures-show-advanced-missile-systems.

32 N. R. Jenzen-Jones, "Mujao Fighters with 9K32 or 9K32M MANPADS in Mali," RogueAdventurer.com, March 22, 2013.

33 March 2014 United Nations Experts report, 42.

34 Ibid., 97–99.

35 Reza Khalili, "October Surprise? Obama Secret Iran deal cut," *WorldNetDaily*, Oct. 2, 2012.

36 Helene Cooper and Mark Landler, "U.S. Officials Say Iran Has Agreed to Nuclear Talks," *New York Times*, Oct. 21, 2012.

37 http://www.commentarymagazine.com/2012/11/05/valerie-jarrett-secret-iran-talks-raise-questions-about-obama-intentions-nuclear.

38 Bradley Klapper, "Lawmakers grill White House advisor on secret Iran talks," Associated Press, Dec. 11, 2013, http://www.timesofisrael.com/lawmakers-grill-white-house-adviser-on-secret-iran-talks.

39 "Krauthammer: WH 'Held Affair Over Petraeus' Head' For Favorable Testimony On Benghazi," Fox News, Nov. 11, 2013.

40 "Sharyl Attkisson on Benghazi Attack Investigation," C-SPAN, May 12, 2013. The short clip of her comments on Petraeus can be viewed here: http://www.c-spanvideo.org/clip/4451039.

41 Terrence Jeffrey, "Sen. Burr: CIA Has 'Flatly Refused' to Give Intel Committee Some Benghazi-Related Documents," CNS News, Feb. 13, 2013.

42 Communications with Kevin Shipp, September 2013. Shipp made similar comments at the first public hearing of the People's Commission on Benghazi, where I also testified, at the Heritage Foundation on Sept. 16, 2013. Video of the public portion of the hearing, including Shipp's presentation and my own, is available at http://www.aim.org/aim-column/citizens-commission-on-benghazi-conference-video-now-online.

43 Trey Gowdy on Greta Van Susteren, Fox News, Aug. 18, 2013, https://www.youtube.com/watch?v=AX5WwWSuL3k.

44 Communication with the author, Feb. 4, 2014.

45 "Rolled her eyes": Confidential source interview, October 2013. "Nonsense": Feinstein on NBC's *Meet the Press*, May 12, 2013, http://www.mediaite.com/tv/meet-the-press-sen-feinstein-says-its-nonsense-that-benghazi-should-preclude-hillary-from-presidency.

46 http://transcripts.cnn.com/TRANSCRIPTS/1312/01/sotu.01.html.

47 House Oversight and Government Reform, Mullen interview, 143. Mullen couldn't remember saying that to Maxwell, but in his interview with the committee he stated that in his view, responsibility for the security decisions in Benghazi "rests inside the State Department at the assistant secretary level." Ibid., 115.

48 Ibid., 132. See also House Oversight and Government Reform Committee, "Benghazi Attacks: Investigative Update, Interim Report on the Accountability Review Board," Staff Report prepared for Chairman Darrell Issa, Sept. 16, 2013, 37. "In the specific case of Benghazi, we were aware that, you know, the villas they had rented would not meet [the department's security standards for overseas facilities]. Everybody was aware of it. Ambassador Stevens was aware of it. The Department as a whole was aware that this did not meet standards. But what we did was put as much effort into it as we could to get as close as possible to the standards," said Eric J. Boswell, the former Assistant Secretary of State for Diplomatic Security. Boswell was one of the four officials put on administrative leave following the ARB report.

49 House Oversight and Government Affairs Committee, "Reviews of the Benghazi Attacks and Unanswered Questions," Sept. 19, 2013, GPO Serial 113–59; exchange with Representative Patrick Meehan (R-PA), 71.

50 Amy Chozick, "Clinton Seeks State Dept. Legacy Beyond That of Globe-Trotter," *New York Times*, Nov. 26, 2013.

51 John McCormick, "Americans don't believe Christie on Jam, Clinton on Libya," Bloomberg, March 12, 2014.

APPENDIX I: QUESTIONS & ANSWERS

1 A more complete list is available through Representative Wolf's website: http://wolf.house.gov/benghazi.

2 Representative Nunes sent his questions in a letter to House Speaker John Boehner on Nov. 6, 2013, http://nunes.house.gov/uploadedfiles/rep._nunes_letter_to_speaker_boehner.pdf.

Index

About the Author

KENNETH R. TIMMERMAN is the *New York Times* bestselling author of *Shadow Warriors, Countdown to Crisis, Preachers of Hate,* and *Death Lobby.* He has written for *USA Today, Time, Newsmax,* the *Wall Street Journal,* the *American Spectator,* and the *New York Times.* He lives in Maryland with his family.